Computer Modelling of Electronic and Atomic Processes in Solids

NATO ASI Series

Advanced Science Institutes Series

A Series presenting the results of activities sponsored by the NATO Science Committee, which aims at the dissemination of advanced scientific and technological knowledge, with a view to strengthening links between scientific communities.

The Series is published by an international board of publishers in conjunction with the NATO Scientific Affairs Division

A	**Life Sciences**	Plenum Publishing Corporation
B	**Physics**	London and New York
C	**Mathematical and Physical Sciences**	Kluwer Academic Publishers
D	**Behavioural and Social Sciences**	Dordrecht, Boston and London
E	**Applied Sciences**	
F	**Computer and Systems Sciences**	Springer-Verlag
G	**Ecological Sciences**	Berlin, Heidelberg, New York, London,
H	**Cell Biology**	Paris and Tokyo
I	**Global Environmental Change**	

PARTNERSHIP SUB-SERIES

1.	**Disarmament Technologies**	Kluwer Academic Publishers
2.	**Environment**	Springer-Verlag / Kluwer Academic Publishers
3.	**High Technology**	Kluwer Academic Publishers
4.	**Science and Technology Policy**	Kluwer Academic Publishers
5.	**Computer Networking**	Kluwer Academic Publishers

The Partnership Sub-Series incorporates activities undertaken in collaboration with NATO's Cooperation Partners, the countries of the CIS and Central and Eastern Europe, in Priority Areas of concern to those countries.

NATO-PCO-DATA BASE

The electronic index to the NATO ASI Series provides full bibliographical references (with keywords and/or abstracts) to more than 50000 contributions from international scientists published in all sections of the NATO ASI Series.
Access to the NATO-PCO-DATA BASE is possible in two ways:

– via online FILE 128 (NATO-PCO-DATA BASE) hosted by ESRIN,
Via Galileo Galilei, I-00044 Frascati, Italy.

– via CD-ROM "NATO-PCO-DATA BASE" with user-friendly retrieval software in English, French and German (© WTV GmbH and DATAWARE Technologies Inc. 1989).

The CD-ROM can be ordered through any member of the Board of Publishers or through NATO-PCO, Overijse, Belgium.

Computer Modelling of Electronic and Atomic Processes in Solids

edited by

Roderick C. Tennyson

University of Toronto,
Institute for Aerospace Studies,
Toronto, Canada

and

Arnold E. Kiv

South Ukrainian Pedagogical University,
Odessa, Ukraine

Springer Science+Business Media, B.V.

Proceedings of the NATO Advanced Research Workshop on
Computer Modelling of Electronic and Atomic Processes in Solids
Wroclaw, Poland
May 20–23, 1996

A C.I.P. Catalogue record for this book is available from the Library of Congress.

ISBN 978-94-010-6387-6   ISBN 978-94-011-5662-2 (eBook)
DOI 10.1007/978-94-011-5662-2

Printed on acid-free paper

Table of Contents

Atomic and Molecular Processes

Electronic Structure and Processes

Structure and Properties

Preface

This publication presents the proceedings of the NATO Advanced Research Workshop (ARW) on Computer Modelling of Electronic and Atomic Processes in Solids. This ARW was held at Szklarska Poreba, Wroclaw, Poland from May 20 - 23, 1996, and brought together scientists from Canada, England, Germany, Israel, Latvia, Poland, Russia, Switzerland, United States, Ukraine and Uzbekistan.

The NATO Advanced Research Workshops program is designed to increase collaboration and exchange of knowledge between the Eastern and Western scientific communities. This particular NATO ARW has already succeeded in that effort, and has spawned collaboration agreements and programs. One joint project in space materials has led to the launch of an experiment to the Russian MIR space station. This NATO ARW was also fortunate to be held concurrently with a workshop of the Wroclaw Technical University, in the same location, which focused on glass materials, thus providing for a larger scientific audience for a number of presentations of both groups.

The primary emphasis of this ARW was on computer models, ranging from fundamental atomic, molecular and electronic structures and processes, through to macroscopic descriptions of materials in terms of their structure and properties. Various elements discussed in these proceedings include environmental effects, predictions of properties, correlations with experiments and material performance parameters. Applications to space and electronics were emphasized.

Several presentations were devoted to fundamental theoretical models in materials science. Much effort involved calculations of the microscopic electronic structure of molecules and solids by the density functional method (DFM). Key speakers developed in their papers combined density-functional and configuration-interaction methods for the electronic structure of solids with impurities, modelling of catalytic materials, and hybrid quantum-mechanical and potential models for studies in solids.

Molecular dynamics (MD) simulations were described following essentially two different approaches, where empirical, classical potentials or quantum mechanical techniques are used. The latter scheme was introduced for investigations of subthreshold mechanisms of defect creation in surface layers of solids. Theoretical studies of atomic emission and defect formation caused by electronic excitation in surface layers of ionic crystals were also presented. Existing experimental results on photo-induced and electron-induced emission from ionic surfaces were discussed on the basis of these calculations.

Another group of presentations included an analysis of radiation processes in high-temperature superconductor structures based on computer simulations, the effect of polymer self-organization, simulations of porous silicon and properties of microporous materials. Interesting fractal models of disordered solids were discussed for dielectric properties of composite materials and defect growth in solids.

Other topics in this Workshop program described the problem of environmental effects on materials. For example, test results on the interaction of composite space materials and micrometeoroids were used to establish data useful for computer modelling and simulation. The Monte Carlo computational technique for pred-iction of atomic oxygen erosion was also demonstrated together with a new phenomenological model for the description of erosion of polymers in the space environment. New results regarding the possibility of improvement of polymer materials by the ion implantation method were also presented. Mechanisms of nonmetal surface destruction caused by multiply-charged ions are described in this section as well.

To better understand the topics covered and their interaction with the above elements, a flow chart is included detailing the contents of these papers and their relationship to these subject areas.

This publication has been divided into three main sections consistent with both the ARW and the flow chart:
- Atomic and Molecular Processes
- Electronic Structure and Processes
- Structure and Properties

Roderick C. Tennyson,
University of Toronto Institute for Aerospace Studies,
Toronto, Canada

Arnold E. Kiv,
South-Ukrainian Pedagogical University,
Odessa, Ukraine

Acknowledgments

We are grateful to a number of organizations for providing the financial assistance that made this Workshop possible. Foremost is the NATO Scientific Affairs Division which provided not only important financial support for the Workshop and this publication, but organizational guidance and leadership in promoting the collaboration between the Eastern and Western scientific communities. In addition, the following institutions also made significant contributions: the South-Ukrainian Pedagogical University; the Technical University of Wroclaw, the University of Toronto Institute for Aerospace Studies and the Integrity Testing Laboratory (Canada).

It is a pleasure to acknowledge the efforts of the two Vice-Directors (and Associate Editors), Dr. Jacob Kleiman (Canada) and Professor Jacob Roizin (Ukraine) who made significant contributions to the organization and scientific content of this Workshop. We are indebted to the other members of the International Organizing Committee and the Local Organizing Committee, particularly Dr. E. Rysiakiewicz-Pasek who made this workshop possible (See membership list in Appendix A). A particular word of thanks must go to the administrative staff from the South-Ukrainian Pedagogical University and the Technical University of Wroclaw who contributed to making the workshop a success. Finally, the energetic efforts of the Managing Editor, Daphne Lavers, of Delta Blue Communications (Canada), ensured that contributions were received and edited in a timely manner to bring this publication to print. The editors are very appreciative of her work.

Roderick C. Tennyson,
University of Toronto Institute for Aerospace Studies
Toronto, Canada

Arnold E. Kiv,
South-Ukrainian Pedagogical University,
Odessa, Ukraine

Flow Chart

This NATO Advanced Research Workshop focused on the development and application of computer models for analyzing various solid materials at the atomic, electronic, molecular and macro-scopic levels. Many of the papers contain environmental effects in the computer models, which were presented as parameters influencing the predicted properties and performance of the materials. Correlations with other predictions and in some cases, with experiments, are also highlighted where appropriate.

The environmental factors that have been considered include space effects such as atomic oxygen, radiation, charged ions and microparticle impacts. Other effects of interest to assessing material behavior include temperature and photo excitation.

To better understand the topics covered in this workshop and their interaction with the various elements presented, the following flow chart details the contents of these papers and their relationship to these subject areas. The text has been divided into three main sections consistent with both the ARW and the flow chart:

- Atomic and Molecular Processes
- Electronic Structure and Processes
- Structure and Properties

The reader can use this flow chart as a supplement to the index, to assess and identify those papers which closely match specific topics of interest. Applications relate to space, electronics and of more general use.

**Overview of NATO
Advanced Research Workshop
on Computer Modelling of Electronic
and Atomic Processes in Solids**

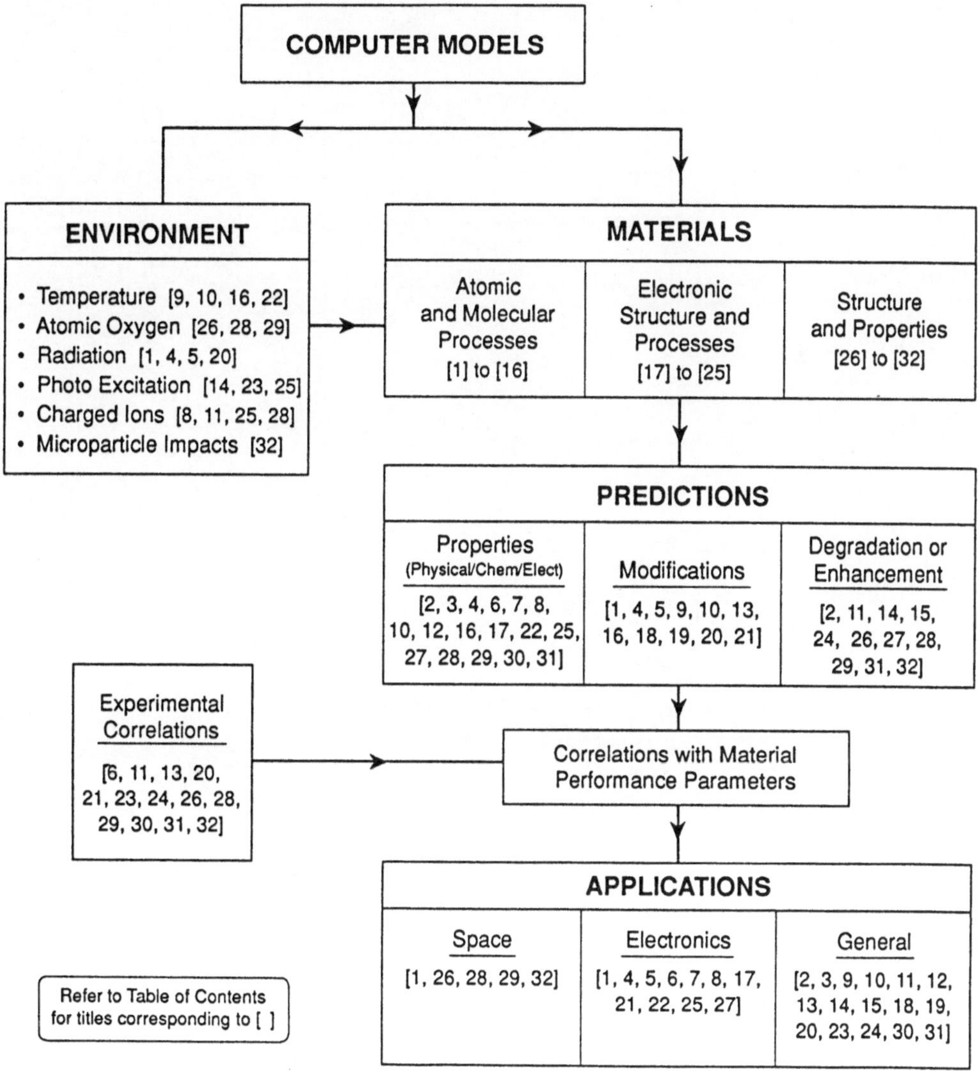

COMPUTER MODELS

ENVIRONMENT

- Temperature [9, 10, 16, 22]
- Atomic Oxygen [26, 28, 29]
- Radiation [1, 4, 5, 20]
- Photo Excitation [14, 23, 25]
- Charged Ions [8, 11, 25, 28]
- Microparticle Impacts [32]

MATERIALS

Atomic and Molecular Processes [1] to [16]	Electronic Structure and Processes [17] to [25]	Structure and Properties [26] to [32]

PREDICTIONS

Properties (Physical/Chem/Elect) [2, 3, 4, 6, 7, 8, 10, 12, 16, 17, 22, 25, 27, 28, 29, 30, 31]	Modifications [1, 4, 5, 9, 10, 13, 16, 18, 19, 20, 21]	Degradation or Enhancement [2, 11, 14, 15, 24, 26, 27, 28, 29, 31, 32]

Experimental Correlations

[6, 11, 13, 20, 21, 23, 24, 26, 28, 29, 30, 31, 32]

Correlations with Material Performance Parameters

APPLICATIONS

Space [1, 26, 28, 29, 32]	Electronics [1, 4, 5, 6, 7, 8, 17, 21, 22, 25, 27]	General [2, 3, 9, 10, 11, 12, 13, 14, 15, 18, 19, 20, 23, 24, 30, 31]

Refer to Table of Contents for titles corresponding to [ ]

Atomic and Molecular Processes

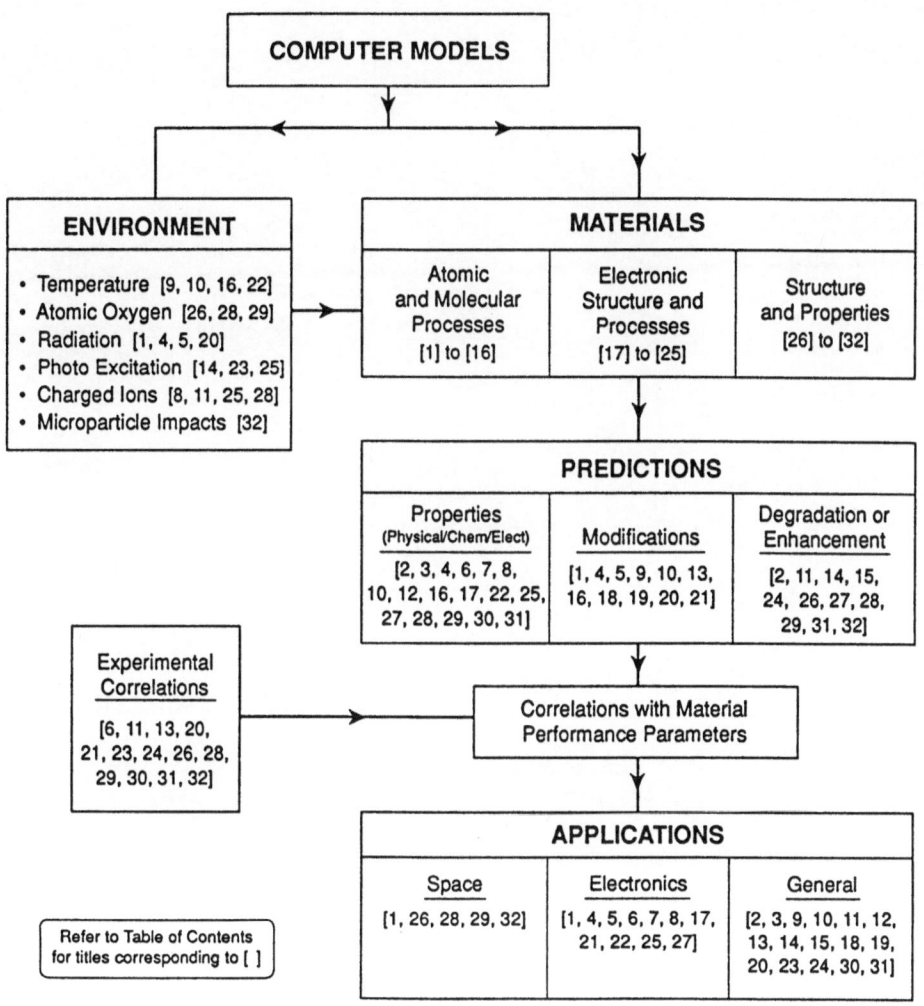

COMPUTER MODELS

ENVIRONMENT

- Temperature [9, 10, 16, 22]
- Atomic Oxygen [26, 28, 29]
- Radiation [1, 4, 5, 20]
- Photo Excitation [14, 23, 25]
- Charged Ions [8, 11, 25, 28]
- Microparticle Impacts [32]

MATERIALS

Atomic and Molecular Processes [1] to [16]	Electronic Structure and Processes [17] to [25]	Structure and Properties [26] to [32]

PREDICTIONS

Properties (Physical/Chem/Elect) [2, 3, 4, 6, 7, 8, 10, 12, 16, 17, 22, 25, 27, 28, 29, 30, 31]	Modifications [1, 4, 5, 9, 10, 13, 16, 18, 19, 20, 21]	Degradation or Enhancement [2, 11, 14, 15, 24, 26, 27, 28, 29, 31, 32]

Experimental Correlations

[6, 11, 13, 20, 21, 23, 24, 26, 28, 29, 30, 31, 32]

Correlations with Material Performance Parameters

Refer to Table of Contents for titles corresponding to [ ]

APPLICATIONS

Space [1, 26, 28, 29, 32]	Electronics [1, 4, 5, 6, 7, 8, 17, 21, 22, 25, 27]	General [2, 3, 9, 10, 11, 12, 13, 14, 15, 18, 19, 20, 23, 24, 30, 31]

NEW MECHANISMS OF RADIATION DEFECT CREATION IN SPACE CONDITIONS

*E.P. BRITAVSKAYA, *V.V. CHISLOV,
**V.I. SHAKHOVTSOV, *I.G. ZAKHARCHENKO
*South-Ukrainian Pedagogical University,
26 Staroportofrankovskaya Str. 270020 Odessa Ukraine
**Institute of Physics, National Academy of Science
46 Prospect Nauki 252628 Kiev Ukraine

1. Introduction

The influence of space radiation on materials is varied. The situation with high energy particles is clear. However, subthreshold radiation effects in space conditions are now of great interest. It is observed the large effects of material destruction when the energy of incident particles is very small (<5 eV) and in some cases destructive effects take place as a result of simple adsorption processes. So it is important to find mechanisms of subthreshold radiation effects when perturbations of electronic and lattice subsystems of solids are extremely small.

2. Model of local electric fields

Irradiation of dielectric materials leads to the appearance of a wide spectra of localized charges. It is well known that localized charges in non-metallic solids and corresponding local electric fields stimulate atom displacements (defects creation, diffusion etc.) [1].

A model of atom displacement as a result of decreasing barriers caused by local fields was investigated in detail. However, in all cases, only the localized charge carriers or the recharged centres were considered as the local field sources [2, 3]. It was shown in [4] that localized vibration modes can activate atom displacements. Below we have shown that very small changes of the

electronic density in non-metallic solids ($\Delta q << e$) can result in destruction processes in a crystal lattice.

In order to solve this problem, we have united two approaches which are given in [4] and [5]. In [4], the influence of local vibrational modes on the

1

R. C. Tennyson and A. E. Kiv (eds.),
Computer Modelling of Electronic and Atomic Processes in Solids, 1–4.
© 1997 Kluwer Academic Publishers.

2

diffusion processes of B in Si is considered. It was shown that in processes of atom displacement, a significant role is played by the distortion of the vibration spectrum of the host crystal due to the presence of the impurities in the interior. For instance, boron in silicon gives rise to localized vibration modes, which impede the penetration of the boron atoms into the interior. Surface vibration modes also affect the value of the distribution coefficient for boron in silicon.

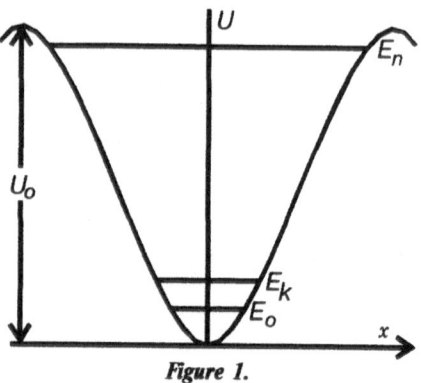

Figure 1.

There are a lot of atom configurations that create local vibrations in the spectra of surface vibration modes. We can consider such an atom configuration as a local oscillator (see Fig.1). The probability of oscillator configuration destruction is

$$\omega_K = \exp\left\{-\frac{U_0 - E_K}{kT}\right\}, \qquad (1)$$

U_0 is the depth of the oscillator pit, E_K is the primary vibration level.

Let us evaluate the probability of the oscillator excitation under the sudden electric field arising by using a standard quantum mechanics procedure [6]. For example, the one-dimensional oscillator is considered. The wave function of the unexcited oscillator is

$$\psi(x, t) = \psi_0(x) \exp\left\{-\frac{i E_0 t}{\hbar}\right\}, \qquad (2)$$

where $\psi_0(x) = \dfrac{\alpha^{1/2}}{\pi^{1/4}} \exp\left\{-\dfrac{1}{2}(\alpha x)^2\right\}$; $\alpha = \sqrt{\dfrac{m\omega}{\hbar}}$; m, ω - the oscillator

effective mass and frequency, respectively, and E_0 - zero energy level.

A local electric field $\Delta\vec{\varepsilon}$ leads to a new wave function:

$$\psi_0(x - x_0) = \frac{\alpha^{1/2}}{\pi^{1/4}} \frac{1}{\sqrt{2^n n!}} \exp\left\{-\frac{\alpha^2 (x - x_0)^2}{2}\right\} H_n[\alpha(x - x_0)] \qquad (3)$$

where $x_0 = \dfrac{e|\Delta\vec{\varepsilon}|}{m\omega^2}$ - the displacement of an oscillator equilibrium position;

H_n - Hermithian polynomials.

The full wave function is:

$$\psi(x, t) = \sum_{n=0}^{\infty} C_n \psi_n(x - x_0) \exp\left\{-\frac{i E_0 t}{\hbar}\right\}, \qquad (4)$$

where C_n - coefficients that determine the probability of an oscillator excitation to the nth level. So according to the Fourier transformations we have:

$$C_n = \int_{-\infty}^{+\infty} \psi_0(x)\psi_n^*(x - x_0)\, dx \qquad (5)$$

The final expression is:

$$\omega_n = \frac{(\alpha x_0)^2 \exp\{-\frac{1}{2}(\alpha x_0)\}}{2^n n!} \qquad (6)$$

The probability of the oscillator configuration destruction caused by an electric field is:

$$\omega_d = \omega_n \exp\left\{-\frac{U_0 - E_n}{kT}\right\}, \qquad (7)$$

ω_n is the probability of the oscillator excitation on the nth level.

If the local electric field leads to the oscillator excitation $E_k \rightarrow E_n$ (where $E_n \approx U_0$) then the condition of the effective realization of the proposed mechanism can be written as:

$$\omega_n \gg \omega_k \qquad (8)$$

Let us evaluate the field value $\Delta\vec{\varepsilon}$ that leads to satisfaction of condition (8).

The expression (6) is a well known Poisson distribution with the mean value

$$\bar{n} = \frac{e^2(\Delta\varepsilon)^2}{2\hbar m\omega^3} \qquad (9)$$

For estimation of $\Delta\vec{\varepsilon}$, we shall use parameters close to the real situation. The depth of the oscillator pit $U_0 \approx 0.1$ eV. Energy of the located phonons is $\hbar\omega \approx 10^{-3} - 5\cdot 10^{-4}$ eV. Therefore, for destruction of an oscillator configuration, an oscillator must be excited to the nth level that corresponds to the relation $U_0 = n\hbar\omega$. So $n = 100 - 200$. If we assume $n = \bar{n}$ (see (9)), then the value of $\Delta\vec{\varepsilon}$ that follows from (9) is unacceptably large. This value is equal to $10^8 - 10^9$ V/m. If the charge source is at the one inter-atomic distance (10^{-10} m) from the oscillator, then the corresponding change of local charge may be in the interval $(0.1-0.01)e$, (e is a single charge).

3. Conclusion

A new mechanism of subthreshold radiation defect creation is proposed. This mechanism is characterized by a wide region of applications. It is shown that very small changes of electronic density near the local oscillators lead to essential modifications of local vibrations. Such local oscillators are always present in real solids, in particular on solids` surfaces. It is shown that the perturbations of oscillator states by local electric fields may be a reason for radical destruction of the atom configuration.

All processes of adsorption and other effects on solid surfaces may lead to atom configuration changes as a result of the mechanism described. It is possible to apply this mechanism for an explanation of electronic devices degradation processes[7], and for solid surface erosion [8], etc.

The research described in this publication was made possible in part by Grant №K5J100 from the Joint Fund of the Government of Ukraine and International Science Foundation.

References

1. Vavilov, V.S., Kiv, A.E. and Niyazova, O.R. (1981) *Mechanisms of Creation and Migration of Defects in Semiconductors*, Nauka, Moscow.
2. Kiv, A., Elango, M., Britavskaya, E. and Zakharchenko, I. (1994) Mechanisms of subthreshold atomic emission from solid surface, *NIM B* **90**, 257-260.
3. Emtsev, V.V. and Mashovets, T.V. (1981) *Impurities and Point Defects in Semiconductors*, Radio i svyaz, Moscow.
4. Bykhovskii, A.D., Ipatova., I.P. and Maradudin, A.A. (1992) The equilibrium doping of silicon with boron, *Solid State Commun.*, **82**, 261-265.
5. Kiv, A.E. and Iskanderova, Z.A. (1975) Mechanisms of radiation-stimulated interstitial diffusion in semiconductors, *Fizika I Tekhnika Poluprovodnikov*, **9**, №2.
6. Migdal, A.B. (1975) *Qualitative Methods in Quantum Theory*, Nauka, Moscow.
7. Vavilov, V.S., Gorin, B.M., Danilin, N.S., Kiv, A.E., Nurov, Yu.L. and Shakhovtsov V.I. (1990) *Radiation Methods in Solids Electronics*, Radio i svyaz, Moscow.
8. Kleiman, J.I., Gudimenko, Yu.I., Iskanderova, Z.A. and Tennyson R.C., (1995) Surface Structure and Properties of Polymers and Pyrolytic Graphite Irradiated with Hyperthermal Atomic Oxygen. *Surf. and Interface Analysis*, **23**, 335-341.

COMPUTER SIMULATION OF CATALYTIC SYSTEMS

C.R.A. CATLOW, L. ACKERMANN, R.G. BELL, D.H. GAY, S.
HOLT, D.W. LEWIS, M.A. NYGREN, G. SASTRE, D.C. SAYLE,
P.E. SINCLAIR
The Royal Institution of Great Britain,
21 Albemarle Street, London, W1X 4BS, UK

1. Introduction

One of the most exciting areas of applications of current computer modelling techniques relates to heterogeneous catalysis. Indeed, computational methods are now increasingly recognised as key tools in the study of structure and reactivity in catalytic materials. As described in several recent articles[1,2,3] they can make vital and unique contributions to the following fundamental problems in catalytic science:

i)   Development of models of the structures at the atomic level of crystalline and amorphous catalysts, including both bulk and surface structures.

ii) Elucidation of the structure of *local* states which provide the active sites in solids or on their surfaces. We note that the active sites in catalysts can in several cases be identified in terms of well defined defect states.

iii) Determination of the mechanisms of molecular diffusion to and docking at active sites.

iv) Understanding of detailed reaction mechanisms of docked molecules at active sites.

Such information is, of course, essential in the development of a molecular understanding of catalysis, which is indeed the main thrust of contemporary catalytic studies.

Modelling of catalysts makes use of the full range of currently available atomistic simulation techniques. These comprise both 'forcefield' (or interatomic potential based) methods and electronic structure techniques. The former include Energy Minimisation (EM), Monte Carlo (MC) and Molecular Dynamics (MD) techniques; while the latter embrace both Hartree–Fock (HF) and Density Function Theory (DFT) methodologies which may be applied to clusters or periodic arrays of atoms. In the case of cluster calculations, the use of accurate embedding techniques is a key technical feature.

5

R. C. Tennyson and A. E. Kiv (eds.),
Computer Modelling of Electronic and Atomic Processes in Solids, 5–29.
© 1997 *Kluwer Academic Publishers.*

Details of currently available methods and their implementation are available in several recent reviews[4,5]. We also note that the development of the field has been assisted by the availability of high quality, general purpose computer codes, and that contemporary molecular graphics provides an essential tool for the analysis and interpretation of the results of computer modelling studies.

Broadly speaking, forcefield methods are most appropriately applied to modelling structures, docking and diffusion, although key configurations may be refined by electronic structure techniques; electronic structure methods are, of course, essential in modelling reaction mechanisms. As we will show (and as shown in, for example, the recent work of van Santen[6] and Norskov[7]), modelling methods are moving in the direction of increasingly accurate prediction of systems of growing complexity.

In the sections which follow we highlight recent applications from our laboratory which demonstrate both the accuracy and increasingly predictive character of modelling studies in this field. The synergy between computational and experimental techniques is also emphasised.

2. Applications

We now describe recent work that illustrates the range and diversity of contemporary modelling studies in catalysis.

2.1 MODELLING STRUCTURES

Knowledge of the structure of the catalysts at the atomic level is of course the first prerequisite for a molecular understanding of catalysis. Possibly the most challenging problems are posed by the structure of microporous materials. We now review three recent applications in the field. The first concerns a subtle problem in the crystallographic (long range) structure of an important material; the other two relate to key local structural properties of direct importance to catalysis.

2.1.1 Structural Features of AlPO$_4$-5

Microporous framework materials such as aluminosilicates (zeolites) and aluminophosphates (AlPOs) are, typically, microcrystalline. Therefore, unlike many other heterogeneous catalysts, their structure often cannot be examined routinely using single crystal diffraction techniques. High resolution powder diffraction now allows us to solve complex structures. Nevertheless, detailed and important aspects of the structures may be obscured in the crystallographic refinement of these systems. An important and topical example concerns T-O-T angles (where T= Si or Al in zeolites and Al or P in AlPOs). Such angles typically take values between 140-165°[8].

However, certain published structures exhibit almost linear (176-179°) angles[9,10]. We have recently investigated this effect in $AlPO_4$-5[11].

We performed lattice energy minimisation calculations on a variety of unit cell models of $AlPO_4$-5 using the pair potential model of Gale and Henson[12]. We investigated not only the previously published structure[10] (hexagonal *P6cc* spacegroup) but also a more recent structure (orthorhombic *Pcc2*) obtained using a combination of X-ray and neutron data[13]. We further constructed supercells of the hexagonal structure.

We find that it is not possible to optimise fully the structure of $AlPO_4$-5 in the *P6cc* spacegroup, as indicated in our calculation by the inability to locate an energy minimum. This result had previously been noted by other workers[14,15]. However, if we reduce the symmetry to a *P6* spacegroup (Table 1), then an energy minimum configuration can be obtained (as also found by Henson *et al.*[15]). This reduction in symmetry removes constraints imposed by the *P6cc* spacegroup on the linear Al-O-P angles and allows these angles to take up more typical values. The effect was further confirmed by calculations on supercells of the hexagonal *P6cc* cell. Only when the symmetry was lowered in such a way as to allow optimisation of the linear Al-O-P angle did we find an energy minimum. We found that the orthorhombic cell[13], which does not contain any such linear angles, is also a minimum energy configuration, albeit slightly higher in energy than the *P6* unit cell; by 0.5kJ mol^{-1} per unit cell. Given that entropic effects are not included in these calculations, we conclude that interconversion between structures with these two spacegroups would be likely.

If we consider the structure of the material in terms of angles and bond lengths (Table 1, Figure 1) it is clear that the structures are very similar, and that it is the selection of the spacegroup in the structure solution which makes them different. Furthermore, our results have more general implications regarding the structure solution of such materials. We have seen that the imposition of a high symmetry spacegroup has resulted in anomalous bond angles and bond lengths. Although the complex structure and microcrystalline nature of these materials often dictates that high symmetries are used, we conclude that refinements which exhibit such anomalous structural features would benefit from re-refinement in a lower symmetry spacegroup. Clearly from this and other examples[9,16,17] computational methods can prove of great value in obtaining accurate structures for such complex materials.

2.1.2 *Modelling Si distribution in SAPO-5*
Our next example shows how 'defect' properties can be of central importance to catalytic materials. It concerns possibly the most important local structural property of microporous aluminophosphates, where it is known that introduction of framework substituting elements in AlPO structures often results in the generation of materials with high catalytic activity in

TABLE 1. Structural parameters for various unit cell models for AlPO$_4$-5.

Structure	Experimental P6cc [10]	Simulated P6cc[a]	Simulated P6	Experimental Pcc2 [13]	Simulated Pcc2
$E_{lattice}$ (eV per AlPO$_4$)	-	-268.0408	-268.0643	-	-268.0587
a (Å)	13.771	13.820	13.759	13.797	13.754
b (Å)	13.771	14.138	13.759	23.899	23.900
c (Å)	8.379	8.678	8.393	8.417	8.417
$\alpha(°)$	90.0	90.0	90.0	90.0	90.0
β (°)	90.0	90.0	90.0	90.0	90.0
γ (°)	120.0	120.0	120.0	90.0	90.4
Al-O (Å)	1.68-1.73	1.696-1.722	1.712-1.734	1.67-1.76	1.693-1.731
P-O (Å)	1.46-1.53	1.507-1.523	1.505-1.530	1.44-1.56	1.504-1.528
⟨Al-O-P⟩ (°)	149-178	147.6-176.4	141.8-154.7	147-156	140.1-156.5

(a) Note this structure is not an energy minimum, but rather a transition state as negative phonon modes are present in lattice dynamical calculations on the structure

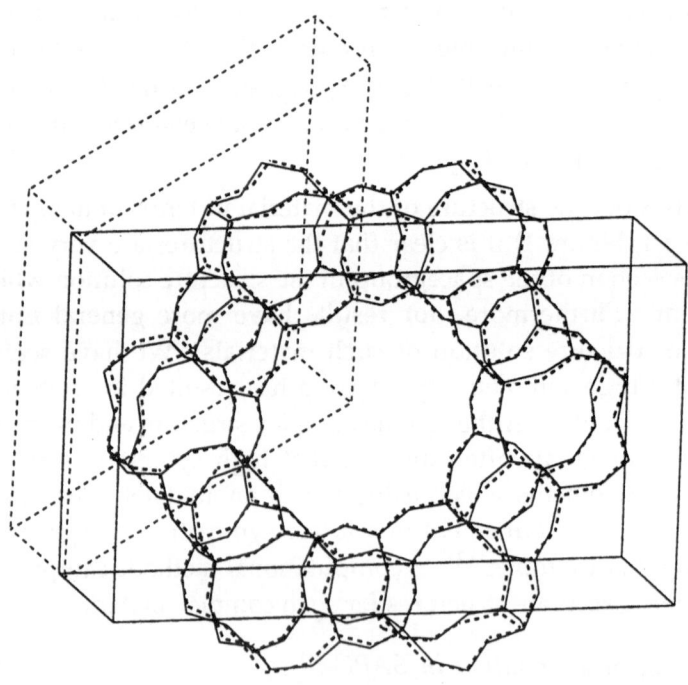

Figure 1. Overlay of the calculated minimum energy structures of AlPO$_4$–5 in both P6 (dotted lines) and Pcc2 spacegroups.

processes requiring medium-high acid strength[18]. For example, when silicon is introduced into a neutral $AlPO_4$ framework, Brønsted acid sites are created, resulting in materials which can be exploited as catalysts. These silico-alumino-phosphates (SAPOs) are effective for the conversion of light olefins to gasoline, catalytic reforming of naphthenes, and for the isomerisation of C_8 aromatics[19]. However there are many questions regarding the mechanism of Si incorporation and their distribution in the framework. For example, it is known that Si can aggregate forming Si islands at certain concentrations, although such behaviour differs in different SAPO structures[20]. We have recently used simulation methods to investigate these effects and now present results pertaining to SAPO-5[21].

Silicon can be incorporated into an AlPO *via* two substitution mechanisms:

	AlPO framework		SAPO framework
SM2:	$[AlPO_4] + Si^{4+} + H^+$	$\longrightarrow$	$[Si,H]_P + P^{5+}$
SM3:	$[AlPO_4] + 2Si^{4+}$	$\longrightarrow$	$[2Si]_{Al,P} + P^{5+} + Al^{3+}$

The notation used shows the species present in the final (in square brackets) and the initial structures (indicated by subscripts). We note that these formal processes are components of a Born-Haber cycle: other reactions are needed as sources of "Si^{4+}" and sinks of "P^{5+}" and "Al^{3+}". The structure of the species is illustrated schematically in Figure 2.

We initially investigated the structural deformations resulting from Si substitution by the mechanism of Brønsted site formation (SM2). These deformations were found only to affect the structure as far as the first neighbouring T site from the substitutent, suggesting that long range effects are small and that there may be little difference in Si substitution energy *via* this mechanism in any other AlPO structure. However, the structural changes at the site where the substitution takes place are quite significant, the resulting bond lengths and angles being summarised below:

	T-O(Å)	O-Al(Å)	TOAl(°)
$AlPO_4$-5 (T=P)	1.52	1.72	149.0
SAPO-5 (T=Si)	1.78	1.82	136.9

We then consider the interactions of two Brønsted sites in SAPO-5. Two different conformations were modelled with a varying topological distance between the two acid sites. We find that a configuration of the type -Si-Al-P-Al-Si- has an energy of $\approx 0.17eV$ lower than that in which two Si substituents are infinitely separated, hence revealing a tendency for the Si to cluster. These findings are in agreement with ^{29}Si NMR measurements[22] in which, at low Si content, a single peak corresponding to Si(4Al)(12P) is observed. As the Si content increases, signals corresponding to Si with some Si in the second or first T environment start to appear. It is also observed that when the Si content in SAPO-5 is >10% of the T atoms, signals corresponding to Si(0Al) appear[23], corresponding to the presence of Si islands[23]; we have therefore performed calculations on such structures.

Many possible Si islands can be simulated in any given SAPO structure depending on the amount of Si atoms present and the topology of its distribution. It appears that even when the silicon content is not very high in SAPO-5, small Si islands are formed. We have therefore determined the stability of islands containing 4,5, and 8 Si atoms. The stability of these islands has been tested in terms of their relative energy with respect to smaller (more dispersed) aggregates. The calculations gave the following results:

	Process	Energy (eV)
4Si island	$[4Si]_{2Al,2P} \rightarrow 2[2Si]_{Al,P}$	+0.68
5Si island	$[5Si,3H]_{Al,4P} \rightarrow [2Si]_{Al,P} + 3[Si,H]_P$	+0.46
8Si island	$[8Si,4H]_{2Al,6P} \rightarrow [4Si]_{2Al,2P} + 4[Si,H]_P$	+1.23

The conformations for all the aggregates showed in the equations are again schematised in Figure 2.

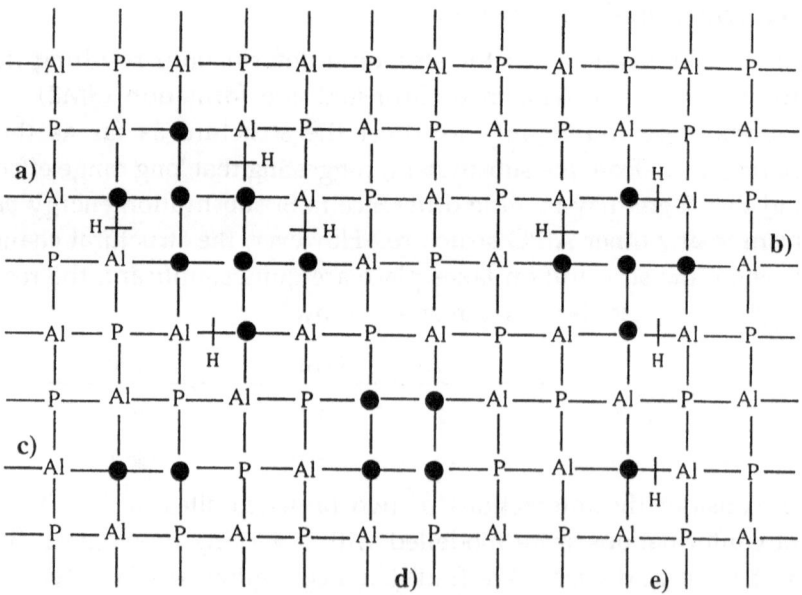

Figure 2. Different Si aggregates produced in SAPO-5 ($\bullet \equiv$ Si). a) $[8Si,4H]_{2aL,6P}$; b) $[5Si,3H]_{Al,4P}$; c) $[2Si]_{Al,P}$; d) $[4Si]_{Al,2P}$; e) $[Si,H]_P$.

The 4Si island is generated by two substitutions described in SM2, while the 5Si and 8Si islands are obtained by means of a combination of SM2 and SM3 substitution. We find that a 4Si island is more stable than two 2Si islands, which can be rationalised in terms of the number of [Si-O-P] linkages present in each of the conformations. Figure 2 shows 4[Si-O-P] linkages in the 4Si islands and 6[Si-O-P] linkages in the two 2Si islands. The instability created by these linkages is in part due to the high electrostatic repulsion generated by the proximity of tetra and pentavalent cations. The absence of [Si-O-P] linkages in SAPO structures has been experimentally established[25] and these calculations provide an energetic rationalisation of this effect. The 5Si island is also found to be more stable than a 2Si island and three Brønsted acid sites. Again, the result can be explained in terms of the number of [Si-O-P] linkages. The calculation on the 8Si island confirms not only the higher stability of an aggregate without [Si-O-P] linkages but also its higher stability with respect to a 4Si island. We can therefore conclude that Si island formation via a combination of SM2 and SM3 processes is the likely mechanism of substitution, a conclusion which is, once more, supported by experimental evidence[24].

2.1.3 Identification of the active site in the NO decomposition catalyst Cu-ZSM-5

The decomposition of NO by Cu-ZSM-5 has stimulated considerable interest in cation exchanged zeolite systems particularly with respect to their potential as vehicle exhaust catalysts[26] a focal point of which is to identify the active site which effects the catalysis.

To identify the active site in Cu-ZSM-5 we have constructed model systems based on the active catalyst preparation of Iwamoto et al.[27] with 96 silicon T-sites per unit cell. The experimental Si/Al ratio of 12 was effected by replacing, at random, 8 of the 96 silicon atoms with aluminium and the 150% copper loading achieved by introducing 4 Cu^+ and 4Cu^{2+} ions at random extra-framework sites within the host zeolite with the excess charge compensated for by 4OH^- species. A simple computer program was written to construct 2000 "trial structures" where the aluminium, copper and hydroxyl species were introduced at various random positions and the energy of each system calculated, with a full energy minimisation applied to the 20 systems with the lowest unrelaxed energy. The procedure should generate all low energy extra-framework copper configurations.

Analysis of the resulting copper clusters from the 20 fully relaxed systems revealed that 41% of the copper species were paired with the most common cluster being a hydroxyl-bridged Cu(I)-Cu(II) copper pair. The high occurrence of such species suggests this structure to be particularly stable and we suggest therefore that the cluster represents a good model for the active site. Further to refine this particular structure, 1000 [Cu(I)-OH-Cu(II)] species were introduced at random into ZSM-5 with a full energy minimisation performed on the 10 lowest energy configurations. The resulting, fully relaxed, lowest energy configuration is illustrated in Figure 3 and Table 2. The figure clearly illustrates both the detailed configuration and location of the cluster within the ZSM-5 zeolite and we suggest is a useful model of the active site within the Cu-ZSM-5 NO decomposition catalyst. We also note the cluster is strongly bound to the host zeolite *via* a framework aluminium.

Figure 3. Illustration of the refined active site structure and location in the Cu-ZSM-5 NO decomposition catalyst. Silicon is represented by light sticks, lattice oxygens by dark sticks; Cu(II), Cu(I), the small and large balls respectively, are bridged by OH (hydrogrn is white).

TABLE 2: Bond distances and coordination number (CN) of the active site within the Cu-ZSM-5 system

	Cu-O	Cu-OH	Cu-Cu	Cu-Al	CN
	(Å)	(Å)	(Å)	(Å)	
Cu(II)	2.00, 2.20, 2.29	1.69	3.10	1.89	4
Cu(I)	2.05, 2.10	1.91		3.24	3

2.2 MODELLING OF SURFACES: α-Fe$_2$O$_3$

To understand how a catalyst functions at the molecular level, we need to understand the interface between the catalyst and reactants, intermediates and products. This requires a detailed knowledge of the surface structure of and products. This requires a detailed knowledge of the surface structure of the catalyst. Here we focus on the modelling of hematite (α-Fe$_2$O$_3$), an oxidation catalyst. Its structure is the same as that of Corundum (α-Al$_2$O$_3$), but the morphology and chemical properties are quite different.

The corundum structure can be described as a hexagonal close packed oxygen anion array with cations in the octahedral sites. In the case of Fe$_2$O$_3$ and Al$_2$O$_3$ the cations are too large to fit into the octahedral cavities without causing distortion.

The surfaces of hematite are usually covered with a layer of Fe$_3$O$_4$, a spinel structure. This certainly affects the chemical properties, but also introduces added complexity for the modeller, and thus the discussion of modelling this new surface will be left for a future publication. This present discussion is concerned with modelling the surface structure and crystal morphology of a purely corundum structured material.

Two different programs were used to develop the models of the surfaces: first GULP[28] (a lattice energy simulation code) to generate the fully equilibrated bulk structure. It is essential that the bulk structure be relaxed, otherwise residual strains will distort the surface structures. The relaxed bulk structure is used as input for MARVIN[29] which models the surfaces. The key features of MARVIN are that the energy is calculated as a 2D infinite lattice summation, with the electrostatic contributions being computed using a 2D Ewald sum. This approach is more general than embedding 2D slabs in a 3D array. We note that MARVIN has been used successfully to model a wide variety of crystal types, including Al$_2$O$_3$ and Zircon[29], molecular/ionic crystals like BaSO$_4$[30, 31] and molecular crystals such as urea[32].

The surface structures pictured in Figure 4 show two of the most stable faces. Their relative stability was determined by the calculated relaxed surface energy and attachment energy. The cleavage plane was selected so that there was no bulk dipole. Figure 4 shows the availability of the different surface sites on the face. The detailed atomistic structure is of great value for understanding catalysis and adsorption processes.

The observed morphologies of hematite (a hexagonal structure), are rhombohedral crystals with the basal plane (0001) and the (10$\bar{1}$1) face being the dominant faces, although the growth conditions can have considerable effects on the final crystal shape. Predicting the morphology can be quite complex since real crystals are seldom of the same quality as theoretical ones. The prediction of crystal morphology is an essential starting point for surface modelling. The morphology provides a good technique for selecting which crystal faces are going to be important for the problems of interest, which could include adsorption, growth modification, defect segregation and catalysis.

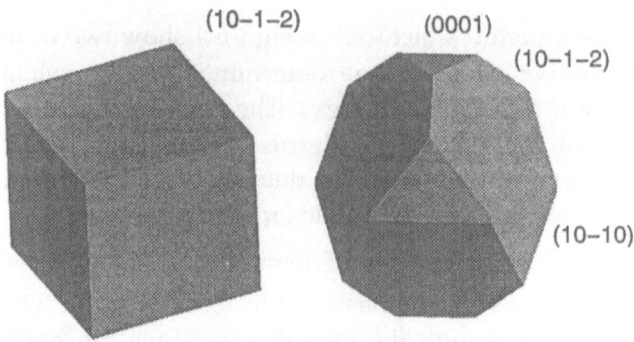

Figure 4. Surface structure of (0001)(a) and the (1012)(b) faces of hematite.

(10–1–2) (0001)

(10–1–2)

(10–10)

Figure 5. Predicted growth and equilibrium morphologies of hematite.

Here we present two predicted morphologies (see Figure 5). The equilibrium morphology is the that which gives the lowest energy for a given volume, and the crystal face with the lowest surface energy will be the dominant face exposed. The growth morphology is based on the attachment energy, which assumes kinetic control of the crystal growth. A crystal will grow in the direction of the face with the highest (most negative) attachment energy which will thereby create the adjacent faces. The morphology will be dominated by the slowest growing face, which will have the lowest (least negative) attachment energy.

TABLE 3:   Surface and attachment energies for α-Fe_2O_3, before and after energy minimisation

Miller plane	Surface Energy (J/m^2)		Attachment Energy (eV)	
	before relaxation	after relaxation	before relaxation	after relaxation
(0001)	5.4	0.8	-107.31	-127.31
$(10\bar{1}4)$	6.0	1.0	-88.24	-107.28
$(10\bar{1}\bar{2})$	2.0	0.5	-20.24	-21.92
$(10\bar{1}\bar{1})$	6.4	1.9	-221.94	-240.70
$(10\bar{1}1)$	4.2	1.2	-96.03	-64.47
$(2\bar{1}\bar{1}0)$	3.0	0.8	-44.09	-40.76
$(10\bar{1}0)$	4.9	0.6	-185.60	-160.77
$(10\bar{1}2)$	6.4	1.1	-256.55	-362.69

From Table 3, we can see that energy minimisation (relaxation) plays an important rôle in the properties used to predict morphologies. Due to the way that surface energies are computed, which is as the difference between the energy of the simulation cell at the surface and the bulk, the surface energy will always decrease during energy minimisation. The same is not true of attachment energies. During minimisation, energy can be exchanged between the growth slice and interplanar interactions, yielding an attachment energy that can either increase or decrease. These two approaches will in general give different predicted morphologies.

In the case of hematite, we get two very different morphologies as shown in Figure 5. Experimentally the (0001) and $(10\bar{1}1)$ faces dominate. From the calculated surface energies, we get the following ordering of morphological importance:

$$(10\bar{1}\bar{2}) > (10\bar{1}0) > (0001) > (2\bar{1}\bar{1}0) > (10\bar{1}4),$$

whilst using attachment energies, the ordering is:

$$(10\bar{1}\bar{2}) > (2\bar{1}\bar{1}0) > (10\bar{1}1) > (10\bar{1}4) > (0001).$$

Neither of these explain the observed crystal morphology. However, as mentioned previously, with hematite the surface is usually covered with layers of Fe_3O_4 (a spinel). Similarly, many metal oxide surfaces are covered with a hydroxide layer, which can change the relative morphological importance of the crystal faces. It would be of interest to investigate whether different morphologies closer to those predicted would be observed under growth conditions under which such surface layers are not formed. Future calculations will also test whether hydroxylated or non–stoichiometric surfaces generate crystal morphologies that are closer to those that are observed.

TABLE 4: Inter-ionic layer (Angstrom) spacing of the α-Fe$_2$O$_3$ (0001) face.

| | Clean Surface | | Hydroxylated Surface | | |
	Unrelaxed	Relaxed	Unrelaxed	Relaxed	
Fe - O	0.838	0.042	0.825	0.759	H - O
O - Fe	0.838	0.986	0.886	1.074	O - Fe
Fe - Fe	0.584	0.282	0.595	0.441	Fe - Fe
Fe - O	0.838	1.046	0.833	0.868	Fe - O
O - Fe	0.838	0.906	0.833	0.796	O - Fe
Fe - Fe	0.584	0.550	0.595	0.653	Fe - Fe
Fe - O	0.838	0.859	0.833	0.797	Fe - O
O - Fe	0.838	0.817	0.833	0.828	O - Fe

During the minimisation of the basal plane, the surface layers collapse (see Table 4), an effect previously noticed with the corundum structure. For corundum, the effect has been repeated with both classical potential models and quantum mechanics calculations. Recent work[33], has shown that a hydroxyl layer reduces this collapse and the resultant surface is more stable than the bare surface. This is also the case for hematite and we estimate the heat of reaction for water removal to be in the range of 5-8eV. The structure of the relaxed surface (shown in Figure 6) demonstates a key feature of a hydroxyl termination of a surface, which is the complete covering of the metal ions. The presence of such hydroxyl groups will certainly have a considerable effect on the catalytic properties of the surface.

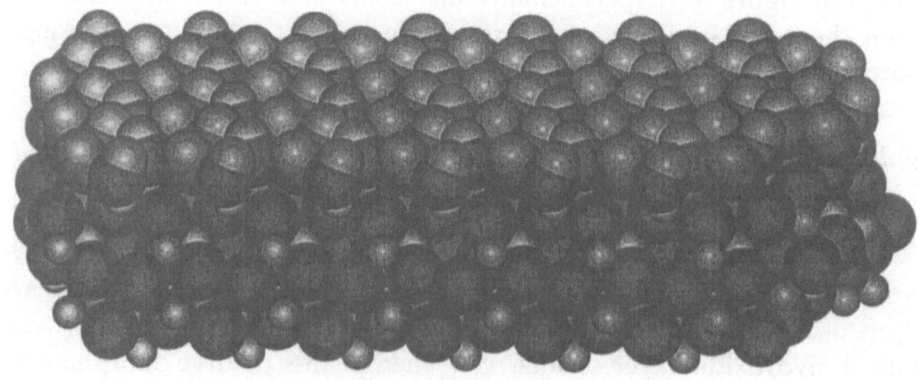

Figure 6: Hydroxylated (0001) surface of hematite.

The calculations described in this section clearly show the critical rôle of the surface chemistry in influencing the morphology of oxide materials.

2.3   MODELLING OF SORPTION

Studying the locations and properties of organic adsorbates, such as catalytic reactants and template molecules, within the internal pores of zeolites poses a number of problems for the experimental scientist. Disorder, both static and dynamic, of the molecules is often encountered. Moreover the lighter elements of which organic molecules are composed, compared to aluminosilicate frameworks, make their detection by x-ray diffraction methods difficult although neutron diffraction techniques have succeeded in several systems in identifying the positions of sorbed molecules.

Probing the sorption sites of organic molecules using molecular modelling methods has provided an increasingly important adjunct to experimental methods. Both molecular dynamics (MD) and Monte Carlo methodologies, often in combination with energy minimisation, have been used successfully in this regard. In particular the automated Monte Carlo docking procedure of Freeman et al. [34], which blends all three methods, has proved especially suitable for locating minimum energy sites in zeolites channel systems. In the procedure, MD is first applied to the guest molecule, as though it were in the gas phase, in order to generate a library of conformations. Each of these conformations is then "docked" into the zeolite framework using a Monte Carlo algorithm. The most favourable site/conformation combinations may subsequently be refined by energy minimisation to yield the overall minimum energy configuration.

Freeman and co-authors[34] originally deployed their automated docking procedure to study the sorption of isomers of butene in silicalite. Other key applications have involved butene in DAF-1[35], butanols in ZSM-5[36] and the modelling of template molecules in a range of zeolites[37,38,39]. Kaszkur and co-workers[40] have also demonstrated how molecular dynamics can be combined with powder x-ray diffraction, in a study of 1,4-dibromobutane in zeolite Y. In this work MD trajectories were obtained for the guest molecules and used to interpret synchrotron powder diffraction data recorded at two different temperatures. The simulations revealed that, at low temperatures, the dibromobutane is sorbed at a position spanning two of the framework supercages, with each Br atom located close to a Na^+ cation. At higher temperatures, one terminal Br remains sorbed while the other exhibits considerable thermal motion.

Recent work by Kaszkur and colleagues[41] used the Monte Carlo docking method to determine the position of *para*-xylene in ferrierite, a medium-pore zeolite which has found wide application as an acid catalyst. Rietveld refinement of high-resolution powder diffraction data, both x-ray and neutron, was complicated by considerable displacements to the framework atoms, observed after the *p*-xylene had been loaded into the sample. This suggested a violation of the *Immm* symmetry of the bare framework. Nevertheless it was possible to obtain a fit within this space group, albeit with large anisotropic temperature factors, the possibility of using lower symmetry models having been discounted due to the excessive numbers of independent parameters. In the light of these problems with symmetry, the accuracy of the *p*-xylene position, as determined from Fourier maps, might justifiably have been regarded with some reservation. Monte Carlo docking calculations were then carried out to attempt to confirm the location of the sorbate. The simulations used a 1 x 1 x 3 supercell of the ferrierite framework, in order to exclude unrealistic molecule-molecule interactions, and involved full energy minimisation of both framework and guest molecule.

The predicted minimum energy location of the molecule was in remarkable agreement with that obtained experimentally (see Figure 7) thus vindicating the approach taken in refining the diffraction data.

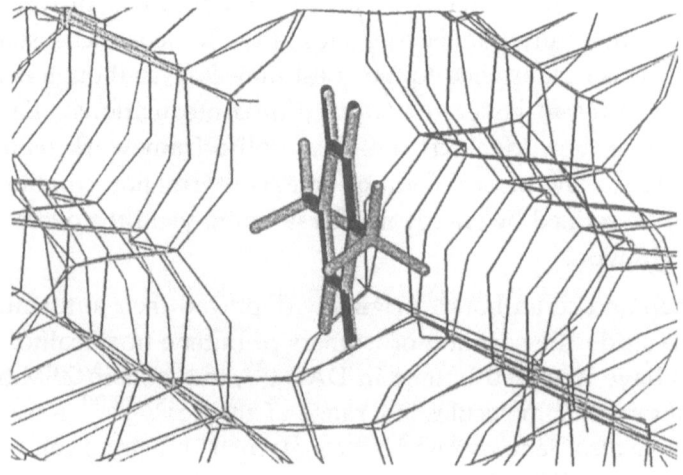

Figure 7: Comparison between the experimental (black) and calculated (grey) sites for *p*–xylene in ferrierite. Methyl hydrogens are omitted from the experimental structure, since their positions are not uniquely defined.

2.4 SIMULATION AND SYNTHESIS: THE COMPETITIVE FORMATION OF CoAlPO$_4$-5 AND CoAlPO$_4$-34 AND THE ROLE OF THE TEMPLATE

Recent work, both computational[35,37,42,43,44] and experimental[45,46] has considered the role of templates in the formation of microprous materials. Much work has concentrated on elucidating how the shape of the template influences the product formed[42,43,45]. Our recent work[47] has expanded on these approaches by considering also the effects of the relative concentrations of both the metal cation (being introduced into the framework) and the template, and any co-operative effect they have, in forming particular phases.

It is well documented that a cobalt-aluminophosphate gel containing triethylamine as a template can form either Co-AlPO$_4$-5 (AFI) or the chabazite (CHA) phase CoAlPO$_4$-34[48-50]. The CoAlPO$_4$-5 phase is typically formed at higher temperatures and low Co concentrations, whilst high Co and template concentrations and lower temperatures favour the CoAlPO$_4$-34 structure. Furthermore, the CHA phase appears as an intermediate phase in certain gel compositions. We have therefore investigated how the template concentration and structure in these two materials can be influenced by the presence of heteroatoms in the gel and resulting crystalline material.

TABLE 5: Binding energies of triethylamine in the CHA and AFI structures

	Template concentration (per unit cell)	T site / Template	E_{inter} / template (kJ mol^{-1})	
			No charges	With Charges
CHA				
triethylamine	1	12	-47.3	-71.6
triethylamine	2	6	-61.1	-83.7
AFI				
triethylamine	0.25	48	-32.7	-62.7
triethylamine	0.5	24	-35.4	-64.1
triethylamine	1.0	12	-37.0	-66.8

$E_{inter} = E_{host} - E_{free}$, where E_{host} is the total energy of the framework/template combination, E_{free} is the energy of an isolated gas phase template molecule. The results are normalised to give a E_{inter} per template. Similar trends are observed for triethylammonium. Note that these calculations are performed in siliceous analogues of the CoAlPO$_4$ systems.

By performing calculations at different template concentrations we show how it is the combination of metal uptake, template-metal interaction and high template concentration which makes the formation of the CoAlPO$_4$-34 structure viable. We see (Table 5) that the interaction energy of the template with the framework does not increase markedly with increasing template concentration in the case of the AFI (CoAlPO$_4$-5) structure. We therefore expect that CoAlPO$_4$-5 formation is relatively insensitive to template concentration. However in CHA, there is a dramatic increase in the interaction energy with template concentration. We therefore propose, that high triethylamine concentrations are essential to stabilise the CHA phase. Consequently, assuming a direct correlation, under synthesis conditions, between the cationic triethylammonium template and the Co content (needed for electroneutrality of the bulk structure), a Co concentration equivalent to a bulk composition of $Co_{0.25}Al_{0.75}PO_4$ is required. We note that CoAlPO$_4$-34 usually forms with such a high Co concentration[48–50]. Furthermore, we note that AlPO-34 has not been reported, further supporting our hypothesis that interaction with anionic framework fragments are required to modify the gel in such a way as to make the formation of the CHA structure viable. Similar arguments limit the Co concentration in CoAlPO$_4$-5 to a composition of $Co_{0.17}Al_{0.83}PO_4$. Again, we note that CoAlPO$_4$-5 forms with Co concentrations of typically <0.1 mol fraction. The inability to attain high purity CoAlPO$_4$-5 with high Co concentrations has been attributed to the formation of the CoAlPO$_4$-34 phase[48], a conclusion supported by these calculations. We would expect that, even at lower Co and template concentrations, the CHA phase can form, due to local concentration effects, but only as a transient and intermediate phase, thus disrupting the formation of phase pure CoAlPO$_4$-5 with an uniform Co concentration; a similar conclusion is drawn from diffuse reflectance spectroscopy[49]. Although the Co-template interactions have not been modelled directly, these calculations allow us to understand their role in the competitive formation of the two microporous structures. These interactions compel the templates to be present in concentrations appropriate for the formation of the CHA framework; without them the template packing is less dense resulting in the AFI structure. We consider that this effect is a general one as similar competitive behaviour is also noted with for example, SAPO-5/SAPO-34, SAPO-5/SAPO-44, CoAlPO$_4$-5/CoAlPO$_4$-35 and EU-1/EU-2, which we believe can all be explained by the interaction of framework cations with the template.

2.5   REACTION MECHANISMS

2.5.1   CH-bond activation on catalytic oxides

Oxidative coupling of methane (OCM) is known to be catalyzed by a variety of oxidic materials[51]. Among these, Li/MgO (lithium doped magnesium oxide) may serve as the prototype. It has attracted a considerable number of investigations, both experimental[51,52] and theoretical[53–59] over the past decades. But although these studies created a large amount of data, many questions concerning the detailed nature of the physical and chemical processes involved are still to be clarified.

The catalytic activation of the process leading from methane to methyl radicals ($\cdot CH_3$, observed experimentally[60,61]) probably involves some kind of surface defect site. Evidence from e.p.r. measurements indicates a correlation between catalytic activity and the presence of O^- surface species[62,63]. In particular, the [Li]^0 centre has been proposed as an active site for OCM catalysis on Li/MgO[52]. The importance of this type of defect, consisting of an electron hole localized on an oxygen ion and stabilized by an adjacent Li (dopant) ion, has been demonstrated[64], but has also recently been questioned by the work of Lunsford and coworkers[52]. In several theoretical approaches [Li]^0 has served as a model of the active surface site[53-59]. Other defects known to be present on MgO surfaces (like dislocations and charged or neutral vacancies) may also be considered (see *eg.* references 58, 65, 66, 67).

Theoretical modeling of reactions at defective MgO surfaces has been undertaken by several workers employing different methods (ranging from EHT to MCSCF)[53-59,65,66,67]. More than 20 years ago, *ab initio* calculations were used to investigate H_2 bond breaking at a cation vacancy site[65] and computational studies of the role of MgO surface defects in catalysis have continued to be performed[67,68,69].

The main challenge lies in choosing an appropriate description of both (a) the surface defect and (b) the reaction. While (a) can be achieved through atomistic modeling (see *e.g.* reference 70), (b) requires a quantum mechanical approach. As will be shown in the present work, however, the level of theory can affect results even qualitatively.

Since the longe–range nature of the Madelung field demands an accurate treatment, to satisfy (a) one has to go beyond the simplest (*i.e.* most localized) approximation. This can be done by using large cluster models (allowing substrate relaxation), embedding of some kind or a (pseudo) two-dimensional representation of the system (slab). Ideally a model should fulfill both (a) and (b) as well as possible, but unfortunately the prohibitively high computational cost involved still restricts the degree to which this goal may be reached. The present work is primarily concerned with (b); the transfer to larger, more realistic models (including surface relaxation) is reported elsewhere[71].

The process here envisaged, $CH_4 + [Li^+O^-]_s \rightarrow \cdot CH_3 + [Li^+(OH)^-]_s$ (index s indicates surface species) involves the homolytic breaking of a CH bond; in the same process an OH bond is formed. We note that open shell systems are present in both reactants and products. Therefore, an electronic structure calculation (and subsequent geometry optimization) ought to be carried out at a sufficiently high level of theory. Starting from Hartree Fock (HF) SCF level this also includes taking into account electron correlation to some extent. Although computationally demanding, a way to do this systematically is by applying post-HF schemes (like MP2, MP4 etc.). Alternatively, and at much lower cost, one can use an approach based upon gradient corrected density functional theory (DFT). Recent improvements in the gradient corrections applied (in this work we use the B-LYP correction[72,73]) have rendered this method very well capable of determining geometries and energies with a high accuracy[74,75]. It seems therefore to be ideally suited for the requirements outlined above.

In previous studies[54,56] very simple models have been used to describe the interaction of methane with a [Li]^0 centre. An efficient, though still crude way of improving the quality of the model is by using a cluster embedded in an array of point charges. In our model (Figure 8) only lithium, oxygen and methane are treated explicitly; the MgO crystal is represented by about 200 point charges (q=±2.0 e) at their bulk positions. The CH and OH distances are varied and for each combination of d(CH) and d(OH) the remaining parameters are optimized. This is important for two reasons: (1) the oxygen ion accepting the hydrogen moves during this transfer proces by several tenths of an Ångstrøm and (2) the methyl radical generated in the process tends towards planarization gaining an energy of about 0.3 eV.

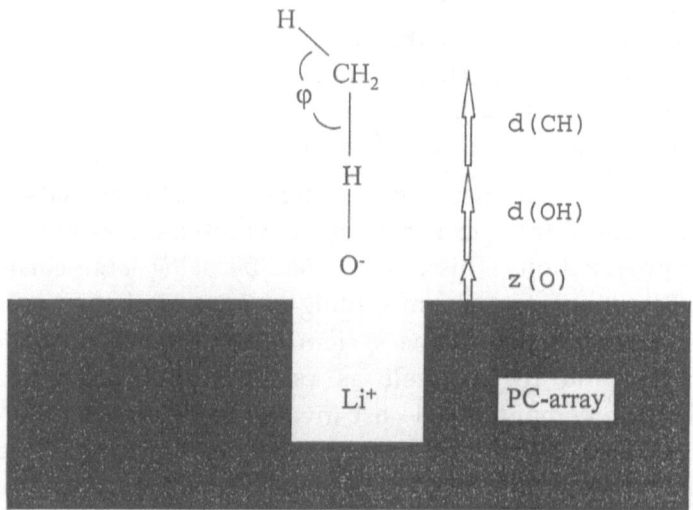

Figure 8: The [Li$^+$O$^-$]-H-CH$_3$ model system. Its geometry is determined by the bond distances d(CH) and d(OH), the position of oxygen relative to the surface z(O) and the HCH angle. Li is kept fixed at 2.11 Å below the surface.

The resulting energy surfaces (Figure 9) display several important features and some significant differences. From HF-calculations (Figure 9a) we find the classic shape of reactants valley (R), transition state (TS) and product valley (P). In contrast to the results of Zicovich-Wilson, however, this is not a late TS[54]; the saddle point is found in a rather symmetric position (at about 0.8 eV above the reactants level). This may be caused by the fact that the planarization accompanying the release of ·CH$_3$ is only very crudely taken into account in reference 54; Their model differs from the one used in this study in the position the dopant atom assumes, but this is not likely to be the cause of such a pronounced, qualitative effect; furthermore, in omitting the ionic crystal field entirely, their description differs from the present one considerably.

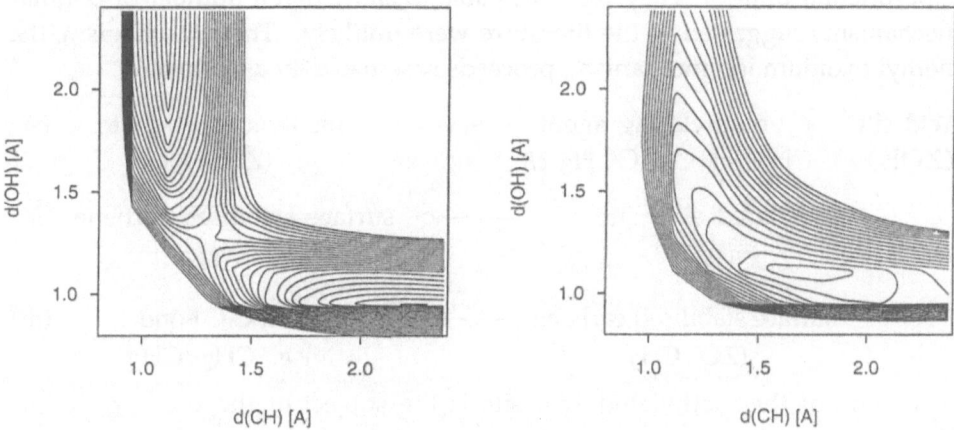

Figure 9: The total energy of [Li$^+$O$^-$]-H-CH$_3$ as a function of the CH and OH distance calculated using (a) HF and (b) B-LYP-DFT. The contour lines are spaced at 54 meV.

The surface resulting from B-LYP-DFT calculations is depicted in Figure 9b. Here no TS is observed. In fact, the transition of hydrogen from carbon to oxygen occurs without activation barrier. However, some energy is needed for the release of the methyl radical fragment. The minimum found at a CH distance of 1.7 Å lies 0.43 eV below the product level (this value is reduced by a factor of two, when substrate relaxation is included[71]). Another difference seen in comparing Figure 9a and 9b is a slightly shorter OH distance at the large d(CH) observed in the HF results. This may be a compensation effect for a slightly smaller z(O) in the DFT calculation, a (very moderate) overbinding effect.

The results presented here clearly demonstrate, that a complete neglect of electron correlation (pure HF) leads to a qualitatively different description of the methane dissociation process than is achived by using an advanced (gradient corrected) DFT scheme. Thus, in further studies employing large clusters it will be essential to use methods that include electron correlation. Considering the computational cost involved, DFT appears to be ideally suited for this challenge.

2.5.2 Reactivity of methanol at alumino-silicate Brønsted acid site

In response to developments of fast computer technology and efficient computer algorithms, computational chemists are now able to probe the very complex nature of organic reactivity at heterogeneous surfaces. Of particular interest has been the prototypical methanol-to-gasoline (MTG) reaction[76] which, even after 20 years of experimental study (and more recently theoretical study), is still poorly understood.

Some years ago, Hutchings and Hunter suggested a mechanism for the MTG process in acidic zeolites that satisfied all of the accumulated experimental data[77]. They were also able to show that a number of popular mechanisms suggested in the literature were unlikely. Their mechanism, the methyl oxonium ion mechanism, proceeds schematically as follows:

Acid site + methylating agent ---------> methylated acid site (i)
(ZOH) CH_3OH, CH_3OCH_3 *etc.*   (ZOCH3)

methylated acid site ---------> surface stabilised carbene (ii)
(ZOCH3)   (ZO⁻..CH2)

surface stabilised carbene ---------> first C-C bond (iii)
(ZO⁻..CH2)   *e.g.*, $CH_2=CH_2$

Formation of the methylated acid site is the subject of the present section. The ongoing debate concerning the nature of the initial complex formed on methanol adsorption at an alumino-silicate Brønsted acid site[78–91] will not be discussed except to state that our current work suggests that at normal temperatures, the dominant adsorbed methanol species in acidic zeolites is a physisorbed species[91]. All reactions reported in the current work involving adsorbed methanol will be with respect to this physisorbed species.

A number of pathways for formation of a methylated acid site are conceivable. The simplest, a S_N2 reaction of a single CH_3OH molecule at a Brønsted acid site, [AlO(H)Si], has been studied by Blaskowskii and van Santen[85] and by Zicovich-Wilson *et al.*[92]. Due to different methodologies, the former reported an activation barrier of 184 kJmol^{-1} whilst the latter reported a barrier of 217 or 171 kJmol^{-1} depending on which oxygen site within their model cluster was methylated.

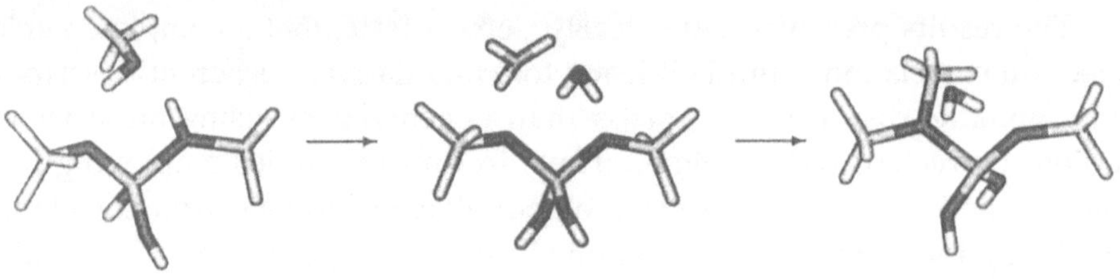

Figure 10: TZVP//DFT(BLYP) optimised structures for the methylation of a model Brönsted acid site by one methanol molecule. The acid site is modelled by a $H_3SiOAl(OH)_2O(H)SiH_3$ cluster and the central structure is the transition state.

Another possibility for formation of the surface methoxyl species is that shown in Figure 10, *i.e.*, a pathway involving a single CH_3OH molecule but with more S_N1 character[91]. As expected on the basis of cluster calculations which omit long range electrostatic effects, no stable, charge separated S_N1 intermediate was observed. The calculated activation barrier for this process of around 230 kJmol^{-1} at the TZVP/DFT(BLYP) level of theory indicates that this pathway is only likely to play a minor rôle in formation of the methylated acid site. We note however, that due to the polar nature of the mechanism, inclusion of long range electrostatic effects into the model is likely to lower the activation barrier.

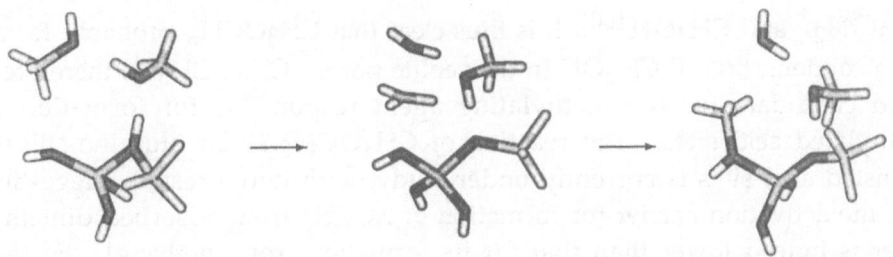

Figure 11: HF/3-21G optimised structures for the methylation of a model Brönsted acid site by two methanol molecules. The acid site is modelled by a (HO)$_2$AlO(H)SiH$_3$ cluster and the central structure is the transition state.

A third possibility for the mechanism of reaction (i) is *via* interaction of two methanol molecules with the acidic Brönsted site, Figure 11. We have been able to show that such a pathway proceeds with an activation barrier in the region of 130-160 kJmol^{-1} at the MP2/6-31G**//HF/3-21G level of theory and using a range of model clusters[93]. The use of different model clusters also enabled us to demonstrate the significant effect of the difference in proton affinities of the two oxygen sites involved on the magnitude of the barrier. This effect was also discussed by Kramer *et al.*[2] who used it to rationalise the different reactivities of methane in MFI and faujasite type zeolites.

Finally, it is possible that dimethyl ether is the primary methylating agent in the initial stages of the MTG process. It is known that dimethyl ether forms almost immediately upon introduction of CH$_3$OH into acidic zeolites and well before the onset of hydrocarbon formation[94]. However, the literature is unclear as to whether dimethyl ether forms from the reaction of methanol with a surface bound methoxyl group (ZOCH$_3$)[95,96,97] or whether it forms from conventional condensation of methanol in the acidic environment of the zeolite pores[99,100].

HF/6-31G** calculations of the formation of (CH$_3$)$_2$OH from CH$_3$OH and ZOCH$_3$ (Figure 12) gave an activation barrier of 150 kJmol^{-1} (desorption of the product was less activated). Assuming that the ZOCH$_3$ species was formed from adsorbed methanol (as above), the effective activation energy for formation of (CH$_3$)$_2$OH^{+} (and from it, CH$_3$OCH$_3$) from methanol at a Brønsted acid site *via* a surface methoxyl species is estimated to be about 120 kJmol^{-1}, *i.e.*, the activation barrier of 150 kJmol^{-1} plus the energy to form

ZOMe (~-5 kJmol^{-1}) plus the energy of adsorption of CH_3OH at the methylated acid site (~-25 kJmol^{-1}). Estimates of the effect of electron correlation (at the MP2 level) suggest that the barrier should probably be lower at about 100 kJmol^{-1}. This value compares with the gas phase activation barrier of ~35 kJmol^{-1} for formation of $(CH_3)_2OH^+$ and H_2O from $CH_3OH_2^+$ and CH_3OH[100]. It is thus clear that CH_3OCH_3 probably forms from condensation of CH_3OH in the zeolite pores. CH_3OCH_3 is therefore a good candidate for the methylating agent responsible for formation of methylated acid sites. The reaction of CH_3OCH_3 at an alumino-silicate Brønsted acid sites is currently under study, with initial results suggesting that the activation barrier for formation of $ZOCH_3$ from adsorbed dimethyl ether is indeed lower than that for its formation from methanol. Further studies of this problem have been reported recently by van Santen[6].

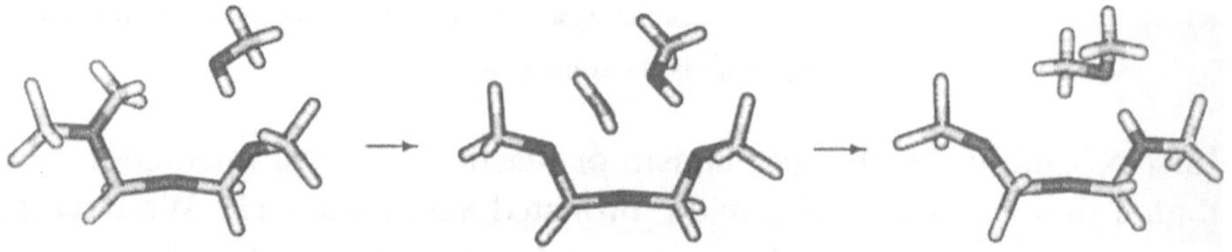

Figure 12: HF/6-31G** optimised structures for formation of dimethyl ether from a methylated acid site and adosrbed methanol. The methylated acid site is modelled by a $H_3SiO(CH_3)Al(H)_2OSi(H)_2OSiH_3$ cluster and the central structure is the transition state.

Although the mechanism of the MTG process is far from being understood, computational quantum chemistry methods are clearly bridging the gap in experimental knowledge.

3.   SUMMARY

The simulation studies described in this paper, together with those summarised in the other theoretical papers in this issue, clearly show the value of computational methods in modelling structure and reactivity in catalytic systems. With the rapid developments taking place in both computer hardware and software, these techniques are poised to play an even greater rôle in the development of a molecular understanding of catalysis.

ACKNOWLEDGEMENT

This chapter is adapted from an article[101] published in a special issue of the *Journal of Molecular Catalysis*, containing the proceedings of the Franqui Colloquium held in honour of Professor E.G. Derouane. We are grateful to Professor Derouane and to the *Journal of Molecular Catalysis* for their permission to use the material in our original paper in this chapter.

REFERENCES

1. C.R.A. Catlow (ed.), (1992) *Modelling of Structures and Reactivity in Zeolites.*, Academic Press, London.

2. Kramer, G.J., van Santen, R.A., Emeis, C.A. and Nowak, A.K., (1993)*Nature*, **363**, 529.

3. Sauer, J., (1994) *Stud.Surf.Sci.Catal.*, **84**, 2039.

4. Catlow, C.R.A.,Bell, R.G. and Gale, J.D., (1994) *J. Mater. Chem.*, **4**(6), 781.

5. Catlow, C.R.A., Bell, R.G., Gale, J.D. and Lewis, D.W., (1995) in L. Bonneviot, S. Kaliaguina, (eds.), *Zeolites: A refined tool for designing catalytic sites*, *Stud.Surf.Sci.Catal.*, **97**, 87, Elsevier, Amsterdam.

6. van Santen, R.A., *J. Molecular Catalysis* —in press

7. Norskov, K., *J. Molecular Catalysis* —in press

8. W.M. Meier and D.H. Olson, (eds.)*Atlas of Zeolite Structure Types*, 3rd Edition, (1992) Butterworth, London.

9. Morris, R. E., Weigel, S. J., Henson, N. J., Bull, L. M., Janicke, M. T., Chmelka, B. F.and Cheetham, A. K., (1994) *J. Am. Chem. Soc.*, **116**, 11849.

10. Bennett, J. M., Cohen, J. P., Flanigen, E.M., Pluth, J. J. and Smith, J. V., (1983) in *Am. Chem. Soc. Symp. Ser.*; American Chemical Society: Washington' Vol. 218; pp 109. See also Qiu, S., Pang, Q. Kessler, H. and Guth, J. L. , (1989)*Zeolites*, **8**, 440.

11. Ruiz-Salvador, A.R., Sastre, G., Lewis, D. W. and Catlow, C. R. A. *J. Mater. Chem.*, submitted

12. Gale, J. D. and Henson, N. J., (1994), *J. Chem. Soc. Faraday Trans.*, **90**, 3175.

13. Cole, M. Fitch, A. N., Goyal, R., Jones, R. H., Mora, A. J., Jobic, H. and Carr, S. W., (1996) *J. Mater. Chem.*, in press

14. de Man, A. J. M., Jacobs, W. P. J. H., Gilson, J. P. W. and van Santen, R. A., (1992) *Zeolites*, **12**, 826.

15. Henson, N. J., Cheetham, A. K. and Gale, J. D. (1996) *Chem. Mater.*, **8**, 664

16. Shannon, M. D., Casci, J.L., Cox, P.A. and Andrews, S. J. (1991) *Nature*, **353**, 417

17. Wright, P.A., Natarajan, S., Thomas, J.M., Bell, R.G., Gai-Boyes, P.L., Jones, R.H. andChen, J. (1992) *Angew. Chem. Intl. Ed. Engl.*, **31**, 1472.

18 Asensi,M.A.,Corma, A., Martinez, A. (1996) *J. Catal.* **158**, 561.

19 Pujado, P.R., Rabo, J.A., Antos, G.J. and Genbicki, S.A., (1992)*Catal. Today*, **13**, 113.

20. D.Barthomeuf, D., (1993) in J. Fraissard, L. Petrakis, Eds., Acidity and Basicity in Solids. Theory, Assessment and Utility, NATO ASI Series C, Vol. 444, 375.

21. Sastre, G., Lewis, D. W. and Catlow, C. R. A., (1996) *J. Phys. Chem.*, **100**, 6722

22. Briend, M. and Barthomeuf, D., (1993) in R. von Ballmoos, J.B. Higgins, M.M.J. Treacy, Eds., *Proceedings of the 9th International Zeolite Conference*; Butterworth-Heinemann, Boston, 635.

23. Ojo, A.F., Dwyer, J., Dewing, J., O'Malley, P.J. and Nabhan, A. (1992) *J. Chem. Soc. Faraday Trans.* **88**, 105.

24. Barthomeuf, D. (1994) Zeolites , **14**, 394.

25. Flanigen, E.M., Patton, R.L. and Wilson, S.T. (1988) In P.J. Grobet, *et al.* Eds. Innovation in Zeolite Material Science; *Stud. Surf. Sci. Catal.* **37**, 13.

26. Burch, R., (1995) Catalysis Today **26**, 97.

27. Iwamoto, M., Yahiro, H., Mizuno, N., Zhang, W.-X., Mine, Y., Furukawa, H. and Kagawa, S., (1992) *J. Phys. Chem.*, **96**, 9360.

28. Gale, J. D., 1992-1996. Royal Institution of Great Britain and Imperial College, London GULP, General Utility Lattice Programme.

29. Gay, D.H. and Rohl, A.L., (1995), *J.Chem.Soc.Faraday Trans.*,**91**(5), 925.

30. Allan, N.L., Rohl, A.L., Gay, D.H.,Catlow, C.R.A., Davey, R.J. and Mackrodt, W.C., (1993)*Faraday Discuss.*, **95**, 273.

31. Rohl, A.L., Gay, D.H., Davey, R.J., Catlow, C.R.A., (1996) *J. Amer. Chem. Soc.*, **118**, 642-648.

32. George, A.R., Harris, K.D.M., Rohl, A.L. and Gay, D.H., (1995) *J. Mater. Chem.* **5**(1), 133-139.

28

33. Nygren, M.A., *et al*, (1996) to be published.

34. Freeman, C.M., Catlow, C.R.A., Thomas, J.M. and Brode, S., (1991) *Chem. Phys. Lett.*, **186**, 137.

35. Bell, R.G., Lewis, D.W., Voigt, P., Freeman, C.M., Thomas, J.M. and Catlow, C.R.A., (1994) *Stud. Surf. Sci. Catal.*, **84**, 2075.

36. Shubin, A.A., Catlow, C.R.A., Thomas, J.M. and Zamaraev, K.I., (1994) *Proc. Roy. Soc. Lond. A*, **446**, 411.

37. Lewis, D.W., Freeman, C.M. and Catlow, C.R.A., (1995) *J. Phys. Chem.*, **99**, 11194.

38. Stevens, A.P. and Cox, P.A., (1995) *J. Chem. Soc. Chem. Commun.*, 343.

39. Weigel, S.J., Gabriel, J.C., Puebla, E.G., Bravo, A.M., Henson, N.J., Bull., L.M. and Cheetham, A.K., (1996) *J. Am. Chem. Soc.*, **118**, 2427.

40. Kaszkur, Z.A., Jones, R.H., Waller, D., Catlow, C.R.A. and Thomas, J.M., (1993) *J.Phys.Chem.*, **97**, 426.

41. Kaszkur, Z.A., Jones, R.H., Bell, R.G., Catlow, C.R.A. and Thomas, J.M., (1996) *Mol.Phys.*, in press.

42. Freeman, C. M., Lewis, D. W., Harris, T.V., Cheetham, A.K., Henson, N.J., Cox, P.A., Gorman, A.M., Levine, S.M., Newsam, J.M., Hernandez, E. and Catlow, C. R. A. (1995) in eds. C.H. Reynolds, M.K. Holloway, H.K. Cox, *Computer–Aided Molecular Design. Applications in Agrochemicals, Materials and Pharmaceuticals* , **326**,(American Chemical Soc, Washington D.C.)

43. Cox, P. A., Stevens, A. P., Banting, L. and Gorman, A. M. (1994) *in* "Zeolites and Related Microporous Materials: State of the Art 1994. Proceedings of the 10th International Zeolite Association Meeting", J. Weitkamp, H. G. Karge, H. Pfeifer, W. Holderich, eds., *Stud. Surf. Sci. Catal.* Vol. 84, p. 2115. Elsevier, Amsterdam.

44. Harris, T. V. and Zones, S. I., (1994) in J. Weitkamp, H. G. Karge, H. Pfeifer, and W. Holderich, eds., "Zeolites and Related Microporous Materials: State of the Art 1994. Proceedings of the 10th International Zeolite Association Meeting", *Stud. Surf. Sci. Catal.* Vol. 84, p. 29. Elsevier, Amsterdam.

45. Gies, H. and Marler, B. (1992) *Zeolites* **12**, 42

46. (a) Burkett, S. L. and Davis, M. E. (1995) *Chem. Mater.* **7**, 1453; (b) Burkett, S. L. and Davis, M. E. (1995)*Chem. Mater.* **7**, 920; (c) Burkett, S. L. and Davis, M. E. (1994) *J. Phys. Chem.* **98**, 4647

47. Lewis, D. W., Catlow, C. R. A. and Thomas, J. M. (1996) *Chem. Mater.*, **8**, 1112.

48. Urbina de Navarro, C., Machado, F., Lopez, M., Maspero, M. and Perez-Pariente, J. (1995) *Zeolites,*, **15**, 157.

49. Uytterhoeven, M. G. and Schoonheydt, R. A. (1994) *Microporous Mater.* **3**, 265.

50. Kraushaar-Czarnetzki, B., Hoogervorst, Q. G. M., Andrea, R. R., Emeis, C. A. and Stork, W. H. J. (1991) *J. Chem. Soc., Faraday Trans.* **87**, 891.

51. Hutchings, G. J., Scurrell, M. S. and Woodhouse, J. R., (1989) *Chem Soc. Rev.*, **18** , 251.

52. Lunsford, J. H. (1995) *Angew. Chem. Int. Ed. Engl.*, **34**, 970.

53. Mehandru, S. P., Anderson, A. B. and Brazdil, J. F. (1988) *J. Am. Chem. Soc.*, **110**, 1715.

54. Zicovich-Wilson, C.M., González-Luque, R. and Viruela-Martín, P. M. (1990) *J. Mol. Sruct. (THEOCHEM)*, **208**, 153.

55. Viruela-Martín, P.M., Viruela-Martín, R., Zicovich-Wilson, C.M. andTomás-Vert, F. (1991) *J. Mol. Catal.*, **64**,, 191.

56. Børve, K. J. and Pettersson, L. G. M. (1991) *J. Phys. Chem.*, **95** , 3214.

57. Børve, K. J. and Pettersson, L. G. M. (1991) *J. Phys. Chem.*, **95** , 7401.

58. Børve, K. J. (1991) *J. Chem. Phys.*, **95** ,4626.

59. Anchell, J. L., Morokuma, K. and Hess, A. C. (1993) *J. Chem. Phys.*, **99**, 6004.

60. Campbell, K. D., Morales, E. and Lunsford, J. H. (1987) *J. Am. Chem. Soc.*, **109** , 7900.

61. Feng, Y. and Gutman, D. (1991)*J. Phys. Chem.*, **95** , 6556; Feng, Y., Niiranen, J. and Gutman, D. *ibid*, 6564.

62. Zhang, H. S., Wang, J. X., Driscoll, D. J. and Lunsford, J. H. (1988) *J. Catal.*, **112**, 366.

63. Lin, C. H., Wang, J. X. and Lunsford, J. H. (1988) *J. Catal.*, **111**, 302.

64. Driscoll, D. J., Martir, W.,Wang, J. X. and Lunsford, J. H. (1985) *J. Am. Chem. Soc.*, **107**, 58.

65. Derouane, E. G., Fripiat, J. G. and André, J. M. (1974) *Chem. Phys. Lett.*, **28**, 445.

66. Pope, S. A., Guest, M. F., Hillier, I. H., Colbourn, E. A., Mackrodt, W. C. andKendrick, J. (1983) *Phys. Rev. B*, **28**, 2191.

67. Kobayashi, H., Salahub, D.R. and Ito, T. (1994) *J. Phys. Chem.*, **98**, 5487.

68. Zhidomirov, G.M., Avdeev, V.I., Zhanpeisov, N.U., Zakharov, I.I. and Yudanov, I.Y. (1995) *Calalysis Today*, **24**, 383.

69. Orlando, R., Corà, F., Millini, R., Perego, G. and Dovesi, R. (1996) *submitted for publication*.

70. Catlow, C.R.A., Jackson, R.A. and Thomas, J.M. (1990) *J. Phys. Chem.*, **94**, 7889.

71. Ackermann, L., Gale, J.D. and Catlow, C.R.A. - to be pblished

72. Becke, A. D. (1988) *J. Chem. Phys.*, **88**, 2547.

73. Lee, C., Yang, W. and Parr, R. G. (1988) *Phys. Rev. B* **37**, 786

74. Ziegler, T. (1991) *Chem. Rev.*, **91**, 651.

75. P. Politzer, J. M. Seminario (Eds.), *Density Functional Theory: A Tool for Chemistry*, Elsevier 1995.

76. S.L.Meisel, S.L., McCullogh, J.P., Lechthaler, C.H. and Weisz, P.B. (1976) *Chemtech*, **6**,86.

77. Hutchings, G.J. and Hunter, R. (1990) *Catalysis Today*, **6**, 279.

78. Vetrivel, R., Catlow, C.R.A. and Colbourn, E.A. (1989) *J. Phys. Chem*, **89**, 4594.

79. Gale, J.D., Catlow, C.R.A. and Cheetham, A.K. (1991) *J. Chem. Soc, Chem. Comm*, 178.

80. Sauer, J., Kölmel, C., Haase, F. and Ahlrichs, R. (1992) *Proc. 9th Int Zeo. Conf*, Montreal, 679.

81. Gale, J.D., Catlow, C.R.A. and Carruthers, J.R. (1993) *Chem.Phys.Lett.*, **216**, 155.

82. Haase, F. and Sauer, J. (1994) *J.Phys.Chem.* **98**, 3083.

83. Bates, S. and Dwyer, J. (1994) *J.Mol.Struct. (Theochem)*, **306**, 57.

84. Haase, F. and Sauer, J. (1995) *J.Am.Chem.Soc.*, **117**, 3780.

85. Blaskowskii, S.R. and van Santen, R.A. (1995) *J.Phys.Chem.*, **99**, 11728.

86. S.P. Greatbanks, Ph.D. thesis University of Manchester (1995)

87. Gale, J.D. (1996)*Topics in Catalysis*, 3(1,2), 169.

88. Shah, R., Payne, M.C., Lee, M.-H. and Gale, J.D. (1996) *Science*, **271**, 1395.

89. Limtrakul, J. (1995) *Chem. Phys.* **193**, 79.

90. Nusterer, E., Blöchl, P.E. and Shwarz, K. (1996) *Angew. Chem. Int. Ed. Engl*, **35**, 175.

91. Sinclair, P.E. and Catlow, C.R.A. — manuscript in preparation.

92. Zicovich-Wilson, C.M., Viruela, P. and Corma, A. (1995) *J.Phys.Chem.*, **99**, 13224.

93. Sinclair, P.E. and Catlow, C.R.A. *J.Chem.Soc., Faraday Trans.*, in press.

94. Changs, C.D. and Silvestri, A.J. (1977) *J. Catal*, **47**, 249.

95. Kubelkova, L.,Novakova, J. and Nedomova, K. (1991) *J. Catal.*, **124**, 441.

96. Bandiera, J. and Naccache, C. (1991) *Appl. Catal.*, **97**, 10732.

97. Bronnimann, C.E. and Maciel, G.E. (1986) *J. Am. Chem. Soc.*, **108**, 7154.

98. Tsiao, C., Corbin, D.R. and Dybowski, C. (1990) *J.Am.Chem.Soc.*, **112**, 7140.

99. Anderson, M.W. and Klinowski, J. (1990) *J.Am.Chem.Soc.*, **112**, 10.

100. Grimsrud, E.P. and Kebarle, P. (1973) *J.Am.Chem.Soc.*, **95**, 7939.

101. Catlow, C.R.A., Ackermann, L., Bell, R.G., Gay, D.H., Holt, S.,Lewis, D.W., Nygren, M.A., Sastre, G., Sayle, D.C. and Sinclair, P.E., (1996), *J. Mol. Catal.*, — in press

APPLICATION OF GREEN-FUNCTION METHOD TO MOLECULAR SYSTEMS

P.W.M.JACOBS,[1] V.A.TELEZHKIN[1,2], A.A.OVODENKO[2]
Centre for Chemical Physics, [1]
The University of Western Ontario, London, Canada

Donetsk Institute of Physics and Technology of the Ukrainian [2]
Academy of Sciences, Donetsk, Ukraine

1 Introduction

Ab initio calculations on molecules and clusters should be carried out with a sufficiently large basis set for the Hartree-Fock limit to be reached. Unfortunately, this is only possible for a limited number of small systems. The use of flexible split-valence basis sets, in which core and valence atomic orbitals are each represented by a few basis functions, or polarization basis sets which, in addition, include functions of higher angular momentum than are occupied in the atomic ground state, do not markedly accelerate convergence to the Hartree-Fock limit. The main reason for such slow convergence is the omission of states that correspond to the continuous spectrum (and which are not considered at all in the standard Hartree-Fock-Roothaan procedure). In the new method a finite set of basis orbitals $\{N\}$ is chosen and eigenvalues ϵ and eigenvectors $|N\rangle$ are found from a one-particle Hamiltonian $\hat{H}$ projected on to this basis set

$$\hat{H}_N = \hat{P}_N \hat{H} \hat{P}_N = \hat{T}_N + \hat{V}_N \tag{1}$$

where $\hat{V}_N$ is the self-consistent potential including Coulomb and exchange interactions and $\hat{P}_N$ is the orthogonal projector on to the subspace $\{N\}$. At the same time, states of the more general Hamiltonian $\hat{H}_1$

$$\hat{H}_1 = \hat{T} + \hat{V}_N \tag{2}$$

31

R. C. Tennyson and A. E. Kiv (eds.),
Computer Modelling of Electronic and Atomic Processes in Solids, 31–42.
© 1997 *Kluwer Academic Publishers.*

may be calculated. If the one-particle spectrum of the Hamiltonian $\hat{H}_N$ is found by application of the Rayleigh-Ritz variational method [1]then the spectrum of $\hat{H}_1$ can be obtained by the degenerate perturbation technique of Lifshits [2]or (what is equivalent) from the Schwinger's or Kohn's variational principles for bound states [3] $\delta\Lambda = 0$ for the functional

$$\Lambda = \langle\psi_N|\hat{V}_N - \hat{V}_N\hat{G}_N^0\hat{V}_N|\psi_N\rangle \tag{3}$$

where $\hat{G}_N^0 = \hat{P}_N\hat{G}^0\hat{P}_N$ is the projection of the free-electron Green function (or resolvent) for the kinetic energy operator $\hat{G}^0 = (\epsilon - \hat{T})^{-1}$. We will refer to this method as the Green-Function (GF) method. Thus, this GF method is analogous to the embedding problem which occurs in the discussion of crystal defects in solid state physics[4], but here the whole Hilbert space plays the role of the host crystal.

A complete exposition of the method, including mathematical details and computational pecularities, such as the solution of the nonlinear eigenvalue problem (3) and the estimation of Green-Function matrix elements over Gaussian functions is given in ref.[5]. Below, we enumerate the main results necessary for the present paper:

a) The discrete spectrum levels (which correspond to bound states) of the Hamiltonian H_N always exceed the corresponding levels of both Hamiltonians $\hat{H}$ and $\hat{H}_1$

$$\epsilon(\hat{H}_N) \geq \epsilon(\hat{H}) \quad , \quad \epsilon(\hat{H}_N) \geq \epsilon(\hat{H}_1)$$

b) The corresponding nonlinear matrix equation for the energy is stable in consequence of $\frac{d(\hat{G}_N^0)^{-1}}{d\epsilon}$ being positive definite. This allows one to build up a precise computational procedure for finding the eigenvalues and eigenvectors of the Hamiltonian $\hat{H}_1$;

c) Although the method gives the possibility of obtaining components of each eigenvector outside the subspace { N } , we will consider only a projected density matrix $\hat{\rho}_N$. Then we can define the "charge" Q_N inside the given basis set $\{N\}$ as

$$Q_N = Tr(\hat{\rho}_N) \tag{4}$$

and consider Q_N as a measure of basis set quality. Notice that Q_N is always less than the number of electrons, in contrast to the Hartree-Fock procedure, because part of the electronic charge is outside $\{N\}$.

In addition to paper [5], we propose to estimate the total energy of a system in the Green-Function method as

$$E_{GF} = E_{HF} + 2\sum_i (\epsilon_i^{GF} - \epsilon_i^{HF}) \tag{5}$$

where the summation is carried out over valence states. It seems that this supposition is reasonable for comparatively good bases, when Q_N converges to the number of electrons. It is possible that similar formulae will be true for methods which take into account electron correlation because the correction (5) concerns only the one-particle kinetic energy operator and does not involve the potential, which can be included in the correlation correction too.

2   Results and Discussion

The basis sets considered in this paper are those containing uncontracted Gaussian functions which were optimized to obtain a minimum total energy for the corresponding atom. The contraction of Gaussians or optimization of individual exponents for some molecules were not used here because the purpose of the present investigation is a comparison of HF and GF methods without obscuring this comparison by further optimization. However, scaling Gaussian exponents of valence state functions and polarization basis functions was carried out in order to satisfy virial theorem conditions.

We will use the notation ns/mp for bases that include n s-type functions and m p-type functions for heavy atoms. The number of Gaussians for hydrogen atoms in hydrogen-containing molecules is chosen equal to m. Exponents for these functions were taken from a handbook [6]. When there were several basis sets in [6] with the same n and m, we chose that which corresponded to the lowest total energy.

For a limited number of molecules, two extended basis sets which include polarization functions have been used. In the first set (which we refer to as $ns/mp^\star$), there are five extra uncontracted d-type Gaussian functions on heavy atoms. The second, more precise set $ns/mp^{\star\star}$, has three p-type Gaussian functions on each hydrogen atom as well. Optimized exponents of polarization functions were taken from [7].

We will compare our HF and GF results with HF limit values which were taken from precise numerical HF calculations for atoms [8, 9] and diatomic molecules [10]. For polyatomic molecules, we will use precise Roothaan- Hartree-Fock calculations [11] for the comparison.

2.1   Atomic systems

Table 1: Comparison of Raleigh-Ritz and Green-Function methods for the Coulomb problem on Gaussian functions. Energies are in atomic units.

n	ref	1s-state			2s-state	
		ϵ_{RR}	ϵ_{GF}	$\epsilon_{2/3}$	ϵ_{RR}	ϵ_{GF}
1	a	-0.424413	-0.519056	-0.487508	-	-
2	b	-0.485813	-0.506900	-0.499871	-	-0.08520
3	c	-0.496979	-0.501852	-0.500228	-	-0.04704
4	b	-0.499278	-0.500323	-0.499975	-	-0.08980
5	b	-0.499810	-0.500096	-0.500001	-	-0.09653
6	d	-0.499946	-0.500023	-0.499997	-0.02474	-0.10687
7	d	-0.499983	-0.500007	-0.499999	-0.05475	-0.11136
8	c	-0.499991	-0.500003	-0.499999	-0.10500	-0.12234
9	d	-0.499998	-0.500001	-0.500000	-0.08731	-0.11775
10	d	-0.499999	-0.500000	-0.500000	-0.02474	-0.10687
exact				-0.500000		-0.12500

[a] C.M.Reeves, J.Chem.Phys., 39 ,1 (1963).
[b] P.G.Mesey, R.E.Kari, I.G.Czizmadia, J.Chem.Phys., 64 ,632 (1976).
[c] S.Huzinaga, J.Chem.Phys., 42, 1293 (1965).
[d] F.B.Van Dujineveldt, IBM Res.J., 16437, 945 (1971).

First, let us consider the one-electron Coulomb problem (or hydrogen atom), where from 1 to 10 s-type Gaussian functions were used for convergence to the exact value of the ground state energy. The results of Table 1 show that the estimated lower bound of the ground state energy from the GF-method gives a better approximation than the standard Raleigh-Ritz method, which estimates an upper bound (despite the use of basis functions optimized especially for that method!). Besides, the GF method describes quite satisfactorily the excited 2s-state(with $n = 8$). Analysis of these ground state results allows us to formulate a semi-empirical correction ("2/3 rule") to the one-particle energies

$$\epsilon = \frac{(2\epsilon_{GF} + \epsilon_{HF})}{3} \qquad (6)$$

Using this correction improves the accuracy of the energy estimate by an order of magnitude. It seems that the correction must have a theoretical foundation which is so far not clear. Coefficients 2/3 and 1/3 could follow from some virial theorem. A similar correction for the total energy can be proposed by the combination of eq.(5) and (6).

Results for the helium and neon atoms are shown in Tables 2, 3, 4. Here, we have a situation analogous to the simple Coulomb problem i.e. bases are optimized and the virial theorem is satisfied. The availability of basis sets with large n makes Ne a favourable test case for studying the convergence of Q and of E (from (6)) as

Table 2: Comparison of Hartree-Fock and Green-Function methods for Helium. Total and one-particle energies are given with respect to their corresponding HF limits [a] , -77.8703 eV and -24.9788 eV.

Basis[b]	Charge	Total energy			1s-state		
		ΔE_{HF}	ΔE_{GF}	$\Delta E_{2/3}$	$\Delta \epsilon_{HF}$	$\Delta \epsilon_{GF}$	$\Delta \epsilon_{2/3}$
2s	1.97429	3.1188	-1.9248	-0.2436	1.6067	-0.9155	-0.0744
3s	1.99062	0.7075	-0.4376	-0.0559	0.3901	-0.1825	0.0084
4s	1.99901	0.1774	-0.0737	0.0100	0.1035	-0.0222	0.0197
5s	1.99953	0.0486	-0.0261	-0.0012	0.0295	-0.0078	0.0046
6s	1.99993	0.0143	-0.0059	0.0008	0.0091	-0.0012	0.0022
7s	1.99997	0.0045	-0.0020	0.0002	0.0029	-0.0004	0.0007
8s	1.99999	0.0015	-0.0006	0.0001	0.0010	-0.0001	0.0003
9s	2.00000	0.0005	-0.0002	0.0000	0.0004	-0.0000	0.0001
10s	2.00000	0.0002	-0.0001	0.0000	0.0001	-0.0000	0.0000

[a] ref.[9]

[b] F.B.Van Dujineveldt, IBM Res.J., 16437, 945 (1971).

well as that of the single particle energies, ϵ. With $n=9$ Q is 9.991 differing by only 0.1% from its proper value. This is a strong indication of the validity of the GF method. E for the same n differs from the value for $14s/9p$ (which must be close to the HF limit) by 0.0022% . Individual results for HF and GF differ from the $14s/9p$ HF value by 0.045% and 0.026% respectively, again showing the more rapid convergence of the GF method and substantiating our claim that eq.(6) improves the estimate to the HF limit by GF by an order of magnitude. The single particle energies illustrate the more rapid convergence of GF than HF with increasing n and show rather effectively how the HF and GF results bracket the limiting value.

GF total energies lie below the HF limit for all atoms considered. This is a very interesting fact because the estimation of lower bounds for energies with sufficient accuracy remains an unsolved problem in quantum mechanics. We propose to investigate this matter of a lower bound in a later publication.

2.2   Diatomic Molecules

The next results are preliminary because the corresponding bases were taken from atomic or ionic calculations, and wave functions do not satisfy the conditions of the virial theorem. Results of HF and GF calculations on the hydrogen molecule are given in Table 5 . Of course, neither method can come close to the experimental

Table 3: Comparison of Hartree-Fock and Green-Function methods for Neon. Total energies are given with respect to the HF limit, -3497.95 eV^a

Basis	Charge	ΔE_{HF}	ΔE_{GF}	$\Delta E_{2/3}$
$4s/2p$	9.586	47.11	-63.83	-26.85
$5s/2p$	9.640	35.05	-40.52	-15.33
$6s/2p$	9.778	30.93	-29.60	-9.42
$5s/3p$	9.772	13.82	-18.85	-7.96
$6s/3p$	9.911	9.60	-7.92	-2.08
$7s/3p$	9.911	7.16	-6.90	-2.21
$8s/3p$	9.919	6.49	-5.02	-1.18
$7s/4p$	9.982	2.42	-2.82	-1.07
$8s/4p$	9.988	1.78	-1.39	-0.33
$9s/4p$	9.991	1.59	-0.90	-0.07

[a] ref. [9]

Table 4: Comparison of Hartree-Fock and Green-Function methods for Neon. One particle energies are given with respect to their corresponding HF limits : $-891.78eV^a(1s)$, $-52.53eV^b(2s)$, $-23.14eV^b(2p)$.

Basis[c]	1s-state		2s-state		2p-state	
	$\Delta\epsilon_{HF}$	$\Delta\epsilon_{GF}$	$\Delta\epsilon_{HF}$	$\Delta\epsilon_{GF}$	$\Delta\epsilon_{HF}$	$\Delta\epsilon_{GF}$
$4s/2p$	5.042	-8.234	3.739	-11.890	6.767	-2.084
$5s/2p$	-0.996	-3.592	3.622	-4.860	6.672	-2.229
$6s/2p$	-1.195	-3.486	1.605	0.294	6.182	-2.705
$5s/3p$	2.027	-0.558	2.969	-5.562	2.321	0.580
$6s/3p$	1.654	-0.623	0.887	-0.422	1.763	0.038
$7s/3p$	0.283	-0.367	0.716	-0.397	1.679	-0.079
$8s/3p$	0.101	-0.068	0.773	-0.367	1.720	-0.005
$7s/4p$	0.661	0.014	0.365	-0.604	0.525	0.190
$8s/4p$	0.272	0.103	0.335	-0.079	0.506	0.171
$9s/4p$	0.188	0.120	0.316	0.142	0.495	0.161

[a] C.F.Bunge, J.A.Barrientos, A.V.Bunge, Atom.Data and Nucl.Data Tables, 53, 113 (1993).

[b] ref.[9].

[c] F.B.Van Duijneveldt, IBM Res.J. 16437, 945 (1971).

Table 5: Comparison of Hartree-Fock and Green-Function methods for H_2 molecule. Energies are given with respect to the HF limit, -30.833 eV[a]. Bond lengths are in Å, energy differences are in eV.

Basis	Atomic basis set				Scaled basis set			
	R_{HF}	ΔE_{HF}	ΔE_{GF}	$\Delta E_{2/3}$	R_{HF}	ΔE_{HF}	ΔE_{GF}	$\Delta E_{2/3}$
$2s$	0.758	1.064	-0.548	-0.010	0.742	0.947	-0.150	0.216
$3s$	0.743	0.317	0.081	0.160	0.734	0.288	0.084	0.152
$4s$	0.737	0.192	0.119	0.143	0.734	0.179	0.123	0.142
$10s$	0.732	0.135	0.135	0.135	0.732	0.135	0.135	0.135
$2s^*$	0.767	0.914	-0.509	-0.034	0.746	0.765	-0.239	0.096
$3s^*$	0.741	0.214	-0.072	0.023	0.737	0.172	-0.095	-0.006
$4s^*$	0.735	0.059	-0.044	-0.010	0.734	0.054	-0.038	-0.008

[a] Estimated Hartree-Fock limit reported by W.Kolos and C.C.J.Roothaan, Rev.Mod.Phys., 32 (1960) 219. Estimated bond length is 0.733Å.

energy or bond length for H_2, but a reliable value for the HF limit is available, and it is the convergence to this value with an increasing size of basis set that is of interest here. One observes that (as anticipated) neither method is close to the limit without the use of atomic p functions to simulate the effect of the second nucleus.

Using functions of different symmetry types is important for representing the self-consistent potential $\hat{V}_N$

$$\hat{V}_N = \sum_n |n> \hat{V}_n <n| \tag{7}$$

because inclusion of a basis function of another symmetry will shift downwards both the HF and GF energies in contrast to the inclusion of a basis function of the same symmetry when the HF energy decreases and GF energy increases. In the expansion (7) index n enumerates basis sets of different symmetry. Inclusion of polarization functions of p-symmetry is essential to improve the energies of H_2 . Then, GF energies of H_2 approach the HF limit from below, the same result obtained for atoms.

The energy calculated from $3s^*$, $4s^*$ scaled basis sets approaches the HF limit satisfactorily, the lower bound from GF being closer to the limit than the upper bound from HF, while E calculated from eq.(6) differs from the limit by less than 0.001% , a rather interesting result.

Results for the fluorine molecule appear in Table 6. Again E_{GF} is a better approximation than E_{HF}, while E from (6) gives a more rapid convergence to the

Table 6: Comparison of Hartree-Fock and Green-Function methods for F_2 Molecule. Total energies are given with respect to the HF limit, -5408.9eV^a . Estimated bond lenghts[b] are in Å, energy differencies are in eV.

Basis	Atomic basis set				Scaled basis set			
	ΔE_{HF}	ΔE_{GF}	$\Delta E_{2/3}$	R_{HF}	ΔE_{HF}	ΔE_{GF}	$\Delta E_{2/3}$	R_{HF}
$4s/2p$	64.4	-86.4	-35.9	1.318	64.4	-82.3	-33.5	1.316
$5s/3p$	19.5	-24.0	-9.3	1.389	19.3	-22.8	-8.8	1.375
$7s/4p$	4.3	-3.4	-1.1	1.413	4.2	-3.3	-0.8	1.410
$4s/2p^*$	63.7	-77.1	-30.2	1.299	61.0	-78.8	-33.5	1.271
$5s/3p^*$	18.8	-18.5	-6.1	1.336	18.7	-18.0	-8.8	1.338
$7s/4p^*$	3.5	-4.2	-1.6	1.329	3.4	-4.2	-1.6	1.330

[a] Numerical Hartree-Fock limit reported by P.Pyykkö et al, Mol.Phys., 60, 597 (1987).
[b] Estimated bond length in HF limit is 1.33 Å(A.C.Wahl, J.Chem.Phys., 41, 2600 (1964).

estimated HF limit, the error being 0.04% for $7s/4p^\star$.

Results for HF appear in Table 7. Again the GF energy for $7s/4p^{\star\star}$ is closer to the HF limit than the HF calculation with the same basis set. The estimated E from eq.(6) is -2722.6 eV which differs from the HF limit by 0.02% . A possible approach is to use a basis set for F that is optimized for the F^- ion but our calculations showed that this is less important than increasing n and m (compare n=4,5,7).

Inclusion of porarization functions of d-symmetry for heavier atoms improves all energies essentially. After this correction, the GF energies again approach the HF limit from below.

2.3   Polyatomic Molecules

Results for hydrogen-containing polyatomic molecules H_2O, NH_3, CH_4 are in Table 8. In all cases, inclusion of polarization functions, i.e. functions of other symmetry, improves the HF and GF energies as well as supporting the validity of the "2/3 rule". This improves the accuracy of the total energy estimate by an order of magnitude. Fulfilment of the "2/3" rule gives the possibility of supplementary checking the basis set quality by simply comparing "2/3" energies. For the water

Table 7: Comparison of Hartree-Fock and Green-Function methods for Hydrogen Fluoride Molecule. Total energies are given with respect to the HF limit, -2721.93 eV^a. Energy differences are in eV, bond length[b] is in Å.

Basis	Atomic basis set				Scaled basis set			
	ΔE_{HF}	E_{GF}	$\Delta E_{2/3}$	R_{HF}	ΔE_{HF}	ΔE_{GF}	$\Delta E_{2/3}$	R_{HF}
$4s/2p$	35.06	-40.29	-15.17	0.953	34.84	-40.21	-15.20	0.955
$5s/3p$	10.77	-11.10	-3.81	0.920	10.68	-10.43	-3.40	0.912
$7s/4p$	2.71	-1.09	0.18	0.918	2.63	-1.24	0.05	0.920
$4s/2p^{**}$	34.00	-33.26	-10.84	0.918	33.80	-34.38	-11.65	0.914
$5s/3p^{**}$	9.92	-6.13	-0.78	0.890	9.82	-7.80	-1.93	0.888
$7s/4p^{**}$	2.01	-1.88	-0.58	0.902	1.94	-1.97	-0.67	0.898

[a] Numerical Hartree-Fock limit reported by D.Sundholm et al., Mol.Phys., 56, 1411 (1985).

[b] Estimated bond length in HF limit is 0.897Å by P.E.Cade and W.M.Huo, J.Chem.Phys., 47 (1968) 614.

molecule, all bases beginning with n=3 lead to $\Delta E_{2/3}$ in the interval -0.2 eV to 0.8 eV, in comparison with ΔE_{HF} estimates of 2 to 6eV.

2.4   Dimers

The GF method is more suitable than the HF one for the description of intermolecular interactions, including hydrogen bonding, because it gives the correct asymptotic behaviour of the tail of a wave function in the intermolecular space [12]. In this space, where the potential is varying slowly, a corrected wave function will really be

$$\psi \sim \int G_N^0(\epsilon, \vec{r} - \vec{r'})\psi_N(\vec{r'})d\vec{r'}$$

where ψ_N is the HF wave function from the basis $\{N\}$. If $r >> r'$ (ψ_N is localized at the origin of one of the molecules) then the asymptotic behaviour of ψ coincides with that of G^0 which corresponds to the correct quantum mechanical value

$$\frac{\exp\left(-\sqrt{-2\epsilon}r\right)}{r}$$

For dimers, there is a basis-set superposition error (BSSE) which arises because the dimer basis set is larger than that of each monomer and this produces an artificial lowering of the dimer energy relative to that of the separated monomers. The

Table 8: Comparison of Hartree-Fock and Green-Function methods for polyatomic molecules. Total energies are given with respect to their corresponding HF limits[a]. Bond lengths are in Å, energy differences are in eV.

Basis	ΔE_{HF}	ΔE_{GF}	$\Delta E_{2/3}$	Θ^o_{HF}	R_{HF}
H_2O-atomic basis set					
$4s/2p$	24.93	-23.60	-7.42	103.3	0.991
$6s/3p^b$	6.92	-2.85	0.41	110.7	0.957
$7s/4p$	2.56	-0.22	0.70	111.6	0.948
H_2O-scaled basis set					
$4s/2p$	24.72	-25.52	-8.77	103.5	0.993
$6s/3p^b$	6.68	-2.17	0.78	110.5	0.953
$6s/3p^{b**}$	5.29	-3.44	-0.53	106.2	0.938
$7s/4p$	2.40	-0.48	0.48	112.2	0.948
$7s/4p^{**}$	1.74	-1.23	-0.24	106.1	0.944
$limit^a$	0	0	0	106.3	0.940
NH_3-scaled basis set					
$7s/4p$	1.78	-0.19	0.47	114.6	0.994
$7s/4p^*$	1.26	-0.80	-0.10	103.9	1.001
$limit^a$	0	0	0	108.1	0.998
CH_4-scaled basis set					
$7s/4p$	1.16	-0.06	0.35	109.5	1.082
$7s/4p^*$	0.60	-0.49	-0.13	109.5	1.082
$limit^a$	0	0	0	109.5	1.0815

[a] Estimated geometries and energies in HF limit reported by R.D.Amos, J.Chem.Soc.Faraday Trans. 2, 83 (1987) 1595. Energies are -2069.043 eV (H_2O, 112 optimized functions), -1529.319 eV (NH_3, 133 functions), -1093.897 eV (CH_4, 154 functions).
[b] Optimized on the O^- ion. Basis set reported in R.A.Poirier, R.Daudel, I.G.Csizmadia, Int.J.Quantum Chem. 19 (1981) 693.

Table 9: Comparison of Hartree-Fock and Green-Function methods for the He-He interatomic potential (in $10^{-6}a.u.$). Basis set includes 4 s-type Gaussians and a polarization function of p-type for each atom.

$R, a.u.$	without CPC		with CPC		Numerical
	HF	GF	HF	GF	HF[10]
1.5	347297	337476	346824	339029	347724
2.0	121752	120247	121835	119971	120749
3.0	13710	13802	13344	13377	13517
4.0	1129	1129	1425	1277	1279
5.6	-9	17	27	19	29

Table 10: Energies of interaction of two neon atoms and two water molecules in *kcal/mol* with and without counterpoise corrections (CPC).

	$R, \overset{\circ}{A}$	*without CPC*		*with CPC*	
		HF	*GF*	*HF*	*GF*
Ne_2	3.17	-0.013	-0.107	0.002	-0.100
$(H_2O)_2$	2.98 [a]	-7.077	-5.506	-6.338	-6.101

[a] Oxygen-oxygen distance fixed at experimental value, oxygen-hydrogen distance varied along the hydrogen bond.

usual procedure is a counterpoise correction (CPC) consisting of the calculation of the energy of each monomer using an extended basis set, which includes the orbitals of the second monomer at the appropriate nuclear separation [13]. The CPC to the GF basis set gives an energy correction in the appropriate direction and decreases the BSSE.

The simplest model system for theoretical consideration is the Helium molecule. As is well known, the HF method does not predict bonding for this molecule. The interatomic potentials for He-He calculated by both methods are given in Table 9 for comparison with numerical HF results [10]. After taking into account the CPC correction, the GF potential lies below HF limit for all distances considered. In the actual region of the bonding near $R_e = 5.6a.u$ the CPC correction does not influence the GF-potential significantly as we supposed above.

The results for the Neon molecule are shown in Table 10. The fitted potential curve for Ne_2 shows a rather flat minimum of - 0.124 kcal/mol at R=3.42Å. On using eq.(6) the value of E is -0.082 kcal/mol which may be compared with the experimental value of -0.083 kcal/mol [14].This agreement is not, however, as good as it seems because of the lack of correlation in the HF and GF calculations.

For water [15],large basis set SCF calculations with CPC give a dimerization energy in the HF limit of -3.73±0.05 kcal/mol [16]. Our GF value with a limited basis set is rather lower than this value, and comparable with the CPC corrected HF 3-21G calculation of -6.2 kcal/mol. The GF calculation after CPC correction seems not to be very different from the HF one, but its superiority before CPC (Table 10) supports our contention that the method gives a better wave function in the intermolecular space when one starts from atomic basis sets without CPC. This might prove to be important in calculations on large molecules and clusters.

This research was supported in part by the Ukranian State Committee for Science and Technology. The collaboration of VAT and PWMJ was made possible by an NSERC Research Grant (to PWMJ).

References

[1] S.H.Gould, *Variational Methods for Eigenvalue Problems*, OUP, London ,1966.

[2] I.M.Lifshits, Uspehi Mat.Nauk (Adv.Math.Sci.,USSR), 7, 171 (1952).

[3] S.R.Singh, A.D.Stauffer, Nuovo Cim. **25B**, 547 (1975).

[4] R.W.Grimes, C.R.A.Catlow and A.L.Shluger (ed.), Quantum Mechanical Cluster Calculations in Solid State Studies, World Scientific, Singapore, 1992.

[5] V.A.Telezhkin and A.A.Rafalovich, Int.J.Quantum Chem. **52**, 1199 (1994).

[6] R.Pourier, R.Kari and I.G.Csizmadia, Handbook of Gaussian Basis Sets, Elsevier, Amsterdam (1985).

[7] P.C.Hariharan and J.A.Pople, Theoret.chim.Acta (Berl.), **28**, 213 (1973).

[8] C.Froese-Fischer, *The Hartree-Fock Method for Atoms* (Wiley, New York,1977).

[9] H.Tatewaka, T.Koga, Y.Sakai, A.J.Thakkar, J.Chem.Phys. **101**, 4945 (1994).

[10] P.Pyykkö, in *Numerical Determination of the Electronic Structure of Atoms, Diatomic and Polyatomic Molecules* (ed.M.Defrancheschi, J.Delhalle) Kluwer Acad.Publ., (1989), p.161.

[11] R.D.Amos, J.Chem.Soc.Faraday Trans. 2, **83**, 1595 (1987).

[12] V.A.Telezhkin, A.A.Rafalovich, A.A.Ovodenko, Mol.Materials. 4, 225 (1994).

[13] P.Hobza, R.Zahradnik, Chem.Rev. **88**, 871 (1988).

[14] K.P.Huber and G.Herzberg, Constants of Diatomic Molecules, vol.4, Van Nostrand Reinhold, New York, 1979.

[15] T.R.Dyke, K.M.Mack and J.S.Muenter, J.Chem.Phys. **66**, 498 (1977).

[16] K.Szalewicz, S.J.Cole, W.Kolos and R.J.Bartlett, J.Chem.Phys. **89**, 3662 (1988); cf. R.J.Vos, R.Hendriks, V.B.van Duijneveldt, J.Comp.Chem.11,1 (1990).

COMPUTER SIMULATION OF HTSC STRUCTURES AND PROCESSES

V. V. KIRSANOV, E. I. SHAMARINA, I. YU. YANOV
Tver State Technical University,
Tver, Russia

1. Introduction

The treatment of materials by the use of various beams of accelerated particles has become a widely-used technology for producing materials with new properties. It has proven useful in changing properties of semiconductors and metals. At present, the first steps have been taken to produce high-temperature superconductors (HTSC) with new properties.

Present-day scientific literature on HTSC physics discusses extensively the problem of increasing an HTSC critical current by creating effective pinning centres. As this takes place, the structural nature of pinning centres plays an important role, the function which, as is known, can be performed by defects of a size comparable to the coherence length. This means that defects can be of nuclear scale, including point defects. Radiation defects (such as point defects, their clusters, post-cascade areas, tracks, etc.) are considered the main possible candidates. Reasons for the detectable increase of HTSC critical current, from ion, neutron and electron irradiation, are vigorously studied.

The use of beams of accelerated particles presents opportunities to modify properties of a new class of semiconductors - doped by alkaline metals of fullerites. In so doing, the formation of an endohedral complex based on "fullerene molecule-implantation," is of special interest. In addition to the influence on the superconducting state (at the expense of phonon spectrum distortion), complexes of this kind can be used as molecular devices for nano-electron technologies.

As is known, one of the main parameters enabling the control of processes for radiation modification of materials is the threshold energy for forming stable displacements in the matrix of the irradiated material. Using this energy, one can calculate the number of defects produced, determine the required energy of bombarding beam particles, etc. Currently, this determination is successfully done by computer simulation methods which take into account both the structure of the irradiated material and the type of inter-atomic or inter-ionic interaction.

Using the molecular dynamics method outlined below, one can determine the displacement threshold energies of HTSC for the system Y-Ba-Cu-O as well as the threshold energies of endohedral complexes formed from the base of fullerene molecules (energies required for dynamic doping of fullerene).

R. C. Tennyson and A. E. Kiv (eds.),
Computer Modelling of Electronic and Atomic Processes in Solids, 43–50.
© 1997 *Kluwer Academic Publishers.*

2. Method

Computations were done using a molecular dynamics method. Simulation details are fully given in [1]. The model microcrystallite, simulating monocrystal $YBa_2Cu_3O_7$ consisted of 1000 ions. A modified potential from [2] was used to describe the interaction between ions of the crystallite. Using the given potential, structural parameters for ortho and tetra modifications $YBa_2Cu_3O_{7-d}$ were obtained. The computed phonon spectrum showed good agreement with experimental data. The specified model was also used to determine the ortho-tetra transition temperature, the value of which (at d = 0, 0.25 and 0.5, the transition temperatures were equal to 1100, 1000 and 900K, respectively) was in good agreement with experimental data (at d = 0.5, -970K) [2]. In addition, the temperature factor of expansion obtained by simulation, reproduces experimental values [2]. This indicates considerable potential for a dynamic method when studying both structural properties of HTSC and, in particular, displacement threshold energies. In our previous work, this was confirmed by good agreement between well-computed separate threshold energies for oxygen displacement and experimentally-determined thresholds.

To simulate the interaction of a fullerene molecule with a bombarding atom of carbon, the traditional variant of a molecular dynamics method was used. First of all, the initial configuration of atoms of a fullerene molecule was formed based on the data obtained experimentally. Then, atoms were allowed to move and classical equations of motion were solved. Difficulty in simulating this molecule lies in the fact that fullerene has no dense package of atoms and the nature of coupling between them is substantially of a covalent character. It results in using many-body potentials which in an explicit form take into account angular directivity of couplings as well as the local environment of the atoms. We have selected a potential that meets such requirements [3].

After obtaining an equilibrium configuration of atoms of the fullerene molecule, one more carbon atom was added to the model, the position of which was determined by a target parameter and the speed was imparted to this atom in the direction of the molecule. The following events were registered: atom repulsion from the molecule; atom penetration into the molecule, and formation of endohedral complexes; passage of an accelerated atom through the molecule.

3. Processes of ion displacements and displacement threshold energy of ions in $YBa_2Cu_3O_7$

Computation results for angular dependencies of the threshold displacement energies (E_d) for oxygen, copper, yttrium and barium ions in monocrystal $YBa_2Cu_3O_7$ are given below and were obtained by the molecular dynamics method. The determination of E_d was done as follows. Energy, sufficient to form a stable vacancy (on the PKA site) as well as the stable interstitial atom, was imparted to one of the ions of the model microcrystallite (PKA = primary-knock-on atom). Thus, the formation of a stable Frenkel pair was simulated. Then, this energy was gradually reduced to a size where the Frenkel pair went from a steady position to an unstable one: the interstitial ion that formed spontaneously returned to its vacancy. The PKA critical energy for this transition was considered as the threshold displacement energy.

The threshold energy for various ions was investigated for a wide set of PKA angles from the site in a crystal lattice. Thus, histograms of angular dependence of displacement threshold energy of an ion were constructed.

The displacement threshold energy of $O(1)$, $O(2)$ and $O(3)$ ions in the ab and ac planes of motion was investigated. In the case of oxygen ion travel $O(1)$ in the ab plane, the PKA was displaced to the nearest oxygen vacancy in all cases. A knock-on angle was measured from the b axis, along which the copper-oxygen chain is located in a base plane. Hence, the E_d is greater at the small knock-on angle than at angles approaching 90°. This is due to the fact that at a knock-on angle close to 90°, PKA directly displaces into the oxygen vacancy, while at small angles it makes intricate movements, bouncing off the ions (which surround it) before occupying a vacant site $O(5)$. With an increase in angle, this process requires less energy, smoothly changing E_d from 18 to 6 eV. The initial direction of ion movement $O(1)$ along the c axis in the ac plane exhibits rather complex behaviour. In this case, a knock-on angle was measured from the axis c. At a small knock-on-angle, the ion $O(1)$ bounces off the next ion of chain $Cu(2)$ and shifts to the oxygen vacancy $O(5)$. At knock-on angles from 10° to 35°, PKA displaces into the vacancy $O(5)$. With the angle increased from 10° to 35°, E_d decreases from 16 to 4 eV, then a sharp jump of E_d takes place that is dependent on the change of the defect formation mechanism in such a collision chain. At angles from 40° to 70°, a substitution chain of the type $O(1)->O(2)$ is formed, creating a vacancy on the site $O(1)$ and ejecting ion $O(2)$ in an interstitial position. Then, for angles from 75° to 90°, the PKA displacement mechanism acquires the former position: the oxygen ion displaces into vacancy $O(5)$.

Minimum, maximum and average values of the threshold energy displacement for the oxygen ions in all directions are presented in Table 1.

Investigation of the angular dependence of E_d for the ion $Cu(1)$ in the ab plane showed that at first, when the knock-on angle increases, E_d falls from 43 to 15 eV as the initial direction of movement departs from 0°, along which the ion $O(4)$ is located. When the initial direction of movement begins nearing the other ion $O(4)$, the E_d begins to increase. When the knock-on angles are close to 90°, E_d again goes down to 10 eV. It is worth noting that the E_d results imply a minimum energy, at which defect forming begins in a copper sub-lattice as the oxygen defect formation began when the E_d values were smaller. The appearance of "copper" defects, as a rule, was accompanied by formation of one or more oxygen defects of the type $O(4)->O(5)$. Typical "copper" defects were a "dumb-bell" form of copper ions in a $Cu(1)$ site, as well as the interstitial ion of copper, located in the centre of a square, which is formed by Cu(1) ions.

The threshold energy displacement for Ba and Y ions was similarly determined. Minimum, maximum and average values of the displacement threshold energy of all ions of $YBa_2Cu_3O_7$ are given in Table 1.

The first column of the table displays the plane of chain distribution and the direction of chain movement along the c axis (if more than one ion is tabulated). If only one ion is indicated, it identifies the chain, extending in the ab plane for this ion. The table also presents the results for Ba and Y collision chains and the block of collision chains taking place in copper and oxygen sub-lattices. The data in the table show that oxygen ions pose the least E_d among all sub-lattices of the system Y-Ba-Cu-O. In this case, O(4) and O(1) ions among oxygen ions have the least energy, which varies from 3 to 28 eV at an average value from 4.5 to 10 eV. Ions O(2) and O(3) have higher threshold energies. It is worth noting here that though thresholds for O(2) and O(3) are almost equal in the ab plane, in the ac plane they are higher for O(3) than thresholds for O(2), which can be explained by the various surroundings for these ions. Ions of barium and yttrium have the highest threshold energies among Y-Ba-Cu-O sub-lattices. They form a skeleton of a lattice Y-Ba-Cu-O. Copper ions occupy an intermediate position. In this case, thresholds for the Cu(1) ions is 1.5 times lower than for the Cu(2) ions.

Thus, it may be inferred that the oxygen and copper ion layers, contiguous to the base plane Y-Ba-Cu-O, have the least resistance to the destruction caused by radiation, and determine material behaviour under irradiation. This should be properly taken into consideration when modifying Y-Ba-Cu-O crystals by accelerated particles, in particular, for the purpose of creating pinning centres.

4. Threshold energy of endohedral complexes formation based on the fullerene molecules

In 1986, Smalley and co-authors presented mass-spectrometric proofs of the existence of endohedral complexes $C_{60}@La$ (where @ designates that the atom La is inside the molecule cavity). Such complexes were obtained in evaporating graphite impregnated with metal salts. Since then, the endohedral complexes formed by atoms of various metals and by various molecules similar to fullerenes were observed. Recently $C_{82}@La$ has been synthesized to the extent of milligrams. The existence of such complexes is also confirmed by experiments on targets from inert gases being bombarded by charge clusters of C_{60}. In this case, atoms He, Ne, Ar and others were observed inside the cavity. Thus, experiments testify unequivocally to the possibility of forming endohedral complexes of C_{60}. However, this process on a molecular level has not been fully investigated.

Table 1

Summary table of minimum, maximum and average displacement threshold energies for investigated chains in $YBa_2Cu_3O_7$

Direction and type PKA	min. E_d,eV	max. E_d,eV	$<E_d>$,eV
Plane *ab*, Ba	25.0	55.0	40.0
Plane *ac*, Ba-Y-Ba-Ba	25.0	50.0	30.0
Plane *ac*, Ba-Ba-Y-Ba	25.0	70.0	45.0
Plane *ab*, Y	28.0	75.0	45.0
Plane *ac*, Y-Ba-Ba-Y	20.0	90.0	40.0
Plane *ac*, Cu(1)-O(1)-Cu(2)	10.0	55.0	25.0
Plane *ab*, Cu(1)	10.0	43.0	24.0
Plane *ac*, Cu(2)-Cu(2)-O(1)	25.0	65.0	38.0
Plane *ab*, Cu(2)	20.0	45.0	34.0
Plane *ab*, O(4)	3.0	9.0	4.5
Plane *ac*, O(1)-Cu(1)-O(1)-Cu(2)	3.0	28.0	7.0
Plane *ac*, O(1)-Cu(2)-Cu(2)-O(1)	4.0	16.0	9.0
Plane *ab*, O(1)	6.0	18.0	10.0
Plane *ac*, O(3)-Ba-Y-O(4)	16.0	52.0	28.0
Plane *ab*, O(3)	16.0	40.0	23.0
Plane *ac*, O(2)-Cu(2)-O(1)-Cu(1)	8.0	26.0	13.0
Plane *ab*, O(2)	16.0	46.0	23.0

Since the direct computation of C atom interaction with a fullerite crystal lattice presents great complexity because of the considerable size of the system, a number of approximations were used. First and foremost, only one - and the most important - stage of the process was considered. It is the direct collision of a bombarding atom with a fullerene molecule. The choice of only two mutual orientations of the fullerene molecule and the direction of the bombarding atom is the second approximation. The origin of the coordinate system was specified in the centre of C_{60}. The z-axis was located at the molecule centre and passed through the centre of one of the faces (pentagonal - orientation 1, hexagonal - orientation 2). The velocity vector of the bombarding C atom runs parallel to axis z, and the initial (x, y) coordinates varied, determining the target parameter.

Lastly, the third simplification lies in the fact that thermal movement of atoms of C_{60} is not taken into account, i.e., their initial velocity is considered to be equal to zero, and the starting configuration of the molecule atoms to be in equilibrium.

Figure 1 presents the dependence of threshold energy to form an endohedral complex $C_{60}@C$ (vertical axis) on the target parameter (x, y) for orientation 1. The top level of the figure displays the chart of isolines of the same surface. Vertical lines correspond to the directions of the atoms forming the pentagonal face of a molecule.

48

Naturally, for target parameter values $R = \sqrt{x^2 + y^2} > R_0$ (where $R_0 = 3.5$ Å - fullerene molecule radius) the bombarding atom is scattered by a molecule at any energy value. Therefore the relief is given in Fig. 1only for values $R < R_0$.

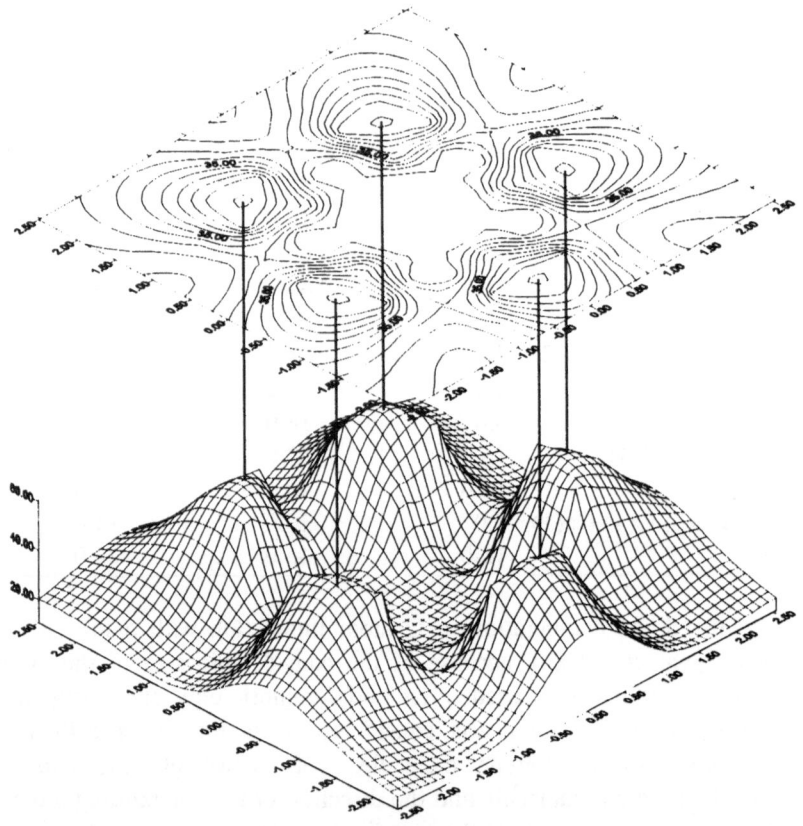

Figure 1.

Dependence of threshold energy of endohedral complex formation (vertical axis) on the target parameter (x, y) for orientation 1. The top level of the figure displays an isoline chart of the same surfaces. Vertical lines correspond to directions onto the atoms forming the pentagonal face of the molecule.

Figure 1 shows that the lowest energy values (about 11 eV) of a bombarding atom, which are necessary to form $C_{60}@C$, are achieved in a pentagonal face centre as well as in directions close to the projections of hexagonal face centres on a plane of a pentagonal face.

Similar results were obtained for the second orientation of the molecule. In this case, the minimum value of energy for forming an endohedral complex makes up 6 eV (as well as for direction through the face centre).

The process of forming an endohedral complex proceeds as follows: the bombarding atom passes through a central part of a pentagonal, or hexagonal face, cluster cavity; then it is reflected from the opposite face of a molecule, returns, is reflected again, etc. In so doing, the essential part of kinetic energy for an ejected atom changes into an oscillatory energy of a molecule.

If the energy becomes greater than 25 eV (for a pentagonal face), atoms having small target parameters stop bouncing off the opposite side of a molecule and pass through. Molecular fragmentation occurs beginning, at the energies ~40 eV, to collide with the target parameter close to directions for the molecule atoms. This process begins as follows: the bombarding atom knocks out two atoms of the molecule on the opposite sides of the sphere, thus forming instabilities resulting in the molecule destruction.

Thus, the computations performed provide values of threshold energies responsible for forming the endohedral complex $C_{60}@C$ at 6 and 11 eV for hexagonal and pentagonal faces of a fullerene molecule, respectively. If the energy of the bombarding atom becomes greater than 16 eV in the first case, and above 25 eV in the second case, the implanted atom passes through the molecule. This means that formation of a complex $C_{60}@C$ is likely to occur at energies of 11-25 eV for directions onto pentagonal faces and at 6-16 eV on the hexagonal faces.

If the energy of a bombarding atom exceeds 40-50 eV, then at some values of the target parameter (approaching those which are close to frontal collisions with molecule atoms), our computation shows that there arises the probability of fullerene molecule destruction. Such a possibility requires further study by stricter quantum-mechanical methods because it offers the prospect for creating pseudopoint defects (for the lack of a site in a molecular lattice) in solid fullerite by implantation methods.

5. Discussion of results

Definition of threshold energies for forming a stable displacement in the system Y-Ba-Cu-O allows one to set a number of methods for upgrading this system. First, it is possible to implant controllable defect formation in different sub-lattices $YBa_2Cu_3O_7$. The latter can be done by gradually increasing the bombarding particles' energy, and it signifies the sequential use of displacement mechanisms first, in an oxygen sub-lattice, then, in copper, barium and yttrium ones. It is evident that the precise computation of the PKA spectrum is required.

Secondly, the knowledge of threshold energies of steady displacement allows more precise use, for example, of the NRT-standard, to determine a radiation-damaging dose for the Y-Ba-Cu-O system, and hence the concentration of radiation defects caused by one or another type of irradiation. Potentially, using more complex cascade computations, one can determine the spatial arrangement of defects that is very important when estimating the originating radiation centres of pinning.

The resulting determination of threshold energies of endohedral complex formation based on the fullerene molecules allows one to more precisely implant additional carbon ions into the fullerene molecules.

An extension of this work is scheduled to determine similar threshold energies for introducing metal impurities into the fullerene molecule in the form of implantations and substitutions, which offers important information to solve the problem of doping these new materials.

6. Acknowledgments

This work is supported by the Russian Fundamental Investigation Fund, Project N96-02-16271-a.

7. References

1. Kirsanov, V. V., Musin, N. N., and Shamarina, E. I. (1992) Displacement threshold energy in high-temperature supercoductors. II. Thresholds for O, Ba and Y in $YBa_2Cu_3O_7$, Phys. Lett. A 171, 2230233.

2. Chaplot, S. (1989) Interatomic potential, phonon spectrum and molecular-dynamics simulation up to 1300 K in $YBa_2Cu_3O_{7-d}$. Phys. Rev. B 42, 2149-2154.

3. Kirsanov, V. V., Yanov, I. Yu. (1993) Modeling of a stable fullerene molecule. Letters in the Journal of Technical Physics 19, 70-73.

SEMI-EMPIRICAL SIMULATION OF RADIATION DEFECTS IN OXIDE MATERIALS

E. A. KOTOMIN

Institute of Solid State Physics, 8 Kengaraga Str.,
The University of Latvia, Riga LV-1063, Latvia

Abstract. Semi-empirical quantum chemical simulations have been undertaken to obtain the self-consistent atomic and electronic structure of the two basic electron defects in MgO and corundum ($\alpha - Al_2O_3$) crystals - F^+ and F centers (one and two electrons trapped by O-vacancy, V_a, respectively). The calculated absorption and luminescence energies agree well with the experimental data, the excited states of both defects are found to be essentially delocalized over nearest-neighbour cations. The activation energy for diffusion of electron defects in MgO crystal is found to increase monotonically in a series $V_a \rightarrow F^+ \rightarrow F$ -center (2.50, 2.72 and 3.13 eV, respectively).

1 Introduction

Oxide materials in general, and MgO and corundum ($\alpha - Al_2O_3$) in particular, are important as catalysts, as ceramics and due to their relevance to microelectronics [1,2]. Point defects arise in oxides naturally, under irradiation and by design, considerably affecting their optical properties. Several kinds of point defects have been identified and studied in oxide materials like MgO and corundum. The two basic electron defects are the so-called F^+ and F-centers (one and two electrons trapped by O-vacancy (V_a), respectively). Substantial theoretical attention has been paid to these centers in MgO in the last few years, when a number of careful theoretical calculations have been performed. However, such calculations are unavoidably restricted to quite small quantum clusters or supercells, thus not allowing the study of *excited* states of these centers. Another technologically important characteristic of the electron centers in oxide crystals which cannot be practically handled by *ab initio* methods is *electronic defect diffusion*. Currently, there is not a single theoretical attempt to calculate the activation energy for F^+ and/or F-center diffusion in MgO crystals.

This demonstrates that a relatively simple theoretical approach is needed, allowing us to study large quantum clusters (about 100-200 atoms) and com-

R. C. Tennyson and A. E. Kiv (eds.),
Computer Modelling of Electronic and Atomic Processes in Solids, 51–59.
© *1997 Kluwer Academic Publishers.*

plex defects (like dimer F_2-centers). On the other hand, a method should be able to optimize the defect geometry (in both the ground and excited states) through a minimization of the total energy, and to reliably calculate the excited states and the relevant absorption and luminescence energies. Such an approach to large-scale simulations of static and dynamic properties of defects in ionic solids has been elaborated in recent years at the University of Latvia, Riga, on the basis of the semi-empirical quantum chemical method of the Intermediate Neglect of the Differential Overlap (INDO)[3], which since then has been applied very successfully to the defect study of many oxide materials, including MgO, SiO_2, Li_2O, ZrO_2, $\alpha - Al_2O_3$.

In this paper, we summarize results of recent INDO calculations of the optical properties and diffusion of F^+ and F-centers in MgO and corundum [4-6]. 125-atom clusters of a cubic shape with high (O_h) point symmetry and modelling nine spheres of atoms around the coordinate origin in the case of MgO, and a 65-atom cluster for defects in corundum, were used to calculate optical properties. These clusters were embedded into the electrostatic field of a non-point infinite crystalline lattice. In the F-type center simulations, the central O-atom has been removed and one or two electrons added to the cluster making no *a priori* assumptions about their localization and the electron density distribution. The ions surrounding the O-vacancy were allowed to relax in order to obtain a minimum of the total energy, as well as the self-consistent electronic and atomic structures of the defect. The relevant INDO computer code SYM-SYM is perfectly suited for such computer simulations, as it is based on a complete treatment of the point symmetry and the automated defect geometry optimization. The absorption and luminescence energies were calculated as the difference of total energies for the relaxed atomic SCF ground and excited states, respectively known as the Δ SCF method.

2 Defects in MgO

The optimized geometry of a bare O-vacancy, and that which trapped one and two electrons, shows that mainly the six nearest-neighbour (NN) cations are displaced in an outward direction relevant to the vacancy. Naturally, for a double-charged V_a defect, the surrounding atomic relaxation is the largest – 6.5% of the Mg-O spacing for the nearest Mg ions, but it is reduced to less than 2% for the neutral F-center. Atoms of three spheres surrounding O-vacancy are found to be noticeably relaxed, even in the ground state of these F-type centers.

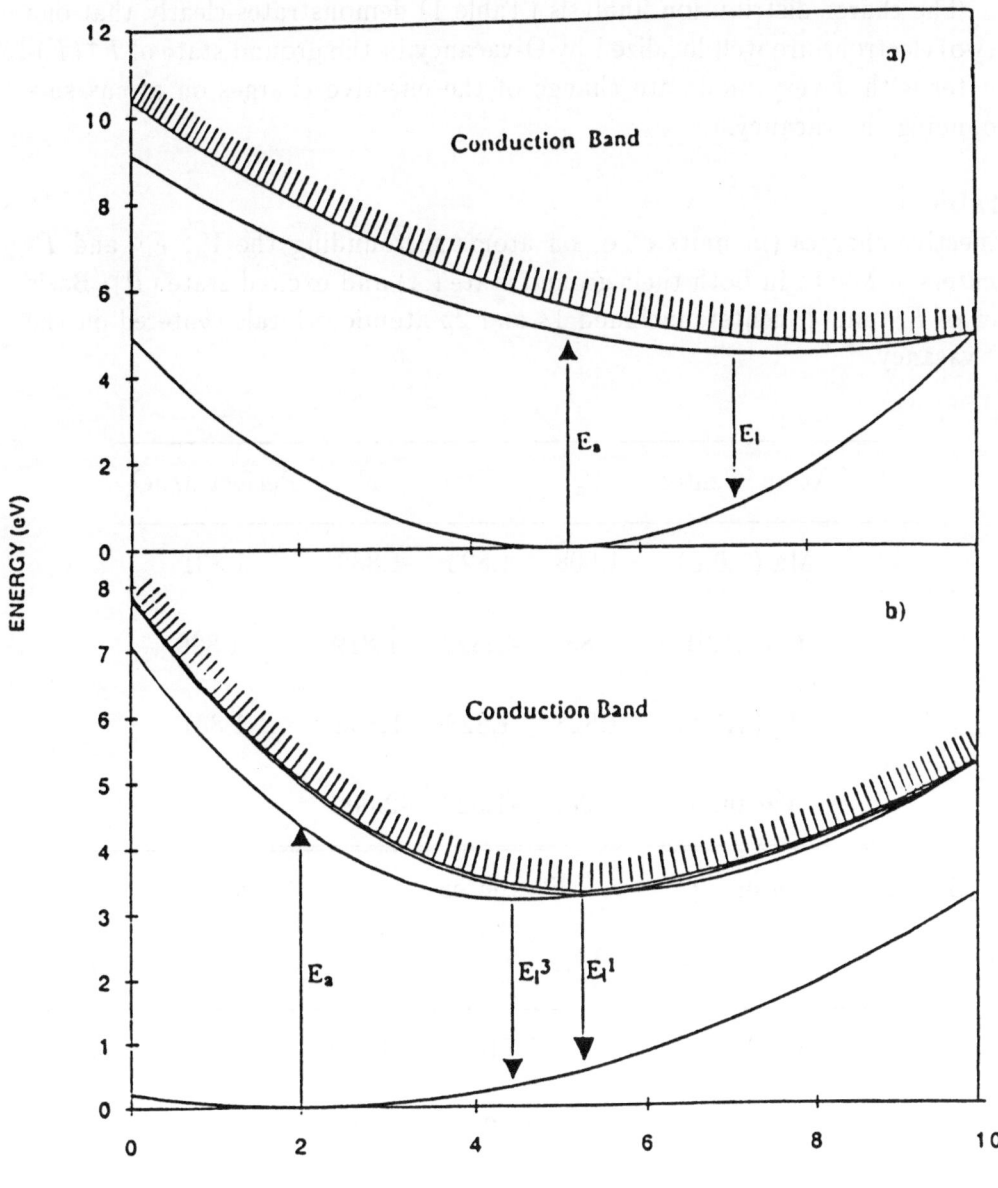

Fig. 1.

A_{1g}-configuration coordinate curves for $F^+(a)$ and $F(b)$ centers in MgO (effect of only the symmetric relaxation of NN cations is shown). E_a is absorption energy. E_l is the luminescence energy of F^+ center whereas E_l^1 and E_l^3 are luminescence triplet and singlet energies for F center, respectively

The charge distribution analysis (Table 1) demonstrates clearly that one (two) electrons are well-localized by O-vacancy in the ground state of $F^+(F)$-center with a very moderate change of the effective charges on atoms surrounding the vacancy.

Table 1

Effective charges (in units of e) on atoms surrounding the V_a, F^+ and F-centers in MgO - in both their ground state (A) and excited state (B). Basis set of F^+ and F centers included $1s$ and $2p$ atomic orbitals centered on the O-vacancy.

A. Atom/Center	V_a	F^+	F	Perfect MgO
Mg (1,0,0)	1.808	1.823	1.838	1.831
O (1,1,0)	-1.835	-1.826	-1.819	-1.829
Mg (1,1,1)	1.823	1.828	1.832	1.831
Vacancy	-0.002	-1.002	-2.002	-

B. Atom/Center	F^+ center	F center*	
		S	T
Mg (1,0,0)	1.715	1.701	1.674
O (1,1,0)	-1.824	-1.817	-1.816
Mg (1,1,1)	1.828	1.832	1.831
Vacancy	-0.476	-1.268	-1.127

*S and T denote the singlet and triplet states, respectively.

This is no longer true, however, for their *excited* states: about $0.5e$ is delocalized from the O-vacancy over the cations surrounding F^+-center, which results in the reduction of their effective charges (by $\sim 0.1e$ on each atom). The singlet excited state of the F center 0.7 e is delocalized. An analysis of the *spin* density distribution for the ground state of F^+-center shows that $\geq$ 90% of the unpaired electron lies inside O-vacancy but only 44% in its excited state.

Our calculations put *both* the ground levels - of F^+ and F-centers- at 3 eV above the top of the valence band - similar to results recently obtained using much more refined super-cell multiple-scattering theory. The potential energy curves as a function of the full-symmetry A_{1g} -relaxation for the F^+ and F centers are shown in Fig. 1. They confirm that the excited states lie very close to the conduction band - in agreement with the experimental data [1], placing the F excited state only 0.06 eV below the bottom of the conduction band. This explains why these excited states are so delocalized.

The luminescence energy for F^+ by 0.4 eV exceeds the experimental one (3.2 eV) (Table 2).

Table 2

Calculated absorption and luminescence energies (in eV); symbols S and T denote theory for the singlet and triplet bands of F-center, respectively.

Defect		F^+	F
Absorption:	theory	4.97	4.98*
	expt [1]	4.95	5.02
Luminescence:	theory	3.6	2.61 (S); 2.76 (T)
	expt [1]	3.2	2.3

*Fitted to the expt. value

Calculations of the excited F-center predict the singlet luminescence peak at 2.61 eV and the triplet one at 2.76 eV. The only observed experimental emission band is at 2.3 eV; probably singlet and triplet bands are too close to be resolved (the width of the experimental emission band is about 0.6 eV). The calculated energies at the minimum of the relaxed singlet and triplet states differ by only 0.04 eV; this probably explains why the optically-detected magnetic resonance (ODMR) experiments indicate that "the emitting state is not predominantly a triplet" [7].

In the simulation of defect diffusion, 224-atom clusters of C_{2v} symmetry were employed [8]. They include several spheres of atoms around *both* a vacancy and its NN O-atom with which it exchanges positions. The use of such a big cluster allows us to avoid any potential complications due to cluster boundary effects. The effective charge of an O-atom in the saddle point (half the distance between two regular O sites along the (110) axis) which is on top of the energy barrier is found to remain practically the same as at the lattice site.

The activation energies for defect diffusion were calculated as the difference of total energies of relaxed atomic configurations corresponding to V_a in its initial (equilibrium) state (the regular lattice site) and the O-atom in the saddle-point as defined above.

The calculated diffusion energies are $E_a = 2.50$ eV (V_a), 2.72 eV (F^+) and 3.13 eV (F-center). The former energy is very close to the experimental value of 2.4 eV. Another two energies are not well established experimentally. However, it has been observed experimentally that in additively coloured MgO crystals, the complex ($F^+ + F$) absorption band begins to decay only at 900°C. Using the standard expression for a hop frequency, $\nu = \nu_o \exp(-E_a/kT)$ with the pre-factor of the order of LO phonons ($\nu_0 \approx 10^{13}s^{-1}$) one can estimate that $E_a \approx 3$ eV. This is again in qualitative agreement with the calculated values.

3   Defects in Corundum

Figure 2.a demonstrates relaxation of the six nearest atoms surrounding the O-vacancy in corundum. It includes inward displacements σ of the two O atoms from the same basic O triangle as V_a, and outward displacements of two *pairs* of Al atoms - Al(1), Al(2) (δ), and Al(3), Al(4) (Δ). In a regular corundum structure, these two kinds of the so-called *long/short* Al-O pairs are characterized by the bond lengths of 1.97 Åand 1.86 Å, respectively. Table 3 shows the calculated absorption and luminescence energies for F^+ and F centers. Due to low symmetry, $1s \to 2p$ absorption band of the former defect in corundum is split into three sub-bands (Fig.2.b) whose energies are reasonably well reproduced in our INDO calculations. Our theory reproduces very well the *luminescence* energies of both F^+ and F centers, and the activation energy for the F^+luminescence quenching, $E_a \approx 0.6$ eV).

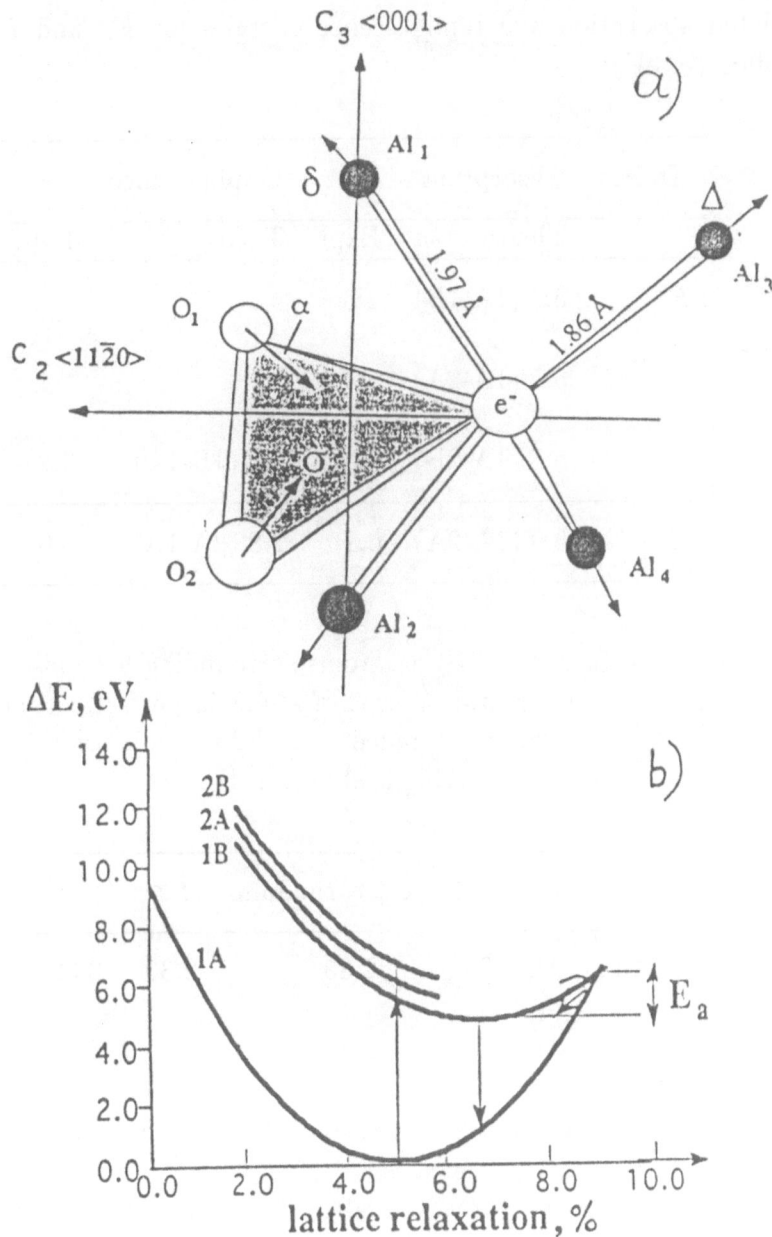

Fig. 2.

(a) Sketch of the F-type center geometry in corundum (6 NN atoms surrounding V_a are shown only).

(b) Potential energy curves for the ground (1A) and split excited states (1B, 2A, 2B) for the F^+-center in corundum. E_a is the thermal quenching energy.

Table 3

Calculated absorption and luminescence energies for F^+ and F centers in corundum (in eV).

Defect	Absorption		Luminescence	
	Theory	Expt	Theory	Expt
F^+	5.2 (1A-1B)	4.8		
	5.5 (1A-2A)	5.4		
	5.8 (1A-2B)	6.0	4.0 (1B-1A)	3.8
F	5.9 (1A-2A)	6.1	2.8 (2A-1A)	3.0

Table 4

Effective charges (in units of **e**) on atoms surrounding ground-state F^+ and F-centers in corundum crystals as well of atoms in perfect corundum.
Basis set of F^+ and F centers included $1s$ and $2p$ atomic orbitals centered on the O-vacancy. Atomic numbering is shown in Fig. 2.

Atoms	Perfect corundum	F^+	F
Al(1), Al(2)	2.35	2.30	2.32
Al(3), Al(4)	2.36	2.29	2.26
O(1), O(2)	-1.57	-1.55	-1.54
Vacancy	—	-0.90	-1.57

The insignificant charge redistribution caused by the O-vacancy at the surrounding atoms is shown in Table 4. Note however, the considerable deviation of the effective charges in the *perfect* corundum from the purely ionic model (Al/O: +3e/-2e), widely used in the modelling of defects in this material. In contrast, a self-consistent quantum chemical calculation indicates the presence of the covalency effects in the corundum chemical bonding.

In conclusion, we believe that the semi-empirical quantum chemical INDO method is a very promising tool for the study of static and dynamic properties of complex defects in oxide materials.

We are greatly indebted to M.M. Kuklja, R. Eglitis, A. Popov, C.R.A. Catlow, P.W.M. Jacobs, A.L. Shluger, M. Stoneham for stimulating discussions.

References

1. Crawford, J.H. Jr (1983) Structure and Properties of MgO and Al_2O_3 Ceramics, In: *Advances in Ceramics, American Cer. Soc., Ohio, Columbus*, **10**, 793;

Crawford, J.H. Jr. Defects in Oxide Crystals (1984) *Nucl. Instr. Meth.* **B 1**, 159-169.

2. Rühle, M., Evans, A.G., Ashby M.F., and Hirsh J.P. (eds.) (1990) *Metal-Ceramic Interfaces, Pergamon Press, Oxford*

3. Evarestov, R.A., Kotomin, E.A., and Ermoshkin, A.N. (1983) *Molecular Models of Defects in Wide-Gap Materials*, Zinatne, Riga.

Kantorovich, L.N., Kotomin, E.A., Kuzovkov, V.N., Tale, I.A., Shluger, A.L., Zakis, Yu.R. (1991) *Models of Defect-Induced Processes in Solids*, Zinatne, Riga.

4. Kotomin, E.A., Kuklja, M.M., Eglitis, R.I., and Popov, A.I. (1996) Quantum Chemical Simulations of the Optical properties and Diffusion in MgO. *Mater. Sci. Eng.*, **B 36**, in press

5. Stashans, A., Kotomin, E.A., and Calais, J.-L. (1994) Calculations of F-type centers in corundum crystals. *Phys. Rev.*, **B49**, 14854-14859.

6. Kotomin, E.A., Stashans, A., Kantorovich, L.N., Livshits, A.I., Popov, A.I., Tale, I.A., and Calais, J.-L. (1995) Calculations of geometry and optical properties of F_{Mg} centers and dimers in corundum. *Phys. Rev.* **B51** 8770-8775.

7. Edel, P., Henderson, B., and Romestain, R. (1982) ODMR study of F centers in MgO. *J. Phys. C: Solid State Phys.*, **15**, 159-168.

8. Popov, A.I., Kotomin, E.A., and Kuklja, M.M. (1996) Quantum Chemical Calculations of the Electron Centers in MgO. *Physica Status Solidi*, **B 195**, 61-66.

CLUSTER MODEL IN SURFACE SCIENCE

*V.V. KOVALCHUK, *L.Yu. KUTSENKO,
*I.A. POLOZOVSKAYA, *G.D. URUM, **V.A. YANCHUK,
***L. ZUNIGA
* South Ukrainian Pedagogical University
26 Staroportofrankovskaya St.270020 Odessa Ukraine
** Institute of Semiconductors, National Academy of Science
45, prospect Nauki 252628 Kiev Ukraine
*** Odessa State University
2, Petra Velykogo St.270000 Odessa Ukraine

1. Introduction

The interaction of halogen atoms (fluorine, chlourine etc.) with a silicon surface has been extensively investigated by both experimental and theoretical methods [1-10]. Recent progress in the combination of visualization with simulation and modelling techniques has demonstrated spectacular results in the investigation of chemical reaction mechanisms [11]. The computer simulation package, called the Surface Molecular Graphics Package (SMGP) for calculations of electronic structure and visualization of geometrical structure is presented. SMGP contains many interfaces with quantum chemical programs such as the ab initio, semi-empirical and molecular surface geometry generation. In this paper, we present several applications of this very useful SMGP to recent models of molecular structures. Besides the traditional quantum chemical "ab initio" SCF methods, based on the Hartree Fock Rutaan scheme, plus a proper treatment of the electron correlation, now the modified semi-empirical approximation techniques have become well established in studies of the electronic structure and geometrical structure of molecular clusters and solids. The Modified Iterative Extended Huckel Theory (IEHT-α [12]) allows the study of rather large systems with reasonable computational efforts and quite good accuracy of results [13]. Furthermore, the IEHT-α calculation, also allows the simulation and visualization of clusters. In this paper, we will present computational results by the IEHT-α method and SCF scheme. The applicability and the accuracy of these methods are illustrated by results of the calculations for a series of some silicon clusters.

61

R. C. Tennyson and A. E. Kiv (eds.),
Computer Modelling of Electronic and Atomic Processes in Solids, 61–68.
© 1997 Kluwer Academic Publishers.

2. Method and realization

2.1. THE SURFACE MOLECULAR GRAPHICS PACKAGE

The SMGP generates detailed, easily interpreted and aesthetically appealing graphics representing models of molecular structures and related properties. This package offers a high level of interactivity through the use of the mouse and via a large set of menus and submenus, organized to enable users to rapidly learn basic operations leading to efficient visualization. For all the menu items, a help facility has been developed. Various representation options and attributes may be selected for adapting the visual output to personal needs and preferences: the molecular structures may be represented as discrete dots, and the global appearance may be modified via attributes such as background appearance, perspective or orthogonal projection, and others.

The purpose of the SMGP is the interactive visual representation of three-dimensional (3D) models of molecular structures and properties for research. Due to the flexibility of the data and program-structure, various chemical systems ranging from small compounds (clusters) to large macromolecules can be investigated and additional interfaces and tools can easily be implemented.

The SMGP contains tools which allow the rapid investigation and visualization of results generated by various external electronic structure calculation program-package such as:

1) IEHT-α is for semi-empirical calculations of one-electron level energies, wave functions and other parameters of electronic structure of quasi-molecular surface systems based on the Iterative Extended Huckel Theory;

2) SCF program package for ab initio quantum chemical calculations with basis sets of contracted Gauss type orbitals;

3) POTENTIAL for calculations based on different types of interaction potentials [15];

4) GP geometrical program based on 3D-representation of molecular structures investigation.

2.2. ABOUT BASIS SETS

The success of the approximate calculation depends on the parameter basis sets (basis electron wave functions) chosen. In order to obtain a compact basis set which still gives a good approximation to atomic energies, we used the radial part of the atomic wave function by Modified Double-Zet Atomic Slater-type Orbitals (MDZ ST AO) [12]. The difference between the values obtained using different basis sets (simple Slater-type Atomic Orbitals (ST AO) and MDZ ST AO) are not small, and illustrate the usefulness of MDZ ST AO (Fig.1). In the latter case, the information must be treated in a qualitative way since small distortions are introduced through the basis set.

The contracted Gaussian basis set for silicon was taken from SCF calculations on the free atom [16]. The primitive s and p basis sets for Si and H were taken from Huzinaga's tables of optimized atomic function [17,18]. The basis set for hydrogen was taken from a cluster study [7]. The basis sets used in the present SCF calculations are little better than MDZ ST AO in the IEHT-α. approximation. This fact is confirmed by the numerical results summarized in Table 1.

TABLE 1.The equilibrium distance (R, Angstroms), binding energy (E_b,electron-volts) of the adsorbing halogen on the cluster Si_4H_9 - X (where X denote the halogen atom)

Clusters	Si_4H_9-F		Si_4H_9-Cl	
Ground State	$_1A^1$		$_1A^1$	
Parameters	R_{Si-F}	E_b	R_{Si-Cl}	E_b
ab initio [7,8]	1.690	3.210	2.200	2.450
ST AO	1.500	3.000	2.011	2.900
MDZ ST AO [12]	1.687	3.194	2.200	2.400
SCF	1.694	3.220	2.200	2.400

The binding energy (E_b) is defined as the difference between the total energy of the absorbate-cluster system and total energies of the isolated fragments. From these results, it is seen that, the use of MDZ ST AO basis enables one to reproduce results which have been obtained by ab initio calculations [7,8], while a difference results from the use of an ST AO basis.

2.3. REPRESENTATION OF ELECTRONIC PROPERTIES

The output of any of these programs (usually huge text files) can be loaded into SMG which retrieves all necessary information such as atomic positions and basis sets (i.e. the electronic wave functions for each atom type), the molecular orbital energies, occupation numbers, the density matrices and other data which cannot directly be represented visually.

The most important properties which can be represented as an isovalue surface are:

1) the molecular orbitals (MO) (orbitals are electronic wave functions which, when squared, yield the electron density distribution of one electron or an electron-pair in the molecular volume), which give significant information on the reactivity of the compound towards various reactions;

2) the electron density distribution (the sum of the squares of all occupied MO's) which indicates for example the type of bonding between atoms and which can be measured by X-ray scattering experiment.

3. Application. Architecture of Silicon Surface

A example of an application of SMGP is the investigation of a silicon surface structure. The Molecular-Cluster Model (MCM) utilizing the semi-empirical MO method of quantum chemistry was introduced twenty years ago to address the problem of deep-level point defects in semiconductors [19]. With the appearance of semi-empirical methods the calculation of the equilibrium geometry and visualization of quite large MCM became possible.

Molecular systems (in particular, surface cluster structures) are made of atoms linked together by chemical bonds. Real surface objects may be built by introducing stereochemistry, i.e., the 3D atomic positions, and it is important to visualize them as molecular models with the usual rendering techniques leading to 3D perception. SMGP visualization allows one to emphasize at length the different aspects of molecular structure of a surface e.g.: chemical topology, conformational details, etc.

We applied the SMGP to the MCM silicon's surface. The single cluster contains ten Si atoms, representing the first four layers of the Si surface. We regard these MCM as hypothetical molecules (quasi-molecules) and attempt to compare the computed results (for example, magic numbers) directly to experimental data for the corresponding impurities in the solids or chemisorbed systems [13].

The mass spectra for charged clusters Si_x^+, or Si_x^- yield the magic numbers and are reported in reference [20]. For example, adsorption processes and chemical reactions on semiconductor surfaces have been extensively investigated and the interaction of halogen-atoms (H, F, Cl, Br, etc.) has become of particular interest because of the role of halogen containing radicals in plasma etching of an Si surface [4-6].

When using an MCM to represent the surface, a choice has to be made about the cluster size, that is, the number of atoms which are treated explicitly in the calculation, and the level of precision of the required computation. Fortunately, the chemisorption of atoms on surfaces seems to be of a local character. This fact is greatly supported by ab initio MCM calculations, and particularly by the calculations for the chemisorption of F and Cl on the Si(111) surface [7-10].

The Fig.2, the MCM Si_4H_9-F and $Si_{10}H_{15}$-F are shown and the distribution of atomic effective charges on each layer and energy characteristics for the ideal silicon clusters are given (approximation IEHT-α MDZ ST AO). The equilibrium cluster geometry parameters were obtained for some small clusters by SMGP. The essential results are presented in Table 2.

TABLE 2. The equilibrium cluster geometric parameters obtained for some small clusters. The bond lengths are in angstroms.

Cluster	bond length	IEHT-α	SCF	Exp[5,6]	ab initio[7-10]
SiH_4	Si-H	1.480	1.480	1.480	1.480
SiF_3	Si-F	1.570	1.570	1.560	1.571
SiF_4	Si-F	1.531	1.552	1.560	1.542
$SiCl_4$	Si-Cl	2.120	2.250	1.950 (0.04)	2.240

4. Conclusion

We have shown that the calculated energy and geometrical characteristics by SMGP are in satisfactory agreement with experimental and other ab initio calculations [7-10]. The present calculations show that the SMGP can be used to obtain a detailed and reasonably accurate description of various aspects of the small halogen-silicon clusters.

In view of the interest of a physicist in visualization of such clusters, one may foresee that data banks representative of the major types of stable systems will soon be available. Therefore, it is important for a physicist to have access the computer tools that allow visualization and generation of computational information.

5. Acknowledgement

We would like to thank Prof. Dr. A.E. Kiv for numerous discussions. Fruitful mathematical comments by Dr. M. Nudelman are gratefully acknowledged.

6. References

1. Winters, H.F. and Coburn, J.W. (1979) The etching of silicon with XeF_2 vapor, *Applied Physics Letters* **34**, N1, 70-73
2. Chuang, T.J. (1980) Electron spectroscopy study of silicon surface exposed to XeF_2 and the chemisorption of SiF_4 on Si, *J.Applied Physics* **51**, N5, 2614 - 2619
3. McFeely, F.R., Morar, J.F. and Skinn, N.D. (1984) Synchrotron photoemisssion investigation of initial stages of fluorine attack of fluorosilylspecies, *Physical Review* **B.30**, N2, 764-770
4. Winters, H.F. and Haarer, D. (1987) Influence of doping on the etching of Si(111), *Physical Review* **B.36**, N12, 6613-6623
5. Kumeda, M., Takayahashi, Y. and Shimizu, T.(1987) Fluorine-incorporation scheme in fluorinated amorphous silicon prepared by various methods, *Physical Review* **B.36**, N5, 2713-2719

66

6. Thornton, G., Wincott, P.L., McGrath, R., McCovern, I.T., Quinn, F.M., Norman, D., and Vvedensky, D.D.(1989) Bonding sites for Cl on Si(100)2-1 and Si(111) 7-7, *Surface Science* **211/212**, 959-968

7. Hermann, K. and Bagus, P.S. (1979) Localized model for hydrogen chemisorption on the silicon (111) surface, *Physical Review* **B.20**, N4, 1603-1610

8. Seel, M and Bagus, P.S. (1983) Ab initio cluster study of the interaction of fluorine and chlorine with the Si(111) surfaces, *Physical Review* **B.28,** 2023- 2038

9. Illas,F., Rubio,J. and Ricart,J.M. (1985) Ab initio cluster-model study of the on- top chemisorption of F and Cl on Si(111) and Ge(111) surfaces, *Physical Review* **B.31**, N12, 8068-8075

10. Ong, C.K. and Tay, L.P. (1989) Chemisorption of fluorine and chlorine on a Si(111) surface, *J.Physics: Condensed Matter* **1**, 1071-1076

11. Weber, J., Deloff, A. and Flukiger, P. (1994) Visualization in Computation Chemistry, *Speedup J.* **8**, N1, 63-70

12. Kovalchuk, V.V. (1995) IEHT-α investigation of silicon surafce, *Ukrainian J. Physics* **40**, N7, 716-719

13. Kovalchuk, V.V., Chislov, V.V. and Yanchuk, V.A.(1995) Cluster model of the real silicon surface, *Physica Status Solidi* (b) **187**, N2 K47-K51

14. Nishida, M. (1978) Cluster model approach for electronic structure of Si and Ge(111) and GaAS(110) surfaces, *Surface Science* **72**, 589-616

15. Gorbachev, Yu.E., Strelchenya, V.F. and Fedotov,V.A.(1991) Models of gas particle-surface interaction potentials, *Surface. Physics, chemistry, mechanics* **2**, 5-20

16. Brode, S., Kolmel, C., Schiffer, H. and Ahlrichs, R. (1987) *Zeitschrift fur Physikalische Chemie Neue Folgl.* **Bd.155**, S.23-28 Ed. by R. Oldenbourg Verlag, Munchen

17. Huzinaga, S. (1971) *Approximate Atomic Functions.* Technical Reports, University of Alberta

18. Ahlrichs, R. and Schart P. (1987) *Ab initio methods in quantum chemistry.* Ed. by K.P. Lawley, Wiley

19. Messmer, R.P. and Watkins,G.D. (1973) Molecular Orbital treatment for deep levels in semiconductors: substitutional nitrogen and the lattice vacancy in diamond, *Physical Review* **B.7**, N6, 2568-2590

20. Bloomfield, L.A., Freeman, R.R. and Brown, W.L. (1985) Photofragmentation of Mass-Resolved Si^+_{2-12} Clusters, *Physical Review Letters* **54,** 2246-2249

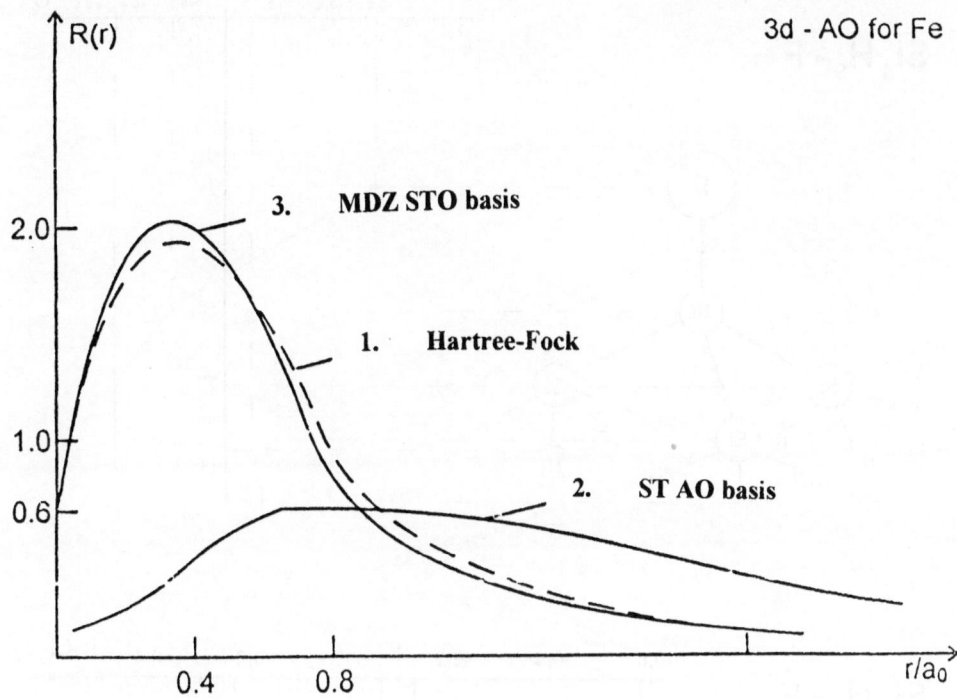

Figure 1

Radial curves (the radial part of atomic wave functions) calculated by
1.   Hartree-Fock-Ruttaan method;
2.   IEHT-α with ST AO basis sets;
3.   IEHT-α with MDZ ST AO basis sets.

68

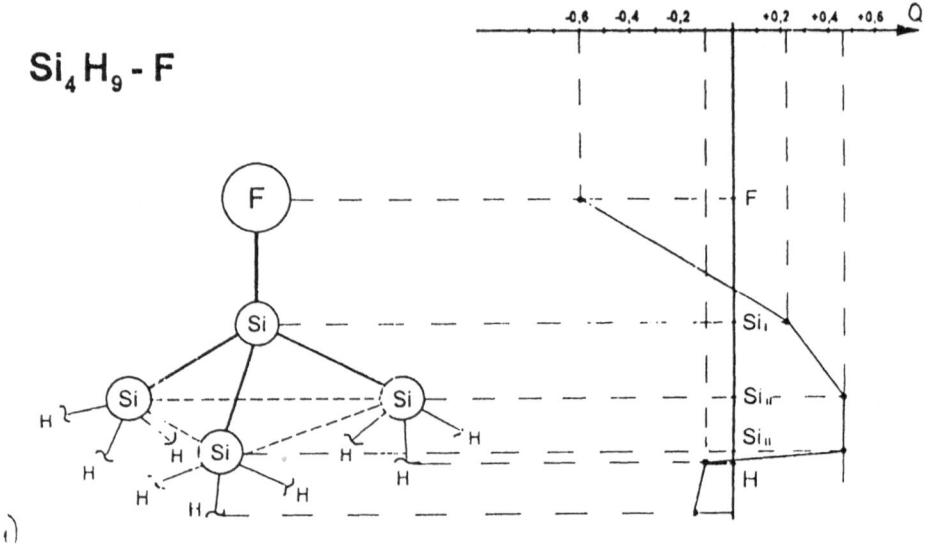

Si₄H₉ - F

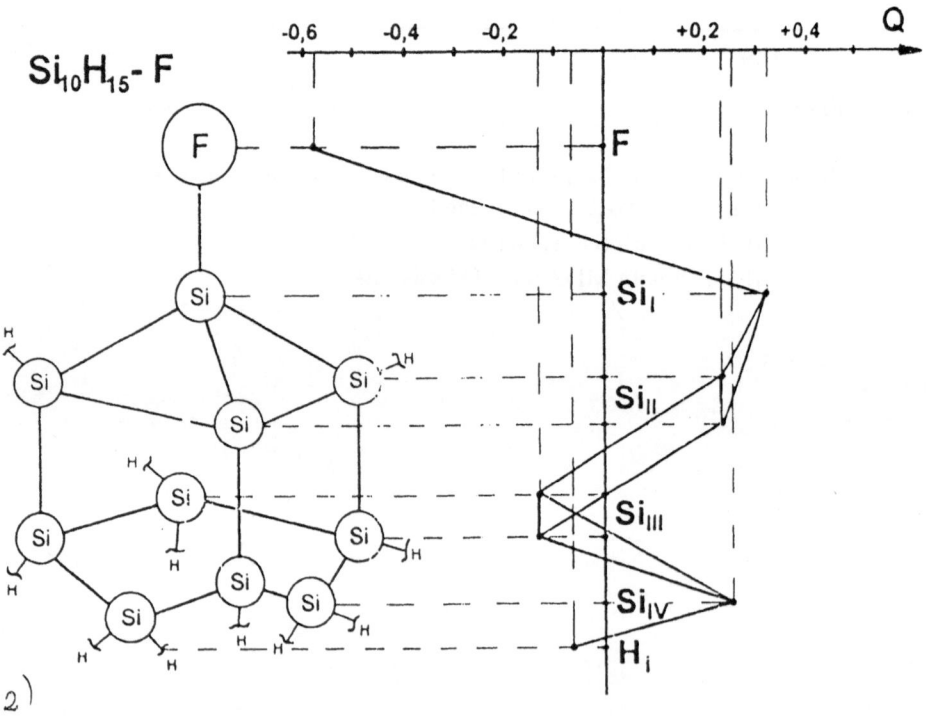

Si₁₀H₁₅- F

Figure 2
Molecular Cluster Model (MCM) and distribution of atomic charges on each layer for:
1. the two-layer Si₄H₉ - F cluster; 2. the four-layer Si₁₀H₁₅ - F cluster.
(Geometry structures for these clusters obtained by the IEHT-α method.)

MODELING OF INHOMOGENEITY IN SOLID COATINGS OBTAINED FROM WATER SUSPENSIONS

D.B. LUKATSKY
Odessa State University, 270000 Odessa, Ukraine
E. RYSIAKIEWICZ-PASEK
Wroclaw Technical University, Wroclaw 50-370, Poland

Abstract

A three-component model is proposed for the investigation of the deposition phenomena from water suspension of charged colloidal particles. The model involves a two-component, charged, hard sphere system (colloidal particles and counter-ions) and neutral hard spheres with dipole moments (water molecules). A system of integral equations for the effective potentials of the model, which is based on hypernetted chain closure, is proposed. Within the approach adopted, effective potentials and structure factors have been found in a linear approximation. Critical behavior is analyzed in terms of singularities of the effective potential.

1. Introduction

Thin film organic and inorganic coatings obtained from the liquid phase find important applications in a variety of electronic devices and materials for aerospace applications [1-3]. Recently, there has been substantial interest in the use of thin polymer films, in particular, as the dielectric in field-effect transistors and integrated circuits [4,5]. One of the main problems that arises in these applications is the homogeneity of coatings obtained. It is well known that heterogeneity leads to the variation of a refractive or absorption index in these media [6,7]. The dimensions of structural inhomogeneities thus play a crucial role in the processes of the interaction of corresponding coatings with the incident radiation. The properties of obtained films and coatings depend strongly on fabrication techniques. Methods of deposition from liquid phases [8-12] are frequently used in the fabrication of electronic devices and coatings. Most research performed in this field is completely empirical. Therefore, simulation of fabrication process is an important problem in the thin and thick film technology.

In this paper, we propose a physical model and numerical analysis that allows the prediction of the properties of coatings obtained from a water suspension. Within the framework of the proposed approach, analysis of inhomogeneities in a deposited phase is performed.

69

R. C. Tennyson and A. E. Kiv (eds.),
Computer Modelling of Electronic and Atomic Processes in Solids, 69–77.
© *1997 Kluwer Academic Publishers.*

2. Model and method

Let us consider a water suspension of charged colloidal particles with a size of order $a \approx 0.1\,\mu m$. In recent experiments [13-15], crystallization and deposition phenomena from such a system were observed. One can expect the appearance of a homogeneous or inhomogeneous deposited phase depending on the parameters of colloidal suspension (temperature, charge and volume fraction of colloidal particles). The size scale of the colloidal particles is some hundred times larger than the size scale of the substrate atoms. Therefore, the interaction between the colloidal suspension and the substrate will be omitted completely in the present model.

In this paper, we consider a three-component model for colloidal suspension, introduced in [17], that involves a two-component charged hard sphere system (colloidal particles and counter-ions) and neutral hard spheres with dipole moments (water molecules).

For a mixture of r species the Ornstein-Zernike equation is

$$G^{\sigma\sigma'}(\vec{R}_1,\vec{R}_2) = C_2^{\sigma\sigma'}(\vec{R}_1,\vec{R}_2) + \sum_{\sigma'=1}^{r} n_{\sigma'} \int d\vec{R}_3 C_2^{\sigma\sigma'}(\vec{R}_1,\vec{R}_3) G^{\sigma'\sigma'}(\vec{R}_3,\vec{R}_2), \qquad (1)$$

where $G^{\sigma\sigma'}(\vec{R}_1,\vec{R}_2)$ is the static correlation function $(G^{\sigma\sigma'}(\vec{R}_1,\vec{R}_2) = F_2^{\sigma\sigma'}(\vec{R}_1,\vec{R}_2) - 1$, here $F_2^{\sigma\sigma'}(\vec{R}_1,\vec{R}_2)$ is the binary distribution function), $C_2^{\sigma\sigma'}(\vec{R}_1,\vec{R}_2)$ is the direct correlation function, n_{σ} is the number density of σ - species, and the sum of integrals represents the indirect contribution to the static correlation function [16]. The superscript σ denotes neutral solvent molecules (σ=1), microions (σ=2) and colloidal particles (σ=3).

Following [17], assume that in the equilibrium state the problem of the calculation of the static correlation function is reduced to the determination of the effective potentials for the particles of species σ and σ' located respectively at the points $\vec{R}_1$ and $\vec{R}_2$,

$$F_2^{\sigma\sigma'}(\vec{R}_1,\vec{R}_2) = e^{-\dfrac{V_{eff}^{\sigma\sigma'}(\vec{R}_1,\vec{R}_2)}{T}} \qquad (2)$$

where T is the absolute temperature in the energy units. Using the hypernetted chain approximation [16] one can obtain:

$$V_{eff}^{\sigma\sigma'}(\vec{R}_1,\vec{R}_2) = V^{\sigma\sigma'}(\vec{R}_1,\vec{R}_2) - \sum_{\sigma'=1}^{r} n_{\sigma'} T \int d\vec{R}_3 (e^{\dfrac{V_{eff}^{\sigma\sigma'}(\vec{R}_1,\vec{R}_3)}{T}} -$$

$$-1 + \frac{1}{T} V_{eff}^{\sigma\sigma'}(\vec{R}_1,\vec{R}_3) - \frac{1}{T} V^{\sigma\sigma'}(\vec{R}_1,\vec{R}_3))(e^{\dfrac{V_{eff}^{\sigma'\sigma'}(\vec{R}_3,\vec{R}_2)}{T}} - 1), \qquad (3)$$

where $V^{\sigma\sigma'}(\vec{R})$ is the direct pair potential.

We assume that $V^{\sigma\sigma'}(\vec{R})$ and $V_{eff}^{\sigma\sigma'}(\vec{R})$ can be divided into a universal short-range part $V_0^{\sigma\sigma'}(\vec{R})$ and long-range parts $v^{\sigma\sigma'}(\vec{R})$ and $\tilde{V}^{\sigma\sigma'}(\vec{R})$ respectively, according to

$$V^{\sigma\sigma'}(\vec{R}) = V_0^{\sigma\sigma'}(\vec{R}) + v^{\sigma\sigma'}(\vec{R})$$

$$V_{\text{eff}}^{\sigma\sigma'}(\vec{R}) = V_0^{\sigma\sigma'}(\vec{R}) + \tilde{V}^{\sigma\sigma'}(\vec{R}), \tag{4}$$

$$V_0^{\sigma\sigma'}(\vec{R}) = \begin{cases} \infty, R < a_{\sigma\sigma'}; \\ 0, R > a_{\sigma\sigma'}; \end{cases} \tag{5}$$

where $a_{\sigma\sigma'} = a_\sigma + a_{\sigma'}$, a_σ being the radius of the solid core for the particles of species σ, $\vec{R} = \vec{R}_1 - \vec{R}_2$.

Substituting Eq. (4) into Eq. (3) and linearizing the result obtained with respect to $\tilde{V}^{\sigma\sigma'}/T$, we find the equation describing the behavior of the effective potentials far from the critical point

$$\tilde{V}^{\sigma\sigma'}(\vec{R}) = v^{\sigma\sigma'}(\vec{R}) - \sum_{\sigma''=1}^{r} n_{\sigma''} T \int d\vec{R}' [\theta(a_{\sigma\sigma''} - R') + \frac{1}{T} v^{\sigma\sigma'}(\vec{R}')\theta(R' - a_{\sigma\sigma''})] \times$$

$$\times [\theta(a_{\sigma''\sigma'} - |\vec{R} - \vec{R}'|) + \frac{1}{T}\tilde{V}^{\sigma''\sigma'}(\vec{R} - \vec{R}')\theta(|\vec{R} - \vec{R}'| - a_{\sigma''\sigma'})], \ R > a_{\sigma\sigma'} \tag{6}$$

Eq. (6) can be solved by the Fourier method. The formal solution of Eq. (6) is the following

$$\tilde{V}_k^{\sigma\sigma'} = \frac{T}{n_{\sigma'}}(\delta_{\sigma\sigma'} - n_\sigma \theta_k^{\sigma\sigma'}) - \frac{T}{n_{\sigma'}}\Pi_{\sigma\sigma'}^{-1}(\vec{k}), \tag{7}$$

where

$$\tilde{V}_k^{\sigma\sigma'} = \int_{R > a_{\sigma\sigma'}} d\vec{R}\, e^{-i k\vec{R}} \tilde{V}^{\sigma\sigma'}(\vec{R}), \tag{8}$$

$$v_k^{\sigma\sigma'} = \int_{R > a_{\sigma\sigma'}} d\vec{R}\, e^{-i k\vec{R}} v^{\sigma\sigma'}(\vec{R}), \tag{9}$$

$$\Lambda^{\sigma\sigma'}(\vec{k}) = \theta_k^{\sigma\sigma'} + \frac{1}{T} v_k^{\sigma\sigma'}, \tag{10}$$

$$\theta_k^{\sigma\sigma'} = \int_{R > a_{\sigma\sigma'}} d\vec{R}\, e^{-i k\vec{R}} = \frac{4\pi}{k^3}(\sin k a_{\sigma\sigma'} - k a_{\sigma\sigma'} \cos k a_{\sigma\sigma'}), \tag{11}$$

$$\Pi_{\sigma\sigma'}(\vec{k}) = \delta_{\sigma\sigma'} + n_\sigma \Lambda^{\sigma\sigma'}(\vec{k}). \tag{12}$$

Static structure factors can be defined by the relations

$$S^{\sigma\sigma'}(\vec{k}) = \delta_{\sigma\sigma'} + \sqrt{n_\sigma n_{\sigma'}} \int d\vec{R}\, e^{-i k\vec{R}} [e^{-\dfrac{V_{\text{eff}}^{\sigma\sigma'}(\vec{R})}{T}} - 1]. \tag{13}$$

Accordingly, for the structure factor $S^{\sigma\sigma'}(\vec{k})$ we have

$$S^{\sigma\sigma'}(\vec{k}) = \sqrt{\frac{n_\sigma}{n_{\sigma'}}}\Pi_{\sigma\sigma'}^{-1}(\vec{k}) \ . \tag{14}$$

Let us specify the long-range parts of the interaction potentials for the model system, namely we have

$$v^{11}(R) = -\frac{3(\vec{d}\vec{l}_R)(\vec{\tilde{d}}\vec{l}_R) - (\vec{d}\vec{\tilde{d}})}{R^3},$$ (15)

$$v^{1\sigma}(R) = \frac{Z_\sigma e(\vec{d}\vec{l}_R)}{R}, \quad \sigma = 2,3,$$ (16)

$$v^{\sigma\sigma'}(R) = \frac{Z_\sigma Z_{\sigma'} e^2}{R}, \quad \sigma,\sigma' = 2,3,$$ (17)

where $\vec{d}$, $\vec{\tilde{d}}$ ($|\vec{d}| = |\vec{\tilde{d}}|$) are the dipole moments of the two water molecules, $\vec{R}$ is the radius vector which connects the centers of mass of the two corresponding particles, $\vec{l}_R$ is the unit vector directed parallel to $\vec{R}$. The interesting structure factor for us $S^{33}(\vec{k}) \equiv S^{33}(\vec{k},\vec{d},\vec{\tilde{d}})$, in this case, takes the form

$$S^{33}(\vec{k},\vec{d},\vec{\tilde{d}}) = \Pi_{33}^{-1}(\vec{k}) = \frac{\Pi_{11}(\vec{k})\Pi_{22}(\vec{k}) - \Pi_{12}(\vec{k})\Pi_{21}(\vec{k})}{\det|\Pi_{\sigma\sigma'}(\vec{k})|},$$ (18)

where the elements of the matrix $\Pi_{\sigma\sigma'}(\vec{k})$ are determined by the Eq. (12). The Fourier transformation of Eq. (15) and Eq. (16) has the form

$$v_k^{1\sigma} = -\frac{4\pi i e Z_\sigma(\vec{d}\vec{k})}{k^2} \frac{\sin ka_{1\sigma}}{ka_{1\sigma}}, \quad \sigma = 2,3$$ (19)

$$v_k^{\sigma 1} = -v_k^{1\sigma},$$

$$v_k^{\sigma\sigma'} = \frac{4\pi Z_\sigma Z_{\sigma'} e^2}{k^2} \cos ka_{\sigma\sigma'}, \quad \sigma,\sigma' = 2,3.$$ (20)

Suppose that the dipole moments of the water molecules have arbitrary chaotic orientations in space. The Fourier transformation of Eq. (15) is defined by

$$v_k^{11} = \frac{4\pi}{(ka_{11})^2} \left(\frac{\sin ka_{11}}{ka_{11}} - \cos ka_{11} \right) \left(\frac{3(\vec{k}\vec{d})(\vec{k}\vec{\tilde{d}})}{k^2} - \vec{d}\vec{\tilde{d}} \right).$$ (21)

The average value of the structure factor $S^{33}(\vec{k},\vec{d},\vec{\tilde{d}})$ is given by

$$\left\langle S^{33}(\vec{k},\vec{d},\vec{\tilde{d}}) \right\rangle = \frac{1}{(4\pi)^3} \int S^{33}(\vec{k},\vec{d},\vec{\tilde{d}}) d\Omega d\tilde{\Omega} d\Omega_0,$$ (22)

where $d\Omega = \sin\theta\, d\theta\, d\varphi$, $d\tilde{\Omega} = \sin\tilde{\theta}\, d\tilde{\theta}\, d\tilde{\varphi}$, $d\Omega_0 = \sin\theta_0\, d\theta_0\, d\varphi_0$ with θ, $\tilde{\theta}$, θ_0 are spherical angles between the vectors $\vec{k}$ and $\vec{d}$, $\vec{k}$ and $\vec{\tilde{d}}$, $\vec{d}$ and $\vec{\tilde{d}}$ respectively. The distribution function of the orientations of the dipole moments in Eq. (22) is assumed to be equal to unity. Numerical analysis of Eq. (22) shows that under the certain values of the wave vector k for the fixed charge number Z_3, the integrand has singularities over a domain of integration. We assume (see [17]) that generation of the critical mode in colloidal suspension is connected with appearance of singularities in the structure factor S^{33} (or in the effective potential $\tilde{V}_k^{33}$). The reciprocal of the wave vector k_0 when such a singularity occurs, will be treated as an estimate for the inter-particle distance R_0 ($R_0 = 2\pi / k_0$) in the present phase. Deposition of the inhomogeneous phase will be attributed to the large magnitudes (some times larger than a_{33}) of the distance R_0.

Carrying out the integration in Eq. (22) by the variables $d\Omega$ and $d\tilde{\Omega}$, and making the numerical analysis of the retaining integral with respect to θ_0, one can obtain the definitive values of the wave vector $\vec{k} \equiv \vec{k}_o$ under which the structure factor $\left\langle S^{33}(\vec{k}, \vec{d}, \tilde{\vec{d}}) \right\rangle$ has singularities. Fig.1 (solid line) represents the dependence

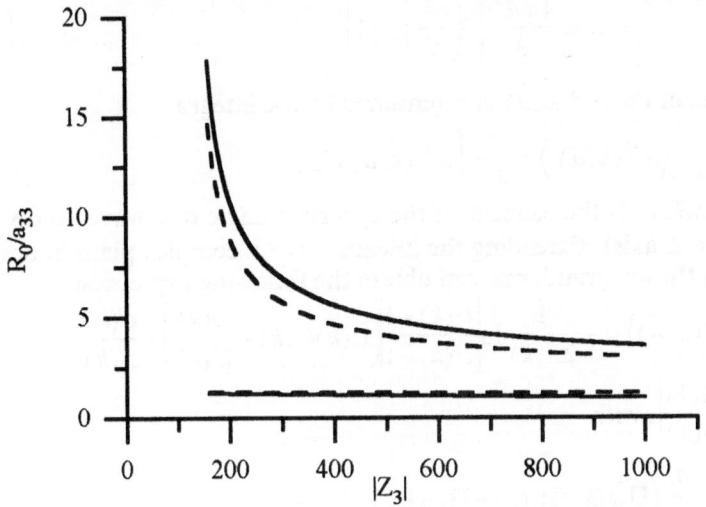

Figure 1. Dimensionless interparticle distance R_0 / a_{33} as a function of the charge number of polystyrene colloidal particles $|Z_3|$ for the three-component model with noncorrelated (solid line) and correlated (dashed line) dipole moments of the water molecules at $n_3 = 3.3 \cdot 10^{12} cm^{-3}$, $a_3 = 5.5 \cdot 10^{-6} cm$, $T = 293 \, ^{\circ}K$ [18] ($Z_{min} = 140$).

of the distance $R_0 = 2\pi / k_0$ on the charge number $|Z_3|$. As the charge of colloidal particles $|Z_3|$ decreases, we observe the effect of the existence of minimal charge at $|Z_3| \equiv |Z_{min}|$. Under the values of the charge number $|Z_3| < |Z_{min}|$ the loss of singularities in Eq. (22) occurs. In vicinity of the minimal charge, one can see the strong increase of the estimated value for the distance R_0 that suggests the appearance of the instabilities in the existing phase. Therefore, we conclude that one could expect deposition of an inhomogeneous phase under the values of the charge number $|Z_3| > |Z_{min}|$. If $|Z_3| < |Z_{min}|$ the homogeneous phase appears.

2.1. STRONGLY-CORRELATED DIPOLE MOMENTS OF THE WATER MOLECULES

Assume that the dipole moments of the water molecules are strongly correlated to each other ($\vec{d} = \tilde{\vec{d}}$). This effect may be caused by the application of an external electromagnetic field or by the influence of the substrate. Using Eq. (9) and Eq. (15), we obtain the Fourier transform of $v^{11}(R)$

$$v_k^{11} = \frac{4\pi d^2}{(ka_{11})^2}\left(\frac{\sin ka_{11}}{ka_{11}} - \cos ka_{11}\right)\left(3\left(\frac{\vec{dk}}{dk}\right)^2 - 1\right), \tag{23}$$

where $d = |\vec{d}|$. Note, that in the case $ka_{11} < 1$ (that is sufficiently satisfied in the types of colloidal suspensions considered) Eq. (23) reduces to the following form:

$$v_k^{11} = \frac{4\pi d^2}{3}\left(3\left(\frac{\vec{dk}}{dk}\right)^2 - 1\right). \tag{24}$$

The average value of the $S^{33}(\vec{k},\vec{d})$ is determined by the integral

$$\left\langle S^{33}(\vec{k},\vec{d})\right\rangle = \frac{1}{4\pi}\int S^{33}(\vec{k},\vec{d})d\Omega, \tag{25}$$

where $d\Omega = \sin\theta\, d\theta\, d\varphi$ is the element of the spherical angle (the wave vector $\vec{k}$ is directed along the Z axis). Extending the integral over a complex plain in domain of singularities of the integrand, one can obtain the following expression

$$\left\langle S^{33}(\vec{k},\vec{d})\right\rangle = -\frac{1}{2A(k)}\ln\left|\frac{U(k)+1}{U(k)-1}\right|\left(C(k)U(k) + \frac{D(k)}{U(k)}\right) + \frac{C(k)}{A(k)}, \tag{26}$$

where $U(k)^2 = \left|\frac{A(k)}{B(k)}\right|$, with

$$A(k) = \frac{4\pi d^2 n_1}{T}\left(\Pi_{22}(k)\Pi_{33}(k) - \Pi_{32}(k)\Pi_{23}(k)\right) +$$
$$+ \left(\frac{4\pi ed}{kT}\right)^2 n_1 n_2 Z_2 \frac{\sin ka_{12}}{ka_{12}}\left(2Z_3 \frac{\sin ka_{13}}{ka_{13}}\Pi_{23}(k) - Z_2 \frac{\sin ka_{12}}{ka_{12}}\Pi_{33}(k)\right), \tag{27}$$

$$B(k) = \left(\Pi_{22}(k)\Pi_{33}(k) - \Pi_{32}(k)\Pi_{23}(k)\right)\left(1 + n_1\theta_k^{11} - \frac{4\pi}{3}\frac{d^2 n_1}{T}\right) +$$
$$+ 2n_1 n_2 \theta_k^{12}\theta_k^{13}\Pi_{23}(k) - n_1\left(n_2\left(\theta_k^{12}\right)^2\Pi_{33}(k) + n_3\left(\theta_k^{13}\right)^2\Pi_{22}(k)\right), \tag{28}$$

$$C(k) = \frac{4\pi d^2 n_1 \Pi_{22}(k)}{T} - \frac{(4\pi)^2 n_1 n_2 Z_2^2 e^2 d^2}{T^2 k^2}\left(\frac{\sin ka_{12}}{ka_{12}}\right)^2, \tag{29}$$

$$D(k) = \left(1 + n_1\theta_k^{11}\right)\Pi_{22}(k) - n_1 n_2\left(\theta_k^{12}\right)^2 - \frac{4\pi}{3}\frac{n_1 d^2}{T}\Pi_{22}(k). \tag{30}$$

In the domain of existence of Eq. (25) the integration leads to the following expression

$$\left\langle S^{33}(\vec{k},\vec{d})\right\rangle = \frac{1}{2}\left(I(1) - I(-1)\right), \tag{31}$$

where

$$I(x) = \frac{C(k)}{A(k)}x + \arctan\left(\sqrt{\frac{A(k)}{B(k)}}x\right)\left(\frac{D(k)}{\sqrt{A(k)B(k)}} - \sqrt{\frac{B(k)}{A(k)^3}}C(k)\right). \tag{32}$$

Singularities of Eq. (26) are determined by the equations
$$U(k) = 1, \; U(k) = 0. \tag{33}$$
Fig.1 (dashed line) represents results of numerical solution of Eq. (33) at different values of the charge number Z_3. As one can see, the model adopted demonstrates qualitatively similar behavior with respect to the tree-component model with noncorrelated dipole moments. The discrepancy lies (see Fig.1) in the magnitudes of the wave vector k_0 under which critical mode generates.

3. Discussion

In summary, we have solved the HNC equation in linear approximation for the three-component charged hard sphere system. The procedure for simulation of the ordering tendency and inhomogeneous deposition effects in such a system is proposed and implemented on the basis of singularities of the structure factor. Applying this approach, the structural properties of the water suspension of charged colloidal particles and deposition phenomena therein are analyzed.

We explored the model when all interactions between all species in the solution, colloidal particles, counter-ions and solvent (water) molecules were explicitly present in the calculations. The effects of the substrate were neglected completely. First, we considered the case where the dipole moments of the water molecules are absolutely noncorrelated to each other. Averaging the structure factor $S^{33}(\vec{k}, \vec{d}, \tilde{d})$ over all orientations of dipoles, we observed that the inhomogeneous phase appear under those values of the charge number of colloidal particles which obey the inequality $|Z_3| > |Z_{min}|$. Note that the effect of the existence of a minimal charge in colloidal suspension is confirmed by experimental [18] and theoretical [19,20] studies. Finally, we considered the case where the dipole moments of the water molecules are strongly correlated. In this case, we found qualitatively-similar behavior of the distance R_0 with respect to the three-component model with noncorrelated dipole moments.

It is interesting to compare the results obtained for the three-component model with those for the one-component model, with the famous DLVO inter-particle potential [21]

$$v^{33}(R) = \frac{Z_{eff}^2 e^2}{\varepsilon R} e^{-\frac{R}{R_D}}, \; Z_{eff} = \frac{Z_3 e^{\frac{a}{R_D}}}{1 + \frac{a}{R_D}}, \; R_D^2 = \frac{T\varepsilon}{4\pi n_c e^2}, \tag{34}$$

where R_D is Debye radius of the counterion cloud, $a \equiv a_{33}$ is the radius of colloidal particle, $n_c \equiv n_2$ is the number density of the counter-ions, ε is the relative permitivity of water. The counter-ions and solvent molecules are treated as a uniform neutralizing background which determines the screening in the system. Assuming $r=1$ in Eq. (1) and implementing the technique adopted, one obtains results represented in Fig.2 (solid line). A two-component model (colloidal particles plus counter-ions), where the solvent is treated as a continuum, leads to the following potentials

$$v^{\sigma\sigma'}(R) = \frac{Z_\sigma Z_{\sigma'} e^2}{\varepsilon R}, \ R > a_{\sigma\sigma'}, \tag{35}$$

where $\sigma, \sigma' = 2,3$. Results obtained for the two-component model are represented

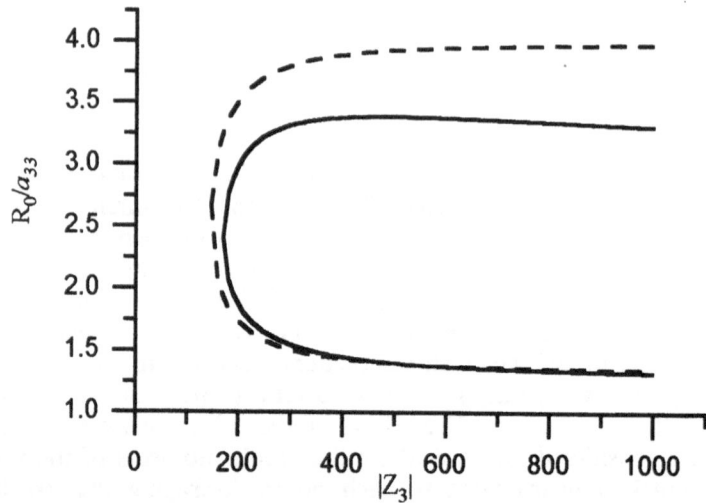

Figure 2. Dependence of R_0 / a_{33} on $|Z_3|$ for the one- (solid line, $Z_{min} = 170$) and two- (dashed line, $Z_{min} = 146$) component models.

in Fig.2 (dashed line). In both cases, the effect of the existence of minimal charge as it was in the tree-component model is observed. The discrepancy lies in the character of singularities in the vicinity of the minimal charge.

4. Conclusion

It was established that phase behavior of the charged colloidal suspension strongly depends on the magnitude of the charge number of colloidal particles. According to our results, one can expect deposition of an inhomogeneous coating from the charged colloidal suspension when the magnitude of the charge number of colloidal particles exceeds a minimal value, and vice versa; homogeneous deposition is more likely to occur in suspensions with low values of the surface charge of colloidal particles. We also conclude that when applying an external field, which makes the dipole moments of the water molecules correlated, no significant change in the structural properties of the deposited phase can be expected.

Our model could be improved by accounting for the interactions between suspension and substrate. The approach that we have used seems suitable also for the study of the deposition phenomena from the suspensions of colloids (a macro-ion) and polymers (a neutral particle) which are the subject of extensive computer study [22,23]. Work in these directions is underway.

4. Acknowledgments

One of the authors (D.B.L.) wishes to express his large sincere gratitude to Professor O.I. Gerasimov for help and stimulating discussions and to Professor A.E. Kiv and Professor Ya.O. Roizin for their attention to the paper presented.

5. References

1. Ohring, M. (1992) *The Materials Science of Thin Films*, Academic, New York.
2. Sze, S.M. (1985) *Semiconductor Devices Physics and Technology*, Wiley, New York.
3. Tu, K.N., Mayer, J.W., and Feldman, L.C. (1992) *Electronic Thin Film Science For Electrical Engineers and Materials Scientists*, Macmillan, New York.
4. Dimitriadis, C.A., Stoemenos, J., Coxon, P.A., Friligkos, S., Antonopoulos, J., and Economou, N.A. (1993) Effect of pressure on the growth of crystallites of low-pressure chemical-vapor-deposited polycrystalline silicon films and the effective electron mobility under high normal field in thin-film transistors, *J.Appl.Phys.* **73**, 8402-8411.
5. Holland, E.R., Bloor, D., Monkman, A.P., Brown, A., De Leeuw, D., Bouman, M.M., and Meijer, E.W. (1994) Effects of order and disorder on field-effect mobilities measured in conjageted polymer thin-film transistors, *J.Appl.Phys.*75, 7954-7957.
6. Kerker, M. (1969) *The Scattering of Light and Other Electromagnetic Radiation*, Academic Press, New York.
7. Van de Hulst, H.C. (1981) *Light Scattering by Small Particles*, Dover, New York.
8. Sherman, A. (1987) *Chemical Vapor Deposition for Microelectronics*, Noyes, New York.
9. Alexandrov, L.N. (1972) *Growth, Kinetics and Structure Formation of Thin Films*, Nauka, Novosibirsk (in Russian).
10. Kasper, E. and Bean, J.C. (eds.) (1988) *Silicon Molecular Beam Epitaxy*, CRC, Boca Raton, Vols. 1 and 2.
11. Baude, P.F., Tamagawa, C.Ye.T., and Polla, D.L. (1993) Fabrication of sol-gel derived ferroelectric optical waveguides, *J.Appl.Phys.* **73**, 7960-7962.
12. Hauser, E.A. (1951) Modern colloidchemical consepts of the phenomenon of coagulation, *J. Phys. and Colloid Chem.* **55**, 605-611.
13. Imhof, A., van Blaaderen, A., Maret, G., Mellema, J., and Dhont, J.K.G. (1994) A comparison between the long-time self-diffusion and low shear viscosity of concentrated dispersions of charged colloidal silica spheres, *J. Chem. Phys.* **100**, 2170-2181.
14. Monovoukas, Y. and Gast, A.P., (1989) The experimental phase diagram of charged colloidal suspensions, *J. Colloid and Interface Sci.* **128**, 533-548.
15. Hoppenbrouwers, M. and van de Water, W. (1995) Charged colloidal systems, The Workshop on Complex Fluids and Plasmas, EUT, Eindhoven, The Netherlands.
16. Hansen, J.P. and McDonald, I.R. (1986) *Theory of Simple Liquids*, Academic, New York.
17. Gerasimov, O.I., Schram, P., Sitenko, A.G., and Zagorodny, A.G. (1996) Critical behavior of effective potentials in colloidal suspensions within the charged hard-sphere model, *Physica A*, in press.
18. Derksen, J. and van de Water, W. (1992) Hydrodynamics of colloidal crystals, *Phys. Rev. A* **45**, 5660-5673.
19. Schram, P. and Trigger, S. (1995) Minimal charge of macroions for crystallization of colloidal suspensions, The Workshop on Complex Fluids and Plasmas, EUT, Eindhoven, The Netherlands.
20. Hastings, R. (1978), On the crystallization of macroionic solutions, *J. Chem. Phys.* **68**, 675-678.
21. Verwey, E.J and Overbeek, J. (1948) *Theory of the Stability of Lyophobic Colloids*, Elsevier, Amsterdam.
22. Dickman, R. and Yethiraj, A. (1994) Polymer-induced forces between colloidal particles. A Monte Carlo simulation, *J. Chem. Phys.* **100**, 4683-4689.
23. Meijer, E.J. and Frenkel, D. (1994) Colloids dispersed in polymer solutions. A computer simulation study, *J. Chem. Phys.* **100**, 6873-6887.

QUANTUM MECHANICAL SIMULATIONS IN SEMICONDUCTOR MATERIALS SCIENCE:

The Tight-Binding Molecular Dynamics Approach

D. MARIC
Swiss Center for Scientific Computing, Switzerland

L. COLOMBO
*Dipartimento di Fisica dell' Universit `a di Milano and
Istituto Nazionale per la Fisica della Materia (INFM), Italy*

1. Introduction

Atomic-scale modeling of materials requires an accurate description of the quantum-mechanical features of chemical bonding. The most fundamental and reliable approach to this problem is first-principles electronic structure calculations, such as the Car-Parrinello molecular dynamics (CPMD) [1, 2]. Unfortunately, CPMD is very computationally-intensive: as a matter of fact, it is usually restricted to short simulation times (~10ps) and to a small number of atoms (~100). Accordingly, the study of *real* materials (i.e. amorphous or glassy materials, extended defects, interfaces, poly- and nano-crystalline materials) and materials science processes (i.e. defect migration, epitaxial growth, microstructural evolution under irradiation, crack propagation, materials fatigue and embrittlement) are too computationally demanding for CPMD.

Very recently, a new simulation scheme has been proposed that has both the accuracy needed to describe complex systems, such as covalent materials characterized by directional chemical bonds, and a reduced computational workload suitable for large-scale simulations. The method is known as tight-binding molecular dynamics (TBMD) [3, 4]. The basic idea of TBMD is to derive the inter-atomic forces from the electronic structure of the simulated system as calculated in the framework of the semi-empirical tight-binding (TB) model. In this way, the quantum mechanical features are properly taken into account, while still keeping the overall computational workload nicely small.

In this work, we briefly review the basic theory of the TBMD scheme and different strategies for its numerical implementation.

R. C. Tennyson and A. E. Kiv (eds.),
Computer Modelling of Electronic and Atomic Processes in Solids, 79–85.
© 1997 *Kluwer Academic Publishers.*

2.   The tight-binding model for simulations: basic formalism

In the TB model [5], the one-electron wavefunctions $|\Psi_n\rangle$ are represented as linear combinations of atomic orbitals $|\varphi_{l\,\alpha}\rangle$ (l is the angular momentum quantum number index and α labels the ions). For large-scale TBMD simulations, the use of an orthogonal basis set (Löwdin orbitals [6]) is numerically-convenient in most cases.

The key objects of the orthogonal TB model are the matrix elements of the one-electron Hamiltonian h, namely $<\varphi_{l'}\,\beta\,|h|\,\varphi_l\,\alpha\,)$. They are usually considered as constants to be fitted. The fitting step is carried out with the following approximations: [7] (*i*) a minimal basis set is used to represent $|\Psi_n\rangle$; (*ii*) first (or second) near neighbour interactions only; (*iii*) two-center approximation.

The TB hopping integrals are evaluated by fitting a suitable data base obtained either from experiments or first principles calculations [5]. Typically, the fitting is applied to the electronic energy bands of the equilibrium crystal phase. This procedure gives the values of the TB hopping integrals for atoms clamped at their equilibrium inter-atomic distance. In order to study finite-temperature systems, we need to rescale the fitted TB matrix elements, with respect to the actual inter-atomic distance, according to suitable scaling functions. The equilibrium hopping integral

$$< \varphi_{l'}\beta\,|h|\,\varphi_l\,\alpha\,) = h_{ll'}^{(0)}$$

is therefore generalized as $h_{ll'}(R_{\alpha\beta}) = h_{ll'}^{(0)}\ f_{u'}(R\alpha\beta)$, where $f_{u'}(R\alpha\beta)$ is known as a scaling function.

Once the single-particle energies are obtained by solving the secular problem

$$h\,|\Psi_n\,\rangle = \varepsilon_n\,|\Psi_n\rangle \tag{1}$$

the total potential energy E_{tot} of a crystalline system of ion cores and valence electrons can be written as:

$$
\begin{aligned}
E_{tot} &= U_{ie} + U_{ii} + U_{ee} \\
&= 2\sum_{k,\,n} f\text{FD}\,(\varepsilon_n, T)\varepsilon_n + U_{ii} - U_{ee}
\end{aligned}
\tag{2}
$$

where $f\text{FD}\,(\varepsilon_n, T)$ is the Fermi-Dirac distribution function and the $-\,U_{ee}$ contribution corrects the double counting of the electron-electron interactions in the first term. The sum over all the single-particle energies is commonly-named band structure energy E_{bs}.

$$E_{bs} = 2\sum_{k,\,n} f\text{FD}\,(\varepsilon_n, T)\,\varepsilon_n\,(k) \tag{3}$$

and contains a factor 2 to take into account the spin degeneracy.

Within the present semi-empirical TB scheme, the last two terms appearing in eq. (2) are usually grouped into an effective repulsive potential $U_{rep} = U_{ii} - U_{ee}$ which is assumed to be short-ranged because of the efficient dielectric screening occurring in semi-conductor materials [5, 8]. This *ansatz* is especially valid for homopolar materials. U_{rep} effectively describes a rich phenomenology: (*i*) steric ion-ion repulsion; (*ii*) overlap interaction due to the non-orthogonality of the basis orbitals; (*iii*) correction of the double counting of electron-electron interactions; (*iv*) possible charge transfer effects. The common fitting procedure adopted to determine U_{rep} and scaling functions is to reproduce the cohesive energy curve of a selected variety of different crystalline phases. In this way, the resulting TB model has the desired transferability and can manage different equilibrium distances, atomic coordinations, and chemical bondings.

As for simulations, during an MD run, we must compute the inter-atomic forces $\mathbf{F}_\alpha$ ($\alpha = 1, 2, \ldots, N_{at}$) to move atoms and to generate trajectories in the phase space. They can be evaluated from the TBMD Hamiltonian H

$$H = \sum_\alpha \frac{p_\alpha^2}{2m_\alpha} + 2 \sum_n^{(occup)} \epsilon_n + U_{rep} \tag{4}$$

(the electronic temperature is assumed to be zero) where the limit (*occup*) indicates that we use just those electron energies ϵ_n belonging to the lower half spectrum of the TB matrix. The force $\mathbf{F}\alpha$ is given by:

$$\mathbf{F}_\alpha = \underbrace{-\frac{\partial}{\partial \mathbf{R}_\alpha} 2 \sum_n^{(occup)} \epsilon_n}_{\text{attractive force}} \underbrace{-\frac{\partial}{\partial \mathbf{R}_\alpha} U_{rep}}_{\text{repulsive force}} \tag{5}$$

where the repulsive force is trivially-computed since U_{rep} is known analytically as a function of the inter-atomic distances. The attractive force, in turn, can be computed only numerically, using the Hellman-Feynman theorem:

$$-\frac{\partial}{\partial \mathbf{R}_\alpha} 2 \sum_n^{(occup)} \epsilon_n = -2 \sum_n^{(occup)} \sum_{l\gamma} \sum_{l'\beta} b_{l\gamma}^n b_{l'\beta}^n h_{ll'}^{(0)} \frac{\partial}{\partial \mathbf{R}_\alpha} f_{ll'}(\mathbf{R}_{\gamma\beta}) \tag{6}$$

It is clear from eqs. 4 and 6 that the spectrum of eigen-values/-vectors of the TB matrix is needed to compute, respectively, the band-structure contribution to the total potential energy and the attractive contribution to the forces. The diagonalization of the TB matrix dominates the overall computational workload of the TBMD scheme.

3. The numerical implementation of TBMD

The most conventional TBMD code is based on the direct diagonalization of the real symmetric TB matrix. This implementation will be hereafter referred to as the standard diagonalization method (SDM).

SDM-TBMD is easy to implement and to port, provided that a standard FORTRAN code is written and a routine from any portable mathematical library is used for the diagonalization. The computational workload of SDM-TBMD scales as the cubic power of N_{at} ($O(N_{at}^3)$) method). Despite such a $O(N_{at}^3)$ bottleneck, the conventional TBMD scheme as outlined above is rather efficient. We obtained a sustained maximum speed of 1.66 Gflops on one processor NEC SX-3 with more than 98% of the operation vectorized for a 512-atom cell.

In order to simulate large systems, an efficient parallel implementation of the conventional scheme has been introduced [9]. The new method employs reordering of the atoms in order to minimize the number of atomic interactions which are cut by processing-element (PE) boundaries. Moreover, the sparsity of the TB Hamiltonian matrix h is exploited such that only non-zero elements are stored and distributed over the PEs in a way that avoids excessive communication when it is referenced. The reordering of the atoms has been implemented through the highly efficient Reverse Cuthill-McKee (RCM) algorithm, while the Lanczos algorithm has been adopted for the diagonalization.

The method outlined above has been benchmarked on a NEC Cenju-3 parallel computer available at SCSC-CSCS. The observed performance in Tab. I illustrates several features of the algorithm, in particular its scalability to large numbers of PEs on the NEC Cenju-3 even for a relatively small $N_{at} = 216$ problem. In addition, at certain problem sizes, a super-linear speedup is observed, e.g., from 16 to 32 PEs on the Cenju-3 for the $N_{at} = 512$ problem size, as opposed to 16 Pes for the same N_{at}. For 32 PEs, the blocks of eigenvectors fit into the secondary cache and effectively avoid main-memory access for the second phase of the diagonalization. On the other hand, the algorithm for a given number of PEs still scales as N_{at}^3, as is inherent to full matrix diagonalization algorithms.

4. Applications

In this section, we review a recent TBMD simulation about the response of a crystalline silicon lattice to ion-beam irradiation [10]. Our main goal is to provide a detailed characterization of both structural and electronic properties of irradiated samples. We have randomly inserted Si atoms into a well equilibrated room temperature sample of c-Si. After that, a careful equilibration at room temperature was performed and finally, the structural and electronic properties have been collected.

Regarding the overall structural features, we observed that the host crystal behaves like a distorted diamond lattice up to a defect concentration of 7 - 8%. Here, the structure of the irradiated sample is very similar to that of c-Si up to the fourth nearest neighbour shell. By increasing the absolute number of defects, we have clearly observed a crystal-to-amorphous transition, as shown in Fig. 1, where the particle-particle correlation function for three samples is reported.

There, we also show the particle-particle correlation function of a-Si obtained by quenching a well equilibrated 216-atom liquid silicon system [11]. We observe that the sample c (bottom panel) is thoroughly disordered by the huge number of defects and the crystal order is clearly lost at distances beyond the second neigbour shell. The good agreement between the solid line and dotted curve indicates that the defect-induced amorphization process has created an amorphous network with structural properties similar to an under-cooled liquid.

The modification of the electronic structure of silicon caused by the insertion of interstitials has been studied by calculating the total electronic density of states (DOS) and the results for the as-implanted samples are shown in Fig. 2. Sample a still presents a close similarity to the DOS of c-Si: in fact, the three main structures close to -12eV, -6eV and -3 eV are in rather good agreement with experimental valence x-ray photoelectron spectroscopy data [12] . By increasing the number of defects, the crystal-like structures in the DOS become less and less evident (sample b, central panel of Fig. 2) and, finally, the amorphous structure is reached in sample c-Si.

An interesting feature common to all of the samples is the sizeable number of states at the Fermi level (0 eV), even if no clear trend is observed for the as-implanted samples. They correspond to the dangling bonds associated with the under-coordinated atoms which are found in any irradiated sample. In the case of sample c, we found 5%, 85.2% and 9.8% of 3-fold, 4-fold and 5-fold coordinated atoms, respectively. The resulting average coordination number is 4.05, in good agreement with the 3.97 experimental value [13].

Performance of TBMD on the NEC Cenju-3								
Problem size		Processing elements						
N_{cell}	N_{at}	1	2	4	8	16	32	64	128
2	64	10.80^2	8.568	5.515	4.067	3.315	2.907	2.897	3.303
3	216	407.40^2	425.6	200.3	98.46	50.90	30.21	20.81	21.23
4	512		6593	3250	1812	883.4	289.7	155.8	113.5
5	1000				10442	5202	2643	1347	632.1
All execution times in seconds								

Table 1

Timing results on the NEC Cenju-3 for the SDM method based on the parallel eigensolver. The original sequential version runs on 1 Cenju-3 PE. The times for the parallel version run on 1 PE are 16.15 secs. and 835.7 secs. (N_{cell} =2,3). Larger N_{cell} values on 1 PE are not possible due to memory restrictions.

84

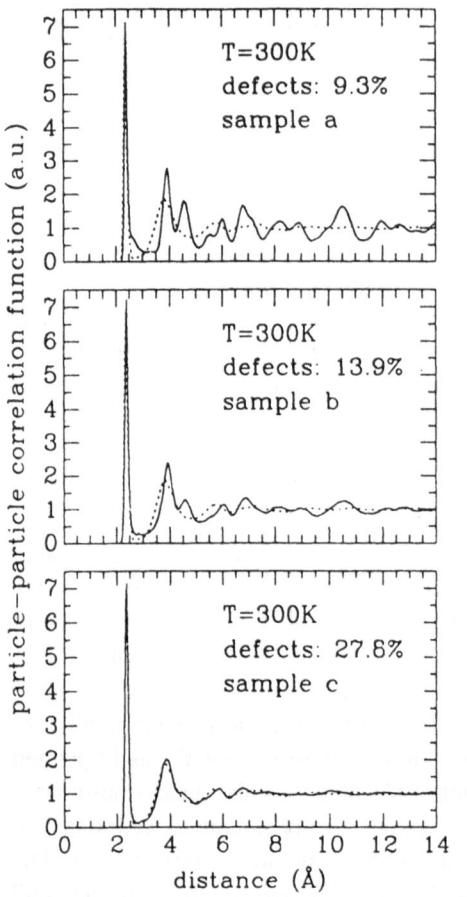

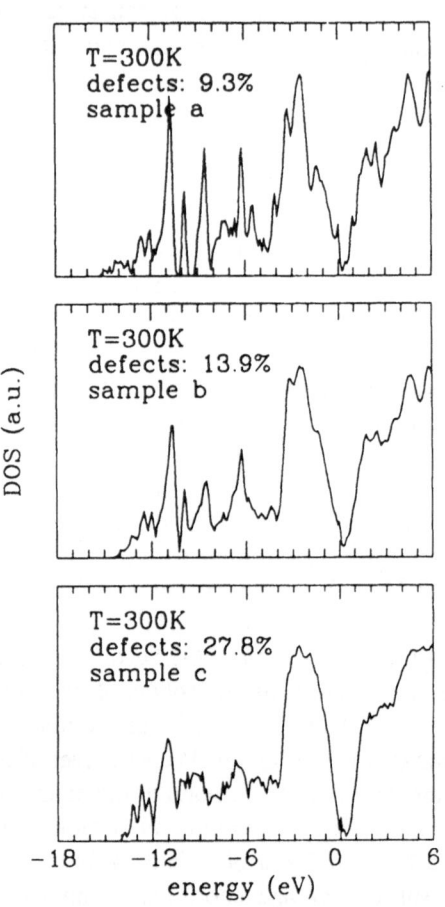

Fig.1

Particle-particle correlation function
for three irradiated samples (solid line).
The absolute number of implanted defects
is 20, 30 and 60 for sample *a*, *b* and *c*,
respectively. Dotted line represents an
undercooled liquid sample (see text).

Fig. 2

Total electronic density of states
(DOS) for three irradiated
samples.

Acknowledgements

The Authors are especially grateful to W. Sawyer for a long-term collaboration. The
parallel eigensolver has been mainly developed by W. Sawyer within the project "Joint
CSCS-ETH ZURICH/NEC Collaboration in Parallel Processing."

References

[1]   R. Car and M. Parrinello, Phys. Rev. Lett.   bf 55 , 2471, (1985)

[2]   G. Galli and F. Mauri, in *Computer Simulation in Chemical Physics* (Kluwer Academic
        Publishers, Amsterdam, 1993), M. Meyer and V. Pontikis Eds., pg.261

[3]   C. Z. Wang, K. M. Ho, in *Advances in Chemical Physics*, I. Prigogine and A. A. Rice Eds.,
                vol. LXXXIX, pg.651

[4]   L. Colombo, Ann. Rev. Comp. Phys., vol. IV (1996), in press

[5]   W.A. Harrison, *Electronic Structure and the Properties of Solids* (Dover Publications,
                New York, 1989)

[6]   P. O. Löwdin , J. Mol. Spectr. 3, 46 (1959)

[7]   J. C. Slater and G.F. Koster, Phys. Rev. 94, 1498 (1954)

[8]   D. J. Chadi, Phys. Rev. Lett. 41, 1062 (1978); Phys. Rev. B 29, 785 (1984)

[9]   L. Colombo and W. Sawyer, Mat. Sci. Eng. B (1996), in press

[10]   D. Maric and L. Colombo, Europhys. Lett. 29, 623 (1995)

[11]   G. Servalli and L. Colombo, Europhys. Lett. 22, 107 (1993)

[12]   *Tetrahedrally bounded amorphous semiconductors* , D. Adler and K. Fritzsche Eds.
                (Plenum, New York, 1985)

[13]   Y. Waseda and S. Suzuki, Z. Phys. B20, 339 (1975)

MOLECULAR DYNAMICS STUDY OF SELF-ORGANIZATION OF POLYMER LIQUID CRYSTALS

A.I.MELKER and A.N.EFLEEV
Department of Metal Physics, Physics and Mechanics Faculty,
St. Petersburg State Technical University,
Polytechnicheskaya str. 29, St.Petersburg, 195251, Russia

In this contribution, we report on a study of the self-organization of extended polymer liquid crystal (PLC) chains into condensed globules using the molecular dynamics method. We have found that a globule structure is formed in several stages: first, is an equilibrium quasi-linear structure; the second stage is associated with dimensional transformation of the quasi-linear structure into two-dimensional liquid-crystal-rich domains; during the third or aggregation stage, a globule forms from the domains, and in the fourth stage, this globule exhibits Brownian motion of its constituents.

1. Introduction

Polymer liquid crystals (PLCs) hold much promise as construction materials. Compared to conventional polymers, PLCs have better stability in vacuum and at elevated temperatures, lower flammability and thermal expansivity, and are stable against visible and ultraviolet light [1]. The simplest type of molecular structure of PLCs, called longitudinal, consists of PLC chains containing approximately rectangular liquid crystals (LC) sequences interconnected with flexible ones. When a system of such chains approaches thermodynamic equilibrium, the LC sequences coalesce and form LC-domains (or islands). Up to now, this phenomenon has not been completely understood, notwithstanding all kinds of plausible phenomenological explanations[2].

To gain a more penetrating insight into the problem, we have chosen molecular-dynamics computer simulation as a tool for investigation because no other method permits the study of structure formation as a self-organization process [3]. The molecular-dynamics models used at present in polymer physics are rather numerous. To make good use of the models, it is well to bear in mind that two-dimensional bead models have the advantage of very simple visualization of the results obtained. Moreover, in any bead model, a monomer is replaced with a bead and behaves as a whole. Both of these traits help to present the main features of self-organization of a system in a state of chaos or other dimension.

R. C. Tennyson and A. E. Kiv (eds.),
Computer Modelling of Electronic and Atomic Processes in Solids, 87–96.
© 1997 *Kluwer Academic Publishers.*

In this paper, we use a two-dimensional molecular-dynamics bead model to study the self-organization of a polymer liquid crystal chain containing rigid and flexible sequences.

2. Computer simulation procedure

In many respects, our computer simulation procedure is identical to that employed previously in [4-7], so we consider only the rest of the procedure developed specially for the problem discussed. In our numerical simulation study, we used a two-dimensional bead model with 360, 1020 and 1080 beads where each bead has two degrees of freedom. Each rigid (R) and flexible (F) sequence of a main chain consists of six beads arranged in a regular way (...RFRFRF...). The bead mass for a rigid sequence is 1.8 times larger than for a flexible one. Interaction between all beads is modelled using the anharmonic Morse potential, but this interaction is assumed to be 10 and 25 times as weak for second neighbours in the rigid and flexible sequences respectively as for all nearest neighbours connected covalently. All other interactions are assumed to be 25 times weaker in comparision with the covalent interaction.

We have studied two different combinations of equilibrium distances between beads:

- Case 1. The equilibrium distance is the same for the interaction between any pair of beads except for the second neighbours in the rigid sequences; the latter is assumed to be two times larger.

- Case 2. By comparision with Case 1 the equilibrium distance for the weak interactions is increased in $\sqrt{2}$ times except for the second neighbours in the rigid sequences where it does not change.

At first, the straightened bead chain is in a state of unstable static equilibrium. To put the chain into motion, the initial velocities of beads are selected at random from the Maxwellian distribution corresponding to a preset temperature that is small, and fulfills the role of a starter.

The Newtonian equations of motion are integrated numerically using the Nordsieck method to the fifth order of accuracy. The optimum time step is equal to 1/22 period of bead vibrations to suit energy and phase trajectories of the beads. All the calculations were made on an IBM PC/AT 486 computer. One evolution history covers 180000 - 220000 time steps, i.e. approximately 800 - 10000 periods of bead vibrations.

3. Results and Discussion

We present and discuss structures for two evolution histories keeping in mind that other histories are similar. *Figure 1a* through *e* shows the time evolution of an initially-straight chain of beads for Case 1. Folding starts at the free ends of the PLC chains. Simultaneously, the straight flexible sequences transform into zigzags. Immediately afterwards, the quasi-linear structure generates large transverse bending waves. This leads to overlap of the rigid sequences and changes the structure of the zigzag flexible ones. Then, these zebra domains are aggregated into domain clusters.

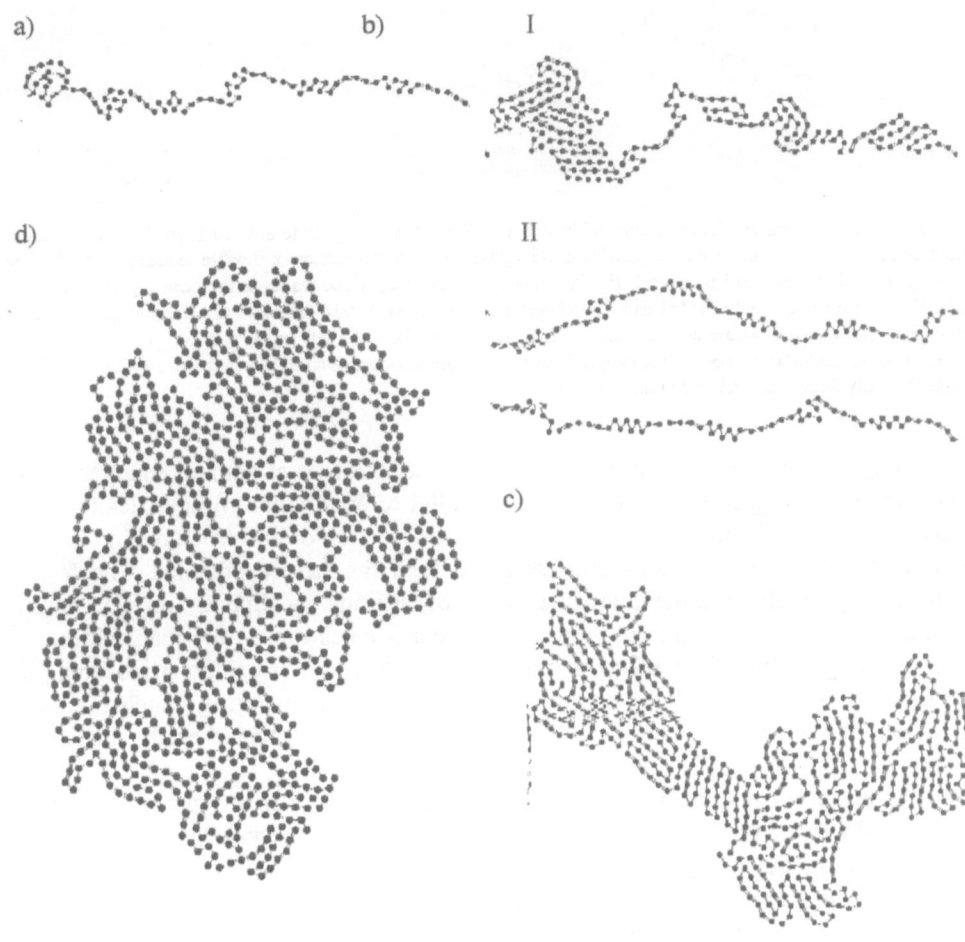

a) b) I

d) II

c)

Figure 1 (a) - (d).

90

e)

Figure 1. Time evolution of spontaneous folding of a PLC chain into a globule in Case 1. *(a)* Folding begins at the free ends of a chain and is accompanied with Straight-Zigzag transformation of flexible sequences: 15000 time steps (here, only the left end is shown). *(b)* Appearance of large transverse waves in the central part of a chain: 45000 time steps (here, only the left end (I) and center (II) are shown). *(c)* Formation and disintegration of zebra domains, appearance of stripe domains: 87000 time steps (here, only the left end is shown). *(d)* Formation of a blend globule: 96000 time steps. *(e)* Reological character of the self-organization of a globule: 145000 time steps. Note that only three small zebra domains are left.

Along with this process but more slowly, the second type of domain emerges. Here, the straight rigid sequences are arranged parallel to each other without flexible ones between them producing a stripe domain. Close inspection of the structures in *Figure 1* shows that the zebra domains are less stable than the striped domains. The former transform gradually into the latter in a variety of fashions: by slipping, rotation, reptile motion, etc. However, unfolding the zebra domains takes more time than folding, so the resulting globule is a blend of the regular zebra and irregular stripe domains. Thereafter, the blend globule exibits the structure evolution during which the zebra domains disappear. It is worth noting that at this stage the globule easily changes its form.

The evolution history in Case 2 (*Figure 2*) is similar in many respects to that of Case 1. However, the zebra domains in Case 2 are still less stable than in Case 1, so that unfolding the zebra domains is practically finished before the stripe domains coalesce into a globule. The further evolution of the globule formed is governed by chaotic motion of rigid and flexible sequences.

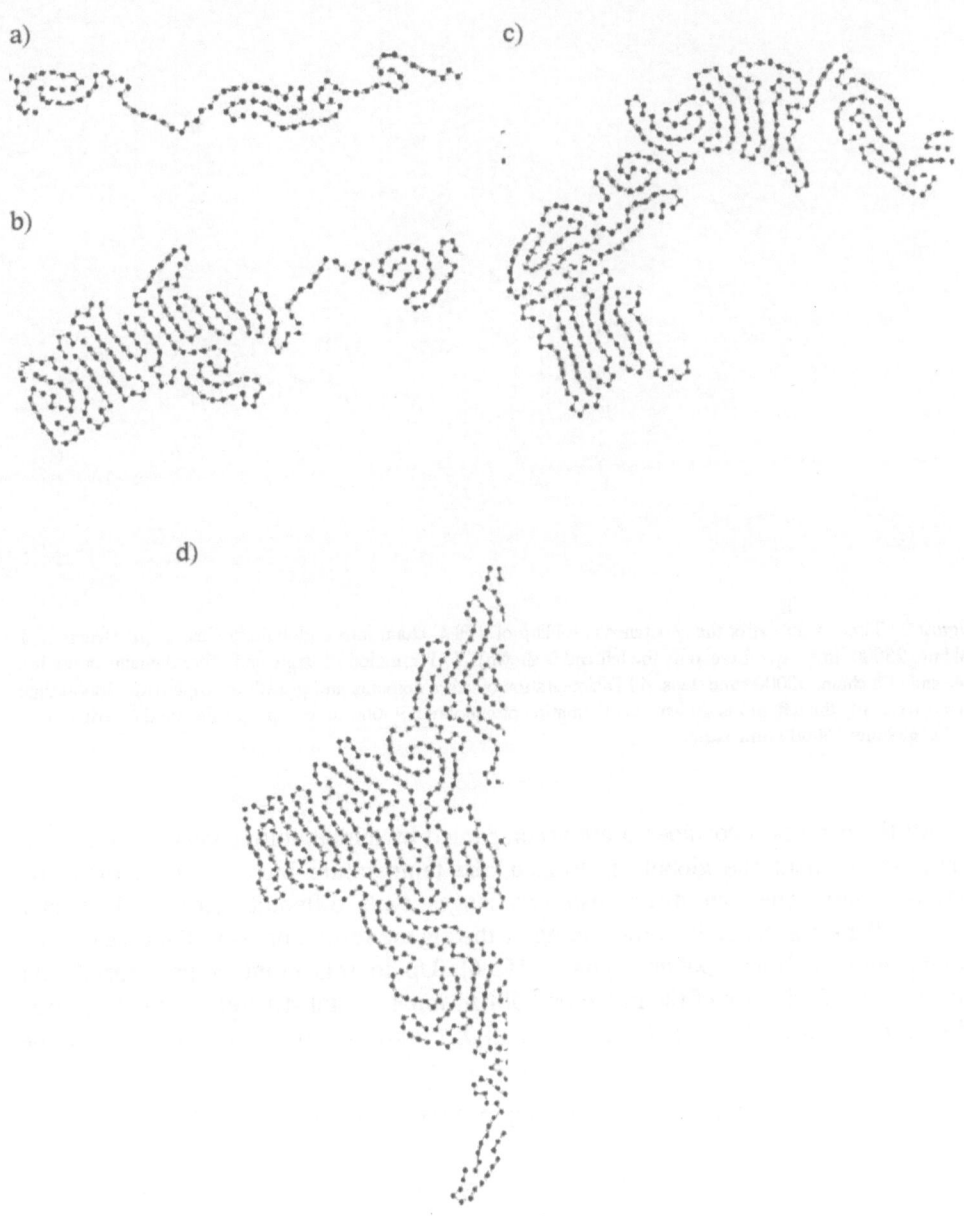

Figure 2 (a) - (d).

e)

Figure 2. Time evolution of the spontaneous folding of a PLC chain into a globule in Case 2. *(a)* Begining of folding: 25000 time steps (here, only the left end is shown). *(b)* Formation of stripe and zebra domains at the left free end of a chain: 55000 time steps. *(c)* Disintegration of zebra domains and growth of stripe ones: 78000 time steps (here, only the left end is shown). *(d)* Formation of a globule: 93000 time steps. *(e)* Fluctuating structure of a PLC globule: 160000 time steps.

All the processes considered are accompanied by temperature changes (*Figure 3*). Until the moment the globule is formed, the temperature increases monotonically. Never-the-less, one can distinguish two stages with different rates. A boundary between them corresponds to the ending of the zigzag formation as in the case of self-organization of linear polymer chains [5, 6]. Up to this point, a pair correlation function gives evidence of the initial one-dimensional crystal structure at a high degree of long range order (*Figure 4*, curves *A* and *a*; *Figure 5*, curve *a'*). The latter can be easily identified with the sharp peaks at n = r / r_0 = 1, 2, 3,... .

It should be emphasized that the zigzag formation presents new peaks:

- Case 1. At $k = \sqrt{(n^2 + n + 1)}$, e.g. $k = \sqrt{3} \approx 1.73$, $\sqrt{7} \approx 2.65$, $\sqrt{13} \approx 3.60$,....
- Case 2. At $m = n\sqrt{2}$, $n + m$, $\sqrt{(n^2 + n\sqrt{2} + 1)}$, $\sqrt{(m^2 + m\sqrt{2} + 1)}$, $\left[(n + m)^2 + \sqrt{2}(n + m) + 1\right]^{1/2}$, e.g., 1.41, 1.84, 2.24, 2.41, 2.83, 3.05, 3.20, 3.41,...

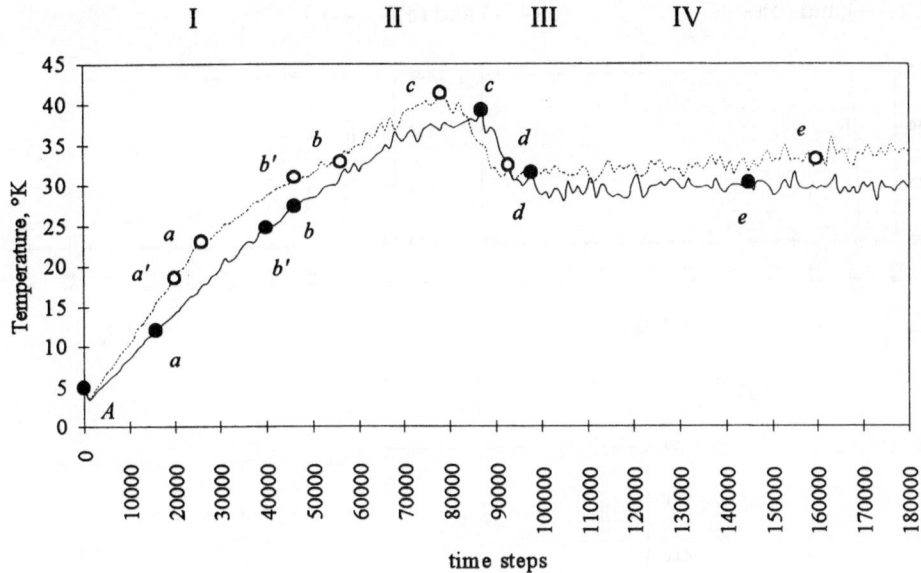

Figure 3. Temperature change during self-organization in Case 1 (————) and 2 (- - - -). Labels (a) - (e) refer to the various conformations shown in *Figure 1* and 2.

The intensity of both old and new peaks has a minimum at 45000 time steps for both cases (*Figure 4*, curve *b*; *Figure 5*, curve *b'*), whereupon it begins to increase. At this moment the long-range order is completely lost and afterwards one can detect only the short-range order (*Figure 4*, curves *c* and *d*; *Figure 5*, curves *b*, *c*, and *d*). For these reasons, the first stage may be named "formation of an equilibrium linear structure" and the second stage, as in [5], "dimensional transformation."

(A)   (a)

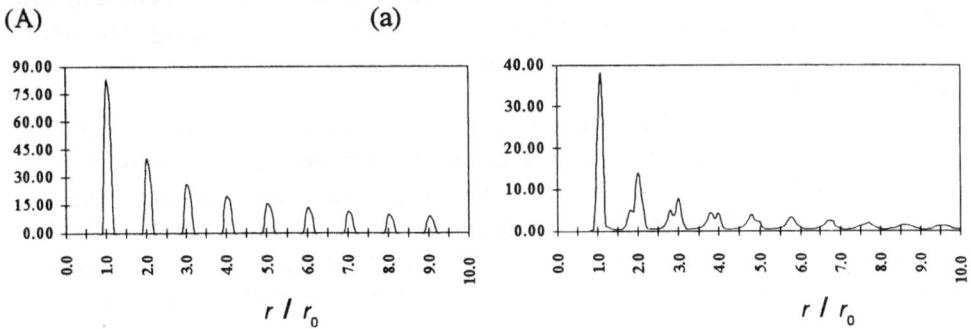

Figure 4 (A) - (a).

94

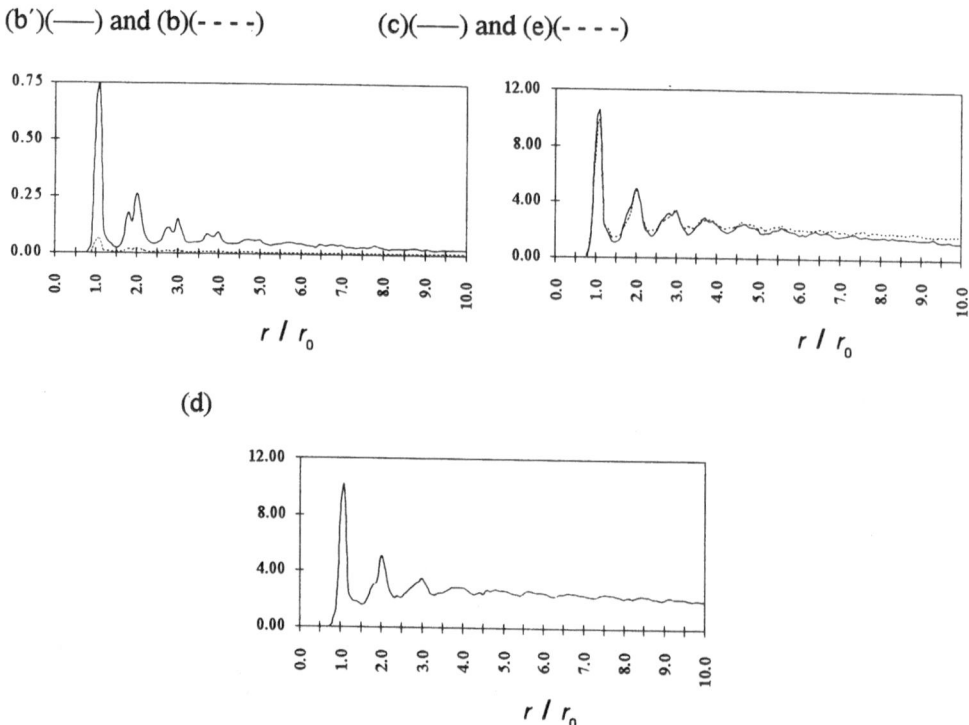

(b′)(——) and (b)(- - - -)   (c)(——) and (e)(- - - -)

(d)

Figure 4. Pair correlation function $\sqrt{s} \cdot r / r_0$ for different stages of folding in Case 1. Here r and r_0 are the distance between beads and the equilibrium distance respectively. Label (A) refers to the initial straight conformation, label (a) - (e) - to the various conformations shown in *Figure 1*.

It is necessary to stress that in Case 2, the boundary between two stages determined with the help of the time dependence of a temperature does not coincide with that found by using a pair correlation function. When the globule forms from domains as a whole, the temperature falls steeply (*Figure 3*) but this third or "aggregation stage" cannot be identified unambiguously with the help of pair correlation functions (*Figure 4* and *5*, curves c and e).

From *Figures 3 - 5* it also follows that during the fourth stage, both temperature and pair correlation function are insensitive to structure changes. However, the structure instability can be detected from the oscillations of an anisotropy factor [8] with time (*Figure 6*). Noteworthy also are the larger oscillations in Case 1 than in Case 2 that may be a connection with the decay of zebra domains. Based on the structure instability, the fourth stage of self-organization may be named the "Brownian motion stage."

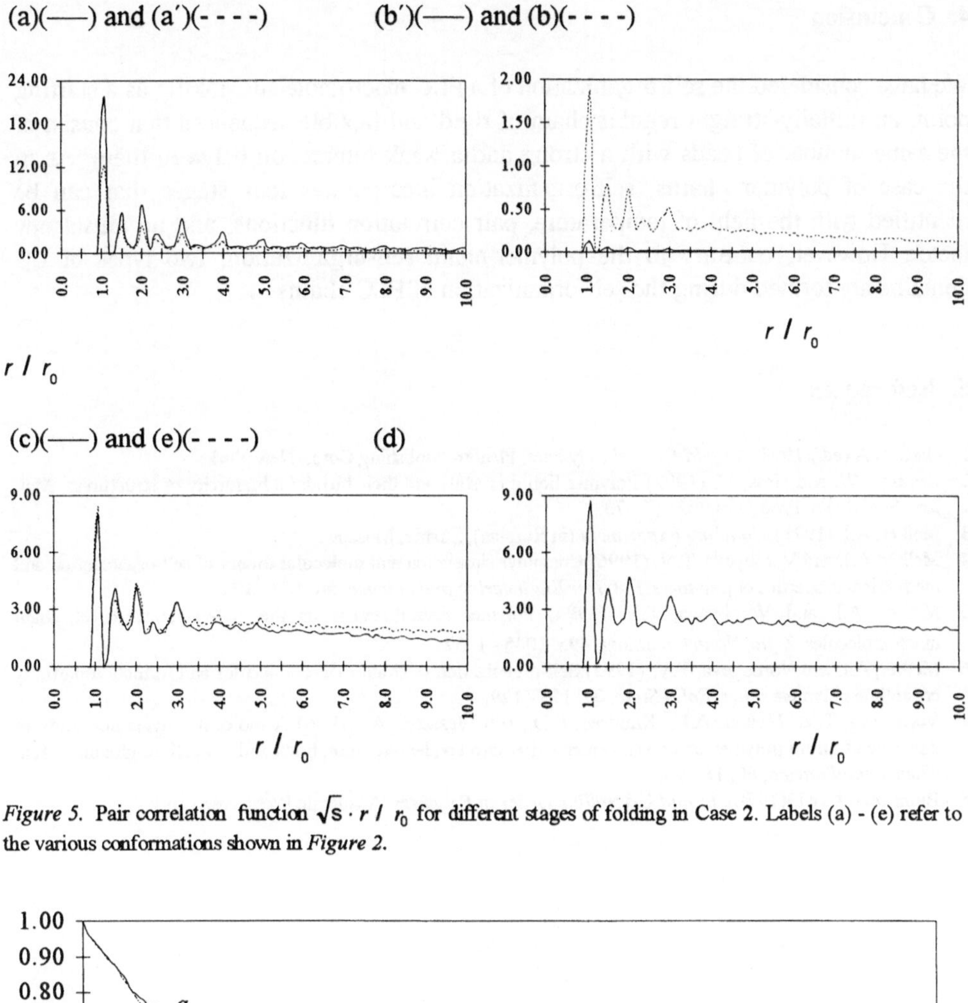

(a)(——) and (a')(- - - -)   (b')(——) and (b)(- - - -)

(c)(——) and (e)(- - - -)   (d)

Figure 5. Pair correlation function $\sqrt{s} \cdot r / r_0$ for different stages of folding in Case 2. Labels (a) - (e) refer to the various conformations shown in *Figure 2.*

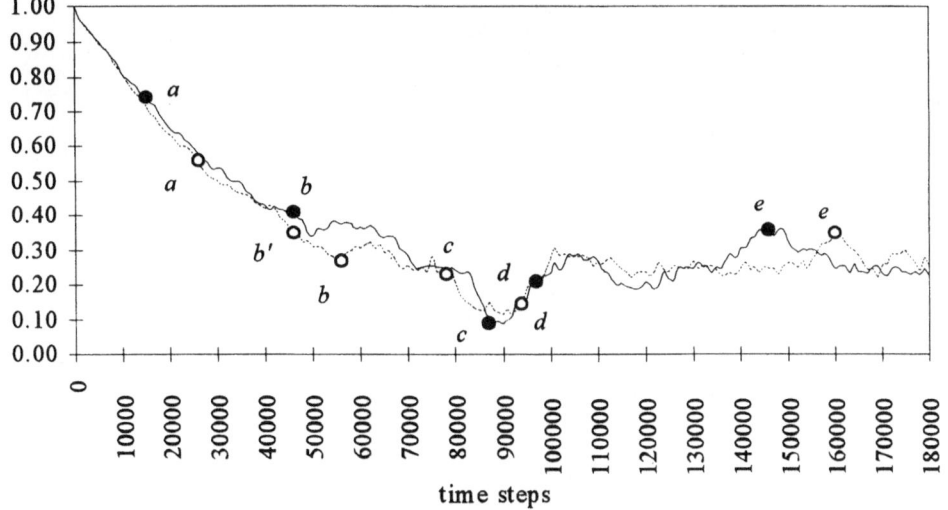

Figure 6. Time dependence of an anisotropy factor in Case 1 (————) and 2 (- - - -). Labels (a) - (e) refer to the various conformations shown in *Figure 1* and 2.

4. Conclusion

We have considered the self-organization of a PLC macromolecule, taking as a starting point, an initially-straight regular chain of rigid and flexible sequences that consist of the same number of beads with a strong and a weak interaction between them. As in the case of polymer chains, self-organization incorporates four stages that can be identified with the help of temperature, pair correlation functions, and an anisotropy factor. However, contrary to the polymer-chain self-organization, two types of LC domains are formed during the self-organization of PLC chains.

5. References

1. Platé, N.A.(ed.)(1993), *Liquid-Crystal Polymers,* Plenum Publishing Corp., New York.
2. Brostow, W. and Hess, M. (1992) Polymer liquid crystals and their blends: a hierarchy of structures, Mat. Res. Soc. Symp. Proc., Vol.255, 57 - 73.
3. Melker, A.I. (1991) *Modeling Experiment* (in Russian), Znanie, Moscow.
4. Melker, A.I. and Vorobyeva, T.V. (1996) Computer simulation and molecular theory of self-organization and mechanical properties of polymers, *Polymer Engineering and Science*, 36, 163 - 174.
5. Melker, A.I. and Vorobyeva, T.V. (1994) Polymer crystallization as the self-organization of chain macromolecules, *Z. für Naturforschung*, 49a, 1045 - 1052.
6. Melker, A.I. and Vorobyeva, T.V. (1995) Self-organization of chain macromolecules and formation partially crystalline polymers, *Phys. Solid State*, 37, 123 - 129.
7. Vorobyeva, T.V., Melker, A.I., Knudsen, K.D., and Elgsaeter, A. (1996) A molecular dynamics study of linear bead-spring polymer chain self-organization into condensed amorphous and crystalline globules, *Acta Chem.Scandinavica*, 50, 18 - 23.
8. Blumstein, A.(ed.)(1978), *Liquid Crystalline Order in Polymers,* Academic Press, New York.

MOLECULAR DYNAMICS STUDY OF SELF-ORGANIZATION AND COMPRESSION OF AN AMORPHOUS POLYMER

A. I. MELKER, A. A. IVANOV and S. N. ROMANOV

Physics and Mechanics Faculty
St. Petersburg State Technical University
Polytekhnicheskaya str. 29, St. Petersburg, 195251, Russia

In this contribution, we report on a study of the compression of an amorphous polymer. This process is computer simulated by a molecular dynamics method using the bead model. The initial amorphous structure is obtained by the self-organization procedure. The structure contains 500 beads that form an amorphous globule. The macroscopic characteristics like stress-strain diagrams, temperature changing during deformation, mesoscopic characteristics like displacements, velocities, and paths of layer-averaged beads as well as microscopic changes are investigated. It is found that the stress-strain diagrams consist of three parts. The first part is due to elastic deformation, the second part is associated with relaxation and the third one is connected with work-hardening. On the basis of the structure changes, a topological model of amorphous polymer deformation is suggested. The model explains many real experiments that have not been understood up to now.

1. Introduction

Polymers are widespread in nature. They are also of considerable current use in industry [1]. It is common knowledge that polymers are highly structured functional materials that easily accommodate themselves to test and environmental conditions. In this connection, two basic questions can be formulated: 1. What causes self-organization in macromolecules that results in the appearance of materials with different shape and structure? 2. What is the relation between the structure and properties?

A large body of information can be obtained with the help of computer simulation to gain a significant qualitative insight into the problems considered [2-8]. Besides, it permits, in some cases, one to develop a microscopic theory [2-5]. In the present paper, we use a molecular-dynamics-bead model [4] to study compression of an amorphous polymer obtained as a result of the self-organization of a straightened bead chain with 500 beads.

2. Computer Simulation Procedure

In many respects, our computer simulation procedure is identical to that employed previously [6-8], so we consider only the rest of the procedure that is developed specially for the problem discussed. The initial two-dimensional amorphous structure was obtained previously in Ref. 6 as a consequence of the self-organization of a

R. C. Tennyson and A. E. Kiv (eds.),
Computer Modelling of Electronic and Atomic Processes in Solids, 97–106.

straightened bead chain. That structure is inserted between two parallel horizontal lines (boundaries) and is divided into five equal-height layers. The layer height is not constant and decreases monotonically during compression. The latter is specified in the following manner. The bottom boundary is fixed and the upper one is moving down at the constant rate $\varepsilon' = 0.001a/\Delta t$ producing compression. Here a is the chain parameter (equilibrium distance for nearest beads) and Δt is the optimum time step equal to $1/22\times$ period of bead vibrations. An outer part of each boundary layer of width equal to one parameter "a" is set off, and the boundaries interact only with the beads of these parts.

The interaction between all beads is modelled by means of the anharmonic Morse potential

$$\varphi(r) = \varphi_0(e^{-2a(r-a)} - 2e^{-a(r-a)}) \tag{1}$$

where φ_0 is the interbead bond, and r is the distance between beads. However, this interaction is assumed to be 25 times as weak for non-nearest neighbours in the chain sequence as for nearest neighbours connected covalently. At the same time, the equilibrium distance for the weak interactions is increased in $\sqrt{2}$ times, i.e., $r_0 = a$, $r_1 = r_2 = r_3 = ... = a\sqrt{2}$, where $r_0, r_1, r_2, ...$, equal the equilibrium distance for nearest neighbours, second neighbours, and so on, in the chain sequence (Case 2 in Ref. 4). The interaction between a bead of the outer part of a boundary layer and an associated boundary is also modelled using the Morse potential (1), where $r = \Delta y$ and $\Delta y \leq a$. Here Δy is the distance between an outer-part bead and an associated boundary.

With the aim to obtain maximum information suitable for subsequent analytic treatment, we calculated the following quantities:

1.   Absolute values of force projections $|F_x|$ and $|F_y|$ acting on each bead. The values are averaged over 1000 time steps (approx. 50 periods of bead vibrations) and over a bead number in a layer or in the whole of the system. Therefore the final values are doubly averaged (over time and over an ensemble). This procedure permits one to eliminate high-frequency fluctuations.

2.   Absolute values of velocity projections $|v_x|$ and $|v_y|$ with kinetic energy $E_k = mv^2/2$ for each bead (the quantities are found in the same manner as above). Here m is the cluster mass and the cluster is equivalent to a monomer CH_2. An average kinetic temperature T for each layer and the whole of a specimen is calculated by means of the expression $E_k = Nk_BT$ where N is the number of beads in a layer and k_B is the Boltzmann constant.

3.   Coordinates x and y for each bead in 200 time step intervals. Thereupon they are averaged to obtain a bead number in a specimen and/or in a layer. Contrary to the quantities considered in items 1 and 2, the coordinates are less susceptible to fluctuations so that there is enough averaging over an ensemble.

It is worth noting that time dependence of $|F_x|$ and $|F_y|$ permits one to obtain stress-strain curves for extension and compression. Previously, we calculated the stress-strain curves for extension through the forces acting only on the beads of the outer parts of the boundary layers. This is convenient, especially in the case of curved boundaries. Besides, the outer part beads constitute only a slight part of the system studied. For this reason, the quantities obtained on the basis of the elements incorporated into the outer parts of the boundary layers are subjected to large fluctuations.

The Newtonian equations of motion are integrated numerically using the Nordsieck method (13) to the fifth order of accuracy. The method is equivalent to a reformulation of the Adams method, but has the advantage of higher stability with minimum degradation of accuracy. The optimum time step is equal to $1/22\times$ period of bead vibrations to meet the requirements of conserving total energy and phase trajectories of the beads. All the calculations were made using IBM computers (model 486).

3. Statistical-Averaged Quantities and Mesostructure

Figure 1 shows the time dependence of the average force acting on the beads along the compression and extension directions. We compress the amorphous polymer globule at a constant rate so the curves can be called the stress-strain diagrams. For different layers, the diagrams are similar to that shown in Fig. 1. At first, the stress increases fast; at the second stage, the stress decreases slowly, and at the third stage, the stress grows exponentially (Fig. 2). All these stages are observed in real experiments on amorphous polymers in a glass state [9]. It is believed that during the first stage, the polymer is deformed like a spring. The second stage has no unified explanation, and the third stage is associated with the work hardening induced by the orientation of chain segments along an extrusion direction.

It is known that the temperature has a dramatic effect on the mechanical properties [9, 10]. For this reason, we have attempted to connect the stress-strain diagrams with the temperature. Figure 3 shows the time dependence of the temperature for different layers during the deformation. One can see that except for

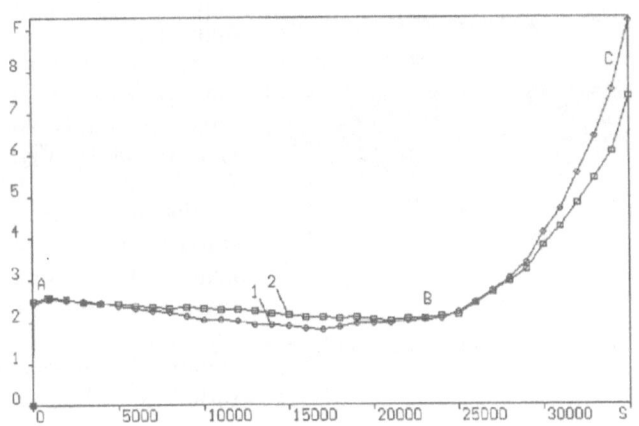

Fig. 1. Stress-strain curves on an amorphous polymer for extending (1) and compressing (2) (here and below time unit corresponds to an integration step. Symbols A, B and C denote different stages of deformation.

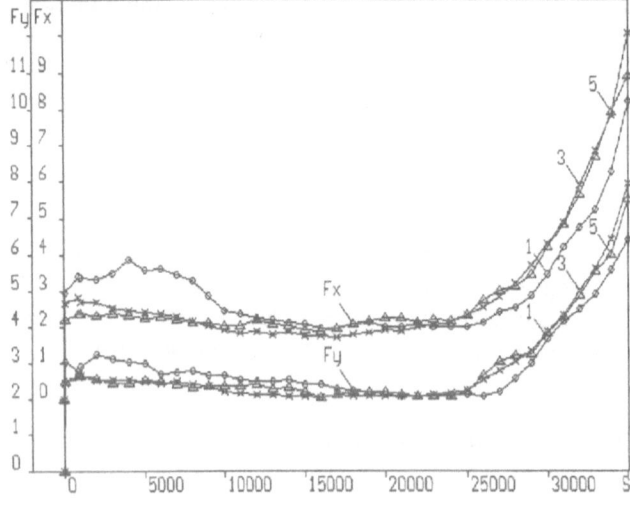

Fig.2. Compression (bottom) and extension (top) curves for different layers of an amophous polymer (here and below a layer number is increasing from a moving upper boundary to a fixed base).

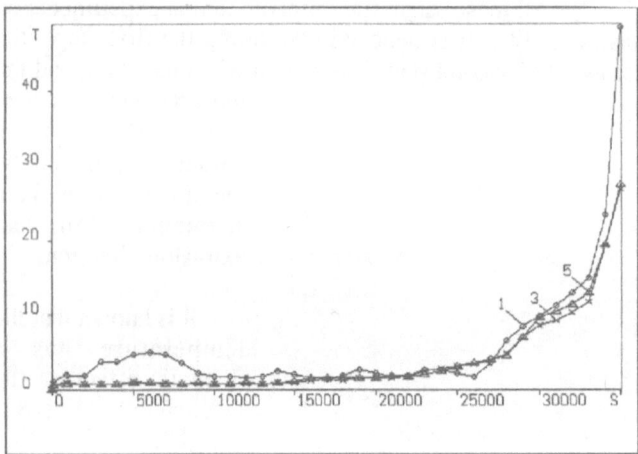

Fig. 3. Time dependence of a temperature for different layers of an amorphous polymer.

the first layer, the temperature is approximately constant during the first and the second stages of the compression, but at the third stage, the time dependencies of the temperature and the stress are similar. This suggests that the shape of the stress-strain diagrams at the third stage is dictated by an entropy constituent of the acting force because this constituent is directly proportional to temperature [9, 10].

In addition to the temperature and traditional mechanical properties (stress-strain) diagrams), we also calculated the quantities which are connected with material flow and are typical for fluids rather than for solids. Figure 4 shows the displacements of the statistically averaged beads of different layers along the extrusion and compression directions. From this figure we notice that the bead displacements have an oscillating character. To better appreciate the physical meaning of this phenomenon, we have drawn the trajectories of the statistical beads (Fig. 5). As can be seen from these diagrams, the trajectories have a highly chaotic character at the second stage of the stress-strain curves.

Let us now examine the velocities of the statistical specimen-bead (Fig. 6). From this figure, we notice that at first, the velocity along the compression direction is higher than along the extrusion one, then the picture is reversed. In statistical physics, the distribution of kinetic energy throughout degrees of freedom is characterised by the projections of an average velocity on coordinate axes, with equal projections for an isotropic system. In the case of small vibrations, average kinetic energy equals the average potential energy. The latter can be written in the form

$$v = k(\delta r)^2 \qquad (2)$$

where k is the elastic modulus, r is the deflection of a coordinate from its equilibrium value. Consequently,

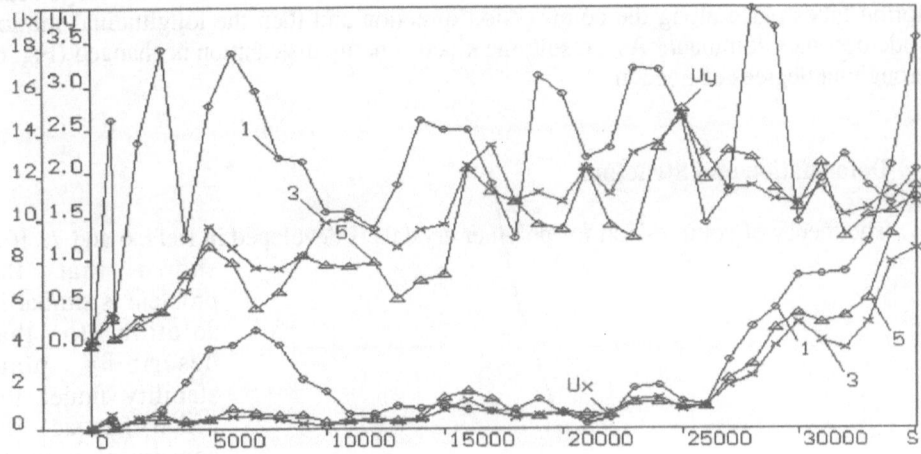

Fig.4. Time dependence of displacements for statistical averaged beads of different layers along an extrusion (top) and compression (bottom) directions.

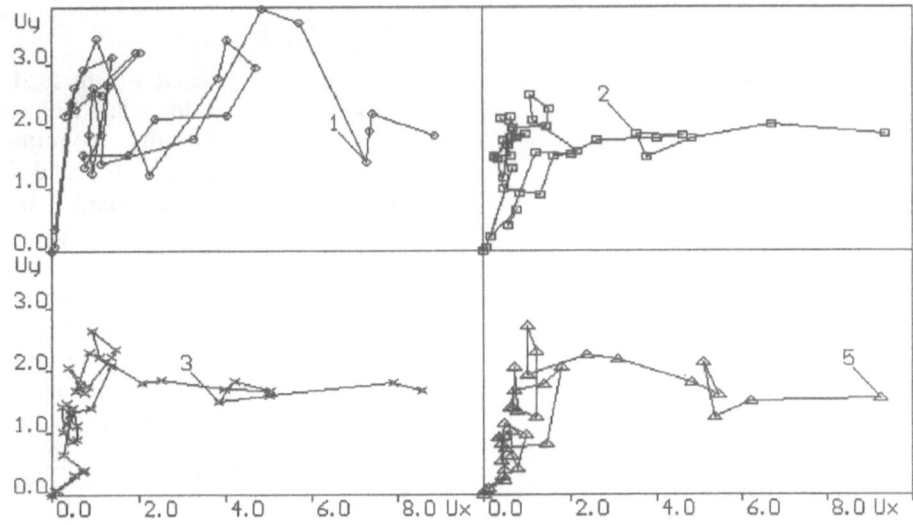

Fig.5. Trajectories of statistical averaged beads of different layers

$$k_x/k_y = ((\langle|v_x|\rangle/\langle|v_y|\rangle))^2 \qquad (3)$$

where v_x and v_y are, respectively, the longitudinal (along the extrusion direction) and transverse (along the compression direction) velocity projections. This relation permits one to estimate the anisotropy of elastic properties for a polymer.

Let us consider the reason for the anisotropy. In hydrodynamics, one differentiates two types of motion: large-scale motion and small fluctuations [11]. An external source of energy is transferred to the large-scale motion that dissipates the energy as heat, forming large-scale fluctuations. From Fig. 3, it follows that at first, the large-scale motion takes place along the compression direction and then the longitudinal motion mode becomes dominant. As a result, the kinetic energy distribution is changed (Fig. 6) throughout degrees of freedom.

4. Deformation and Structure

The theory of compression for polymer crystals is developed in Refs. 6 and 7. It is shown that the problem is similar in solution to that describing plate stability under the action of a load. The constituent equation is of the form

$$f = C_1/\lambda^2 + C_2/\lambda^2 \qquad (4)$$

where f is the load, λ is the relative change of the specimen length, C_1 and C_2 are constant. It is worth noting that the equation contains two constants of different nature: the first constant refers to a material and the second takes into account the rigidity of a boundary interacting with the material.

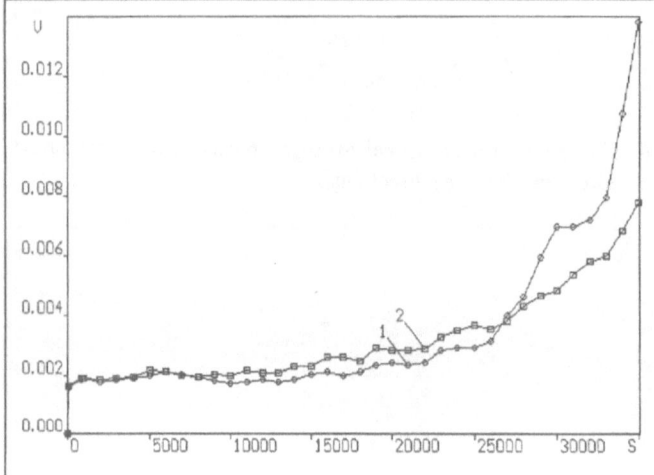

Fig. 6. Time dependence of velocities for a statistically averaged bead along extrusion (1) and compression (2) direction. Contrary to the previous figures, here a bead is equivalent to the whole of a specimen.

This equation is also applicable for compression of amorphous polymers. From the extreme condition $df/d\lambda = 0$, it follows that

$$\lambda_{min} = (C_2/C_1)^{1/4} \tag{5}$$

since $\lambda \geq 0$, it is evident that in the case of softening at the second stage, $C_1 < C_2$. By contrast, the corresponding stage for polymer crystals shows work-hardening [7, 8]. Both results are in good agreement with those observed in real experiments on polymer crystals and amorphous solid polymers in the glassy state [9].

Softening solid polymers means that an amorphous solid polymer under deformation behaves itself not only as a solid but also as a fluid. With the aim to understand this implicit liquid-like behaviour, we have thoroughly examined microstructural changes in the polymer model during compression. The initial folded structure of a macromolecule chain is fully amorphous and resembles a cactus or algae plant (Fig. 7a). In contrast to polymer crystals [4, 7, 8], the loose amorphous structure is better adapted to compression and transforms in such a way as though the algae were in the flow of an ebb-tide along the extrusion direction (Fig. 7a through d). Further compression leads to drastic consequences: a major part of the folds straightens and becomes normal to a compressing force. The final structure consists of alternating sequences of amorphous and quasi-crystalline constituents (Fig. 7e). The algae change not only their shape, but also their structure as a result of decreasing the number of branches.

Visually, the two-coloured structures shown in Fig. 7 resemble a complicated system of winding paths, i.e., a labyrinth. Let us consider a geometrical equivalent of the labyrinth (a labyrinth graph) that will be referred to in the subsequent discussion as a 'bush'. We can bring any irregular fold into one-to-one correspondence with a relevant bush and as a consequence, we obtain the following picture (Fig. 8) that shows a topological structure of the folds observed. From Fig. 8, it also follows that the number of elementary folds incorporated into any complex irregular folds remains constant at the first and second state. This means that one can consider the line (arc) number of the labyrinth graph as a topological invariant.

Now we can define any amorphous structure as a set of all the possible bushes that form a topological space. One can see that this space is homeomorphic to that constructed to analyze branchings in atomic collision cascades [12, 13]. Therefore, the enumeration method developed elsewhere [12, 13] enables all the possible bushes to be found for any irregular fold. The transition from the amorphous structure to a mixture of amorphous and quasi-crystalline regions during the third state is accompanied with work-hardening and unfolding. It must be emphasized that unfolding takes place only for fold elements which are normal to an extension force and are in contact with compressing boundaries. Topologically [14], this is equivalent to splitting a bush root. Therefore, we can define the decomposition of an amorphous structure of a polymer under the action of mechanical stresses as a set of local transitions in a topological space of all the possible bushes.

104

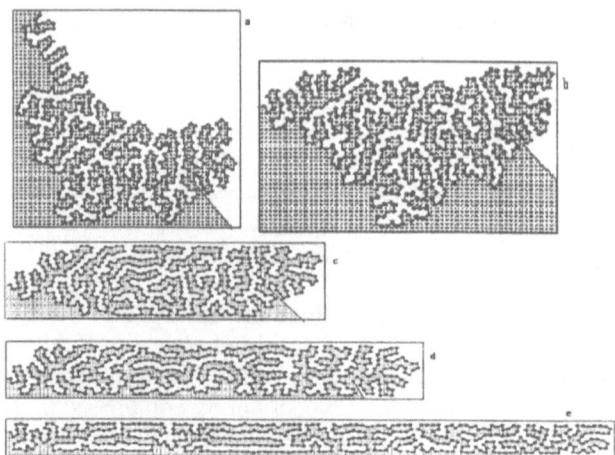

Fig. 7. Structural evolution of an amorphous polymer globule during deformation.

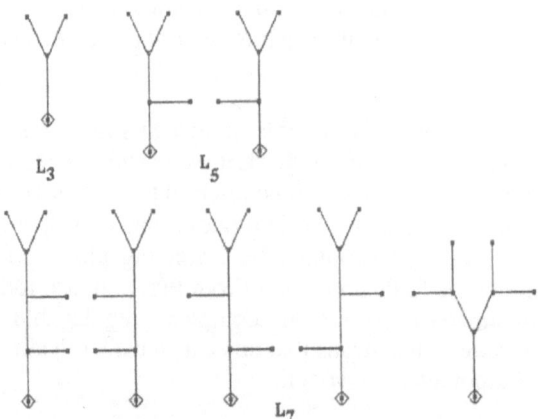

Fig. 8. Topological structure of irregular folds: only bushes of 3, 5 and 7 vertices are shown. Note that the roots denoted by double circles are not incorporated into the vertex enumeration. Topological structure of bushes of 9 vertices is given elsewhere [12].

Such an approach permits one to explain not only our own computer simulation results, but also many real experiments that have not been understood up to now. Indeed, it is known that in a glassy state, polymer materials behave in such a way as if they were in various substates with their own unknown mechanisms of relaxation and fracture [9, 15]. Our topological model enables one to gain greater insight into the nature of this phenomenon.

5. Conclusion

Polymers can be regarded simultaneously as gas, liquid, and solid [16]. Any reliable model for the prediction of polymer properties must take into consideration such a manyfold nature. We have developed a molecular-dynamics-computer-simulation procedure that is perfectly suited to this requirement. The procedure is based on a concept of three types of particles and permits one to study polymer properties simultaneously at micro, meso, and macroscopic levels. At the microscopic level, the particle is a bead, i.e., a small group of atoms corresponding to a monomer that moves as a whole. At the mesoscopic or hydrodynamical level, the particle is a unit volume of the continuum corresponding, in our case, to a layer. At the macroscopic level, the particle is the whole of a specimen.

Using all the approximation levels, we have studied the general features of structural transitions, rheological behaviour, and mechanical properties of an amorphous polymer. The large body of information obtained permitted us to construct a topological model of an amorphous polymer structure. In its turn, with the help of this model we have gained greater insight into the actual rheological and mechanical behaviour of real polymers in a glassy state.

6. References

1.	Elias, H. G. (1987), *Mega Molecules*, Springer-Verlag, Berlin.
2.	Melker, A. I., and Vorobyeva, T. V. (1994), Polymer crystallization as the self-organization of chain macromolecules, *Z. F. Naturforschung*, **49a**, 1045-1052.
3.	Melker, A. I., and Vorobyeva, T. V. (1995), Self-organization of chain macromolecules and formation of partially crystalline polymers, *Phys. Solid State*, **37**, 123-129.
4.	Melker, A. I., and Vorobyeva, T. V. (1996), Computer simulation and molecular theory of self-organization and mechanical properties of polymers, *Polymer Engineering and Science*, **36**.
5.	Vorobyeva, T. V., Melker, A. I., Knudsen, K. D., and Elgsaeter, A. (1996), A molecular dynamics study of linear polymer chain self-organization into condensed amorphous and crystalline globules, *Acta Chemica Scandinavica*, **50**, 18-23.
6.	Melker, A. I., and Vorobyeva, T. V. (accepted), Parametric resonance of the transverse waves in polymer macromolecules: a reason for folding, *Nanobiology*.
7.	Melker, A. I., Ivanov, A. A., Vorobyeva, T. V., and Romanov, S. N. (accepted), A molecular dynamics study of compression of a polymer crystal, *Phys. Solid State*.
8.	Melker, A. I., Ivanov, A. A., and Romanov, S. N. (submitted), Polymer deformation. Computer simulation and theory of compression of a crystalline polymer globule, *Polymer Engineering and Science*.
9.	Askadskii, A. A. (1973), *Deformation of Polymers (in Russian)*, Khimia, Moscow.

10. Ward, I. M. (1972), *Mechanical Properties of Solid Polymers*, Wiley Interscience, London.
11. Landau, L. D., and Lifshits, E. M. (1988), *Hydrodynamics (in Russian)*, Nauka, Moscow.
12. Melker, A. I., and Romanov, S. N. (1984), Group representation of the formation probabilities of athermal vacancy clusters in atomic collision cascades, *Physica Status solidi (b)*, **126**, 133-140.
13. Melker, A. I., Romanov, S. N., and Tarasenko, N. L. (1986), Defect clusters and subcascades. II. The splitting threshold and subcascade dimensions, *Physica Status Solidi (b)*, **133**, 111-118.
14. Harary, F. (1969), *Graph Theory*, Addison-Wesley Reading.
15. Bartenev, G. M. (1984), *Strength and Mechanism of Polymer Fracture (in Russian)*, Khimia, Moscow.
16. Isihara, A. (1971), *Statistical Physics*, Academic Press, New York.

MODELING NON-METAL SURFACE DAMAGE CREATED BY MULTIPLY-CHARGED IONS

E.Parilis
California Institute of Technology,
200-36, Pasadena, CA , 91125, USA

Abstract

A model based on partial Auger neutralization of a highly-charged ion outside the solid surface with consequent positive charge deposition in surface layers and their expand due to Coulomb repulsion has been applied to explain the creation of very shallow blisters and craters on mica crystal surface under slow, highly-charged xenon ion impacts.

1. Introduction

When a slow multiply-charged ion approaches a solid surface, an intense capture of electrons causes several interesting phenomena including secondary electron emission, light emission, secondary neutral and ion emission, radiation damage and surface erosion.

The investigation of sputtering and secondary ion emission under multiply-charged ion bombardment has a relatively short history compared with the efforts spent to study sputtering by single-charged ions. In the first publications [1,2] some multiply-charged ions with charge not exceeding + 7 were used. The highly charged ions became available [3] only with development of new techniques.

The theory is based on the mechanism of Coulomb explosion of a domain on a non-metal surface positively-charged during step-by-step Auger neutralization of the multiply-charged ions [4,5]. This work proposes a simple theoretical model to describe the surface damage produced by highly-charged ion impacts on an insulating, atomically flat surface of single non-isotropic mica crystal.

R. C. Tennyson and A. E. Kiv (eds.),
Computer Modelling of Electronic and Atomic Processes in Solids, 107–113.

2. Summary of Experimental Data

The model is based on experimental findings by a joint group from Sandia National Laboratories, Livermore, California, Kansas State University, Manhattan, Kansas [6] and Lawrence Livermore National Laboratory [7] of some nanometer-size shallow craters and blisters on a mica surface exposed to slow highly-charged xenon ion bombardment.

Experimental data :
- the charge of the projectile Xe ions $q = + 44$
- their energy $E_0 = 0.1 - 20$ KeV• $q = 4.4 - 880$ KeV
- the ion flux $I = 10^7$ ions / cm^2• s
- the total ion fluence $F = 1 - 4•10^{10}$ ions / cm^2
- the incidence is normal
- the crater or blister diameter $D = 18 - 20$ nm
- their depth or height $h = 0.3$ nm
- the crater or blister volume $V = 40$ nm^3
- the number of atoms emitted or displaced $N = 100 - 200$
- both D and h were found to be independent of E_0

Some relevant parameters of Muscovite mica:

- the composition: $KAl_2 [Al Si_3O_{10}] [OH]_2$
- the structure: pseudohexagonal alumino-oxigen-silicate layers formed by double $[Si_2O_5]$ tetragonal structures in which one Si ion is replaced by Al ion, the layers being connected by Al $[OH]_3$ forming negatively charged sheets with positively-charged potassium K^+ ions in between
- the atomic density is $\delta = 4$ atoms / nm^3
- the elasticity of elongation $\mu = 22\ 133$ Kg / mm^2
- the electrical resistance $\rho = 10^{15} - 2•10^{17}$ Ω•cm
- the breakdown electrical force $F_b = 300-400$ KV / mm = 0.3 - 0.4 V / nm
- the work function $\phi = 4$ eV

The main conclusion made by authors of the experiments is that the shallow, circular, blister-like and crater-like features on a mica surface produced by single slow- moving highly-charged xenon ions are due to local lattice disorder and layer delamination caused by local charge depletion via Coulomb explosion during the near surface highly charged ion Auger neutralization process.

3. The Model

All the geometry of the phenomenon - the very- thin- pancake shape of the features
(the thickness-to- width ratio $h / D = 0.015$ with the thickness of the damaged layer
$h = 0.3$ nm being comparable with the distance between mica atomic layers)- leaves no
doubt that the damage comes from outside the surface rather than from the ion track
inside the solid (fig.1). There are no secondary agents (electrons, recoil atom cascades ,
shock waves etc.) connected to the nuclear or electronic stopping power in the solid that
could be responsible for such a geometry.

In the meantime for $q = 44$ the electric field of the highly-charged ion $F=qe / x^2$
becomes equal to the electric breakdown force of the best Muscovite mica F_b at a
distance $x = 12$ nm from the surface , a length that gives an immediate estimate for the
radius of the damaged spot, which indeed equals $D / 2 = 9$ - 10 nm. Therefore $x_b \propto$
$q^{1/2}$ and the damaged area $\pi x_b^2 / 4 \propto q$, i.e. is linear with the ion charge.

This gives an estimate of the charge dependence of the blister volume because the
thickness is the same for all blisters: 1-2 atomic layers. It agrees with the recent
observation by M.A.Briere et al.[7] and rather contradicts the earlier finding by
D.Schneider[11] because the total ionization energy (actually the outer shell part of it) is
proportional to q^2.

The time $t = (2E_0 / m)^{-1/2} x = 1.2 \cdot 10^{-14}$ - $1.5 \cdot 10^{-13}$ sec that the ion needs to cover
this distance is large enough to develop an avalanche-type breakdown in the upper layer
of the solid releasing free electrons to participate in Auger neutralization and leaving a
charge depleted layer that screens deeper layers from the ion electric field. The electrons
with typical energy $E_e \cong 5$-10 eV moving toward the ion, cover this distance
approximately $(mE_e / m_eE_0)^{1/2} = 1.2$ -7.8 times faster than the ion does.

The current density is $j = 4(q + \gamma) e / t\pi D^2 = 2 \cdot 10^8$ A / cm^2 for secondary electron
emission equal to

$\gamma = 10^2$ el / ion .

The Auger neutralization of the 18 upper O vacancies in Xe^{+44} , releasing just 2-3
KeV of the total amount of 50 KeV neutralization energy, is quite enough to provide the
observed surface damage and secondary electron emission. The typical time t for Auger
neutralization by capturing electrons direct from the solid (S) into the vacant xenon
shells via O - SS Auger transitions or by prior filling of the upper, including Rydberg,
ion states with subsequent autoionization of the hollow atom equals
$\tau \cong 10^{-15}$ - 10^{-14} sec , i.e. $\tau < t$. Filling of the remaining inner M and N
vacancies is accompanied by emission of high-energy Auger electrons and (with the
fluorescence yield > 50 %) photons during $\tau = 2 \cdot 10^{-13}$ sec $> t$, i.e. inside the solid.
They add nothing or very little toward creation of the surface damage structures under
discussion.

Therefore, the estimates show that all the events described above are likely to be accomplished *before* the highly-charged ion hits the surface. The experimental findings of the velocity independence of the damaged sites' dimensions within the velocity range $v = 8 \cdot 10^{6} - 1 \cdot 10^{8}$ cm /sec [6,7] is the most convincing evidence of the situation.

Of course, the ordinary sputtering of a solid due to development of collision cascades under heavy atom bombardment, known to increase with the projectile energy, does contribute to the creation of the damaged sites, but the sputtering coefficients , typical for this energy [8] equal to $S \cong 1$-10 atoms / ion, i.e. corresponds to the amount of matter sputtered, which is just about 1-10 % of the damaged volume. This quantity, which is within the error bars on the experimental plots [6], cannot determine the energy dependence of the phenomenon.

The main source of the surface damage is the ion neutralization energy W_q , equal to the sum of the ionization energies of the highly charged ion

$$W_q = \Sigma_{i=1}^{i=q} \; W_i \tag{1}$$

which is released in a series of step-by-step Auger neutralization processes and is shared among the Coulomb repulsion energy W_c of a charged domain containing N_q positive charges, appeared after double ionization of N_q anions, emission of $(N_q - q)$ secondary electrons and $N_{a,p}$ energetic Auger electrons and photons. The relevant energy balance equation is

$$W_q - q\phi = W_c + (N_q - q) E_e + N_q I_2 + N_{a,p}. \tag{2}$$

For an extremely flat disc with diameter D and thickness $h << D$ containing N_q positive charges

$$W_c = 8 N_q^2 e^2 (h - h_o) / D^2 . \tag{3}$$

This formula reflects the non-uniform distribution of the charges due to specific anisotropic structure of the mica surface layers. After stripping electrons from the anions and neutral atoms, the energy I_2 being spent to create each of N_q charges , two layers of positively-charged ions, together with a layer of cations K^+ in between will be created to form a three - layer charged structure. The initial distance between anion and cation layers is h_o. The Coulomb repulsion in this structure would cause a blister, if the dome maintains its integrity, or a crater, if it is fragmented (fig. 2).

In the earlier investigations [4,5,9] the number of electrons N_q stripped from the atoms of the surface layers in partial outside- the- surface Auger neutralization of a q-charged ion, was believed to be proportional to $W_q \sim q^2$.

This assumption based on a model of step-by -step Auger neutralization with energy steps equal to 15 - 25 eV gives for Xe^{+44} the numbers N_q = 2000 - 3000. The corresponding secondary electron emission should be equal to $\gamma = k (N_q - q) \cong$ 600-1000 el / ion . Actually, the model works only up to q = 10- 20. For higher charges, the dependence $N_q (W_q)$ is nonlinear due to emission of energetic Auger electrons. The emission γ = 70 - 90 el / ion has been found for Xe^{+30} and Xe^{+44} on Cu and Au [10]. This corresponds to $(N_q - q) E_e$ = 2000 - 3000 eV and $N_q \cong$ 200 - 300. This amount relates to the slow electrons emitted from the surface.

The delayed high-energy Auger electrons emitted inside the solid carry away a major part of the total W_q and spend it on some processes in bulk. Yet there is no clear experimental evidence of the energetic Auger electrons emitted outside the surface. Some simultaneous measurement of secondary electron emission and surface damage under highly-charged ions would be very useful.

The linear increase of the blister volume with total ionization energy W_q observed by D.Schneider et al. [11] does not mean that all this huge amount of energy (up to 180 KeV for Th^{+74}) is spent to create the blisters with volume V =200 - 800 nm^3 containing $N = \delta V \cong$ 800- 2400 atoms, i.e. to create some hot domains with energy deposition 75 eV/atom. Actually, the main part of total W_q is carried out by the high energy Auger electrons or is deposited in bulk. The proportionality $V \propto W_q$ itself is a manifestation that the electron capture in just the upper levels does play role in the phenomenon of creating blisters.

Returning to the energy balance, we can use the value N_q = 250 to estimate W_c. If the final thickness of the damaged site h exceeds the initial distance between charged anion and cation layers h_0 by a factor 2 , i.e.

$h_0 = a\, h$, with a =1/2 , then we get

$$W_c = 8 N_q^2\, e^2\, h\, (1 - a) /D^2 \cong 255 \text{ eV} \qquad (4)$$

Even if we take into account the pre-existing charge of the K^+ cations to get $W_c \cong$ 1020 eV, still this amount of energy will remain a very small part of the total neutralization energy W_q = 50 KeV. Nevertheless, it is enough to provide the elastic deformation of both the upper and the lower layers (fig. 2) after the binding forces between cations and former anions disappear due to stripping of electrons. This deformation consisting in a 10^{-4} increase of the area of the layer bent to form a dome needs in Muscovite mica (μ = 22 133 Kg / mm^2) an amount of energy equal to 21 eV.

The irreversible deformation includes some changes in the crystal structure. There are no bonds to be broken while the dome is integer. The fragmentation of a part of damage features could be connected with point structure defects (vacancies, dislocations etc.) in the upper layer.

4. Conclusion

A model based on partial Auger neutralization of a highly-charged ion outside the surface of a mica crystal with consequent positive charge deposition in surface layers and their expantion due to Coulomb repulsion enables some estimates that could explain the creation of very shallow blisters and craters on a surface.

A full description of events displayed in fig.2 needs a more detailed knowledge of the mica crystal structure and ionization energies of the anions and cations in the solid as well as the elastic deformation constants and binding energies of the bonds. The dome fragmentation and consequent formation of a crater depends on concentration of point structure defects (vacancies, dislocations etc.) in the upper layer.

Acknowledgment

This work was a part of a research project for the United States Department of Energy under Contract DE-AC04 - 94AL85000. The corresponding SANDIA report SAND95-8206 • UC 411 January 1995, 188 pages is available from:
National Technical Information Service U.S. Department of Commerce
5285 Port Royal Rd. Springfield, VA 22161 U.S.A.

References

1. U.A.Arifov et al., Izv.AN SSSR , ser.fiz. 40(1976)2621
2. S.T. De Zwart et al. Surf.Sci.177(1986)L939
3. R.W.Schmieder and R.J.Bastasz, NIM B 43 (1989) 318
4. E.S.Parilis, Atomic Collisions in Solids, Ed. by Palmer,North-Holland,1970, p.342
5. I.S.Bitensky and E.S.Parilis, Journal de Physique, C2, t. 50, 1989,
6. D.C.Parks, R.Bastasz, R.W.Schmieder, and M.Stockli , Journ.Vac.Sci.Tech. B 13 (1995) 941-8
7. M.A.Briere, D.Schneider, J.McDonald, M.Reaves, C.Rühlicke, G.Weinberg, and D.Knapp, NIM B 90 (1994) 231-236
8. R.Berish, ed., Sputtering by particle bombardment I and II, Topics in Applied Physics V.57 (1981) & V.52 (1983) , Springer -Verlag, Berlin.
9. M.Delaunay, M.Fehringer, R.Geller, D.Hitz, P.Varga and H.Winter, Phys. Rev. B 35(1987) 4232
10. J.W.McDonald et al. , Phys. Rev. Letters 68 (1992) 2297
11. D.Schneider, M.A.Briere, J.McDonald and W.Siekhaus, NIM B 87 (1994) 156

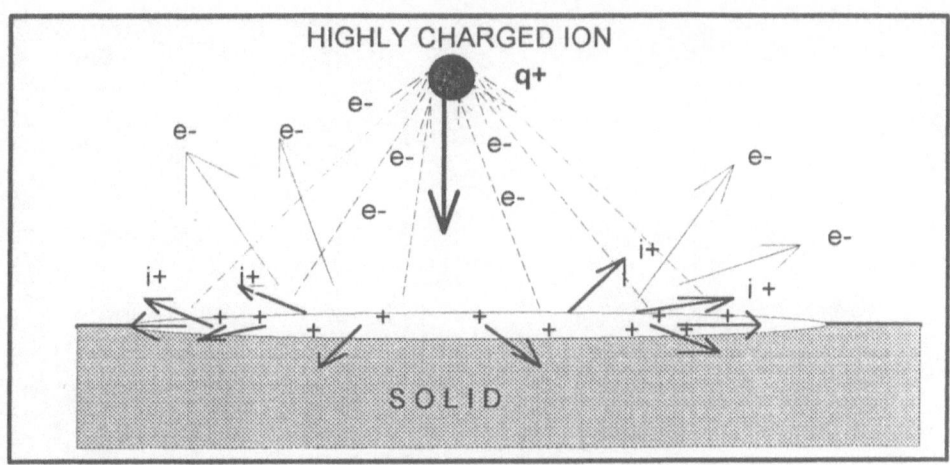

Fig . 1

General scheme of highly-charged ion - surface interactions

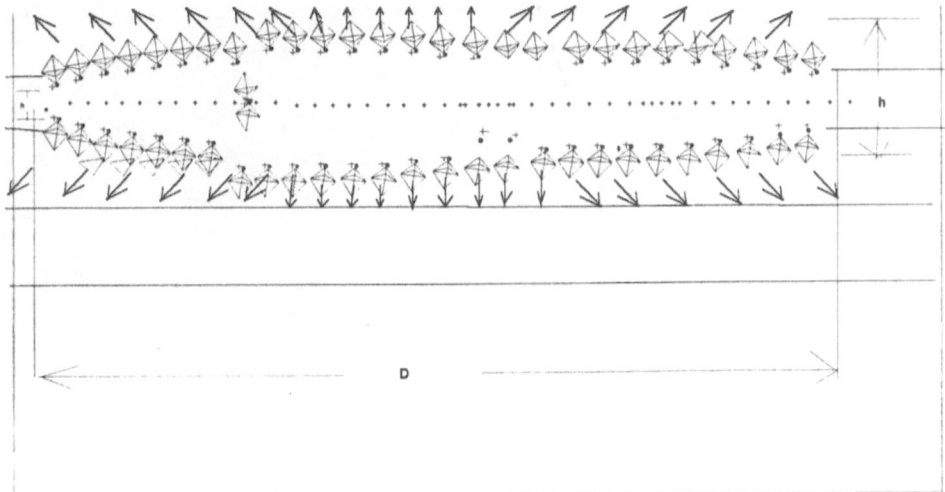

Fig . 2

Mica surface layers deformed due to Coulomb repulsion after multiple stripping of electrons in Auger neutralization of a highly-charged ion outside the surface.

MODELLING OF LOCAL CENTRES OF ICOSAHEDRAL SYMMETRY IN SOLIDS AND FULLERENES

A.B.ROITSIN, A.A.KLIMOV, L.V.ARTAMONOV
Institute of Semiconductor Physics,
Prospekt Nauki, 45, 252650, Kiev-28, Ukraine

Abstract

This paper provides the group-theoretical basic data for investigation of icosahedral symmetry (IS) centres. The table of characters of irreducible representations (IR) of the icosahedral group (IG) is presented. The method of obtaining the matrix representation for all elements and all IR of all IG is developed and all these matrices are obtained. On their ground the Koster-Slatz perturbation matrices of various operators for all possible pairs of IR of IG are obtained. The group-theoretical analysis of energy structure, transition rules and resonance properties, is given for central and noncentral impurity ions in a crystal field of IS. The Generalized Spin-Hamiltonian for IS paramagnetic centers is deduced. The spectra of magnetic and paraelectric resonances of local centres of IS are predicted.

1. Introduction

The symmetry figures which contain pentagons have not been the focus of attention for a long time, as it was believed that the corresponding structures were not present in nature. No proper attention has been given to such symmetries (as C_5, C_{5V}, D_5, D_{5h}, Y, Y_h especially to latest two) as it has been done to the known 32 point groups [1-3]. However, during the last few years there appeared data testifying that such symmetries are not only of academic interest and reflect real structures but they also are promising for applications [4]. So, it has been noted [5] that small ($\leq 10^2$ nm) metal particles may take the shape of a rectilinear pentagon, or have a more complicated structure (icosahedron, pentagonal pyramid) possessing five-fold axes. The so-called quasi-crystals are also distinguished by their specific symmetry [6]. In their grains, it is possible to observe clearly the pentagonal dodecahedra. IS are also revealed in some molecules (such as B_{12}, H_{12}^{-2}, $C_{20}H_{20}$) and in more complicated formations (of virus type) [7]. For almost forty years, the model structure of specific paramagnetic centres (so-called dangling bonds) in diamond-like crystals has been discussed [8]. Recently a new model based on the five-coordinated silicon has been proposed [9].

R. C. Tennyson and A. E. Kiv (eds.),
Computer Modelling of Electronic and Atomic Processes in Solids, 115–124.
© 1997 *Kluwer Academic Publishers.*

But, apparently, the greatest interest was aroused in the discovery of fullerenes, C_{60} molecules having IS [10], and of fullerites, i.e. crystals created on their base. Of interest was not only the peculiar structure and symmetry of this new carbon modification, but also the unusual properties of such substances. So, the introduction of atoms of other elements has been shown to lead to the formation of semiconductive, metallic (including superconducting) properties [11]. In view of the possibility of introducing the elements of different groups of the periodic system (up to lanthanides and even including uranium) into fullerenes [12] and possibly in other atomic formations with the same symmerty investigations of the structure and various physical properties of local centres of IS become urgend. The article is devoted to such investigations.

2. The Group of Icosahedron

The simple IS group Y consist of 60 elements-rotations around the symmetry axes. There are 6 five-fold, 10 three-fold and 15 two-fold axes. The direction of the axes depends on the figure in question (see Fig.1). The full group Y_h is obtained from Y by the addition of the inversion centre ($Y_h = Y \times C_i$, where C_i is inversion group). It contains 120 elements. We shall also distinguish the double groups Y' and Y'_h. The introduction of them is necessary to describe the half-integer value of angular momentum J. The group is obtained from Y by the addition of the rotation Q through the angle 2π. So, $Y' = Y \times Q$, $Y'_h = Y' \times C_i$. Groups Y' and Y'_h contain 120 and 240 elements respectively. The existence of a great number of elements in the IS groups leads to new peculiarities in the properties of various phenomena.

Table1.IRCharacters of groupY'

IR	1E	1Q	$6C_5^1$ 2-7 $6QC_5^4$	$6C_5^4$ 8-13 $6QC_5^1$	$6C_5^2$ 14-19 $6QC_5^3$	$6C_5^3$ 20-25 $6QC_5^2$	$10C_3^1$ 26-35 $10QC_3^2$	$10C_3^2$ 36-45 $10QC_3^1$	$15C_2^1$ 46-60 $15QC_2^1$
	1	2	3	4	5	6	7	8	9
A	1	1	1	1	1	1	1	1	1
F_1	3	3	ε_+	ε_+	ε_-	ε_-	0	0	-1
F_2	3	3	ε_-	ε_-	ε_+	ε_+	0	0	-1
G	4	4	-1	-1	-1	-1	1	1	0
H	5	5	0	0	0	0	-1	-1	1
E_1'	2	-2	ε_+	$-\varepsilon_+$	$-\varepsilon_-$	ε_-	1	-1	0
E_2'	2	-2	ε_-	$-\varepsilon_-$	$-\varepsilon_+$	ε_+	1	-1	0
G'	4	-4	1	-1	-1	1	-1	1	0
I'	6	-6	-1	1	1	-1	0	0	0

C_n^m - m-fold rotation around the axes of order n. The figure before elements is their number. $\varepsilon_\pm = \left(1 \pm \sqrt{5}\right)/2$

The IR characters of group Y' are presented in table 1 [13]. By multiplying them into characters (±1) of group C_i it is easy to obtain the characters of group Y'_h [14]. The number of IR of the group Y'_h is doubled due to the addition of the indices "g" (even) and "u" (odd). In the table, all 60 elements of group Y are numbered. From geometrical analysis it can be established, that all these elements can be obtained in strict sequence from two generating elements. If we choose two rotations through 72 degrees around the axes b→a (element 2) and d→c (element 5) as generating elements (see Fig.1.1), then other elements can be obtained by products of elements as shown in table 2.

Table2.The rules for obtaining all elements of the group $Y^{1)}$

1	2	3	4
14=2×2	10=7×9	35=2×23	48=27×35
20=2×14	18=6×6	30=2×24	49=31×35
8=2×20	24=6×18	33=2×25	50=26×29
17=5×5	12=6×24	26=36×36	51=27×29
23=5×17	22=10×10	37=27×27	52=28×30
11=5×23	4=22×22	38=28×28	53=31×32
3=2×11	16=4×4	29=39×39	54=32×33
7=11×2	1=2×8	40=30×30	55=26×28
15=3×3	27=2×3	41=31×31	56=30×35
21=3×15	28=2×4	42=32×32	57=26×31
9=3×21	32=2×5	43=33×33	58=27×32
19=7×7	44=2×6	34=44×44	59=28×33
25=7×19	31=2×7	45=35×35	60=29×34
13=7×25	36=2×21	46=27×30	
6=2×9	39=2×22	47=26×34	

[1)] The multiplication should be realized from the first column from top to bottom.

To investigate the properties of IS centers it is also necessary to know the matrix representation of group elements. For IR A, E'_1 , F_1, G', H and I' they can be obtained from the matrices $D_J(\alpha,\beta,\gamma)$ of the IR of the rotation group by substituting for J respectively 0,1/2,...,2,5/2 and for Euler angles α,β,γ definite values. So, for element N2 $\alpha=72°$, $\beta=\gamma=0°$ and for element N5 $\alpha=\gamma=18°$, $\cos\beta = 1/\sqrt{5}$. The choice of the coordinate system is shown in Fig.2. The matrices of the rest IR (E'_2 , F_2, G) can be obtained by means of other methods [15]. The form of the matrices of IR G, for example, is

$$N2 = \begin{pmatrix} -c_3^* & 0 & 0 & 0 \\ 0 & c_4^* & 0 & 0 \\ 0 & 0 & c_4 & 0 \\ 0 & 0 & 0 & -c_3 \end{pmatrix} ; \quad N5 = \begin{pmatrix} -c_6 & -c_1^* & c_2^* & a \\ -c_1^* & -c_5^* & -a & c_2 \\ c_2^* & -a & -c_5 & -c_1 \\ a & c_2 & -c_1 & -c_6^* \end{pmatrix} ;$$

$$a = \frac{1}{p}, \quad c_{1,2} = \frac{1}{2p} + i\frac{N^\pm}{2p}, \quad c_{1,2} = \frac{1}{2p} + i\frac{N^\pm}{2p}, \quad c_{3,4} = \frac{(p\pm1)}{4} + i\frac{R^\mp}{2q}, \quad c_{5,6} = \frac{(p\pm1)}{4p} + i\frac{R^\mp}{2pq},$$

$i = \sqrt{-1}$, $p = \sqrt{5}$, $q = \sqrt{2}$, $R^\pm = \sqrt{5\pm p}$, $N^\pm = \sqrt{5\pm2p}$. Using matrices of generating elements and taking into account table 2, we have obtained the matrices of all elements for every IR of the IG.[1]

---

[1] Because of page-limited article we intend to publish all matrices separately.

3. Properties of Central Impurity Ions

3.1. SPLITTING OF ATOMIC LEVELS.

Table3. Splitting of atomic levels.

J	IR	J	IR
0	A	1/2	E_1'
1	F_1	3/2	G'
2	H	5/2	I'
3	F_2+G	7/2	$E_2'+I'$
4	G+H	9/2	$G'+I'$
5	F_1+F_2+H	11/2	$E_1'+G'+I'$
6	$A+F_1+G+H$		

After expansion of the reducible representation (RR) of the rotation group D_J into IR of the IG, we obtain the rule for splitting of the atomic levels in the fields of IS (table 3). As seen from the data, the splitting takes place if $J \geq 3$ (as compared with cubic symmetry, where it is necessary that $J \geq 2$, and lower symmetries, where $J \geq 1$). It means, that for $J \leq 3$ the IS can not be revealed and investigated directly.

3.2. TRANSITION RULES.

The non-vanishing matrix elements (ME) of any perturbation operator is determined by means of decomposition of the direct product of IR characterized levels in question. Such decomposition for any pair of IR is given in table 4. If decomposition

Table4. The transition rules.

Product	A	F1	F2	G	H
[F1×F1]	1	0	0	0	1
{F1×F1}	0	1	0	0	0
F1×F2	0	0	0	1	1
F1×G	0	0	1	1	1
F1×H	0	1	1	1	1
[F2×F2]	1	0	0	0	1
{F2×F2}	0	0	1	0	0
F2×G	0	1	0	1	1
F2×H	0	1	1	1	1
[G×G]	1	0	0	1	1
{G×G}	0	1	1	0	0
G×H	0	1	1	1	2
[H×H]	1	0	0	1	2
{H×H}	0	1	1	1	0

Product	A	F_1	F_2	G	H
$[E_1' \times E_1']$	0	1	0	0	0
$\{E_1' \times E_1'\}$	1	0	0	0	0
$E_1' \times E_2'$	0	0	0	1	0
$E_1' \times G'$	0	1	0	0	1
$E_1' \times I'$	0	0	1	1	1
$[E_2' \times E_2']$	0	1	0	0	0
$\{E_2' \times E_2'\}$	1	0	0	0	0
$E_2' \times G'$	0	0	1	0	1
$E_2' \times I'$	0	1	0	1	1
$[G' \times G']$	0	1	1	1	0
$\{G' \times G'\}$	1	0	0	0	1
$G' \times I'$	0	1	1	2	2
$[I' \times I']$	0	2	2	1	1
$\{I' \times I'\}$	1	0	0	1	2

contains IR which characterizes perturbation operator then the ME differs from zero and the transition is allowed.

3.3. SYMMETRIZED FUNCTIONS.

For any quantum chemical modeling calculations it is convenient to deal with the functions which transform under the action of the group elements in accordance with the matrices of IR. These functions can be obtained by means of the projection operator methods [1]. It follows from the table 3 that for J=0, 1/2, ..., 2, 5/2 the symmetrized functions φ_i^α (α means IR) coincide with the atomic ones $\psi_M^J (M = -J,...,+J)$. For J=3 and J=7/2 we have:

$$\varphi_{1,3}^{F_2} = \pm\sqrt{\frac{3}{5}}\psi_{\mp2}^3 - \sqrt{\frac{2}{5}}\psi_{\pm3}^3 ; \qquad \varphi_2^{F_2} = \psi_0^3 . \qquad \varphi_{1,4}^{G} = \sqrt{\frac{3}{5}}\psi_{\mp3}^3 \mp \sqrt{\frac{2}{5}}\psi_{\pm2}^3 ; \qquad \varphi_{2,3}^{G} = \psi_{\mp1}^3 .$$

$$\varphi_{1,2}^{E_2'} = \sqrt{\frac{3}{10}}\psi_{\mp\frac{7}{2}}^{\frac{7}{2}} \mp \sqrt{\frac{7}{10}}\psi_{\pm\frac{3}{2}}^{\frac{7}{2}} .$$

$$\varphi_{1,6}^{I'} = \pm\frac{1}{\sqrt{50}}\psi_{\mp\frac{5}{2}}^{\frac{7}{2}} - \frac{7}{\sqrt{50}}\psi_{\pm\frac{5}{2}}^{\frac{7}{2}} ; \qquad \varphi_{3,4}^{I'} = \pm\psi_{\mp\frac{1}{2}}^{\frac{7}{2}} ; \qquad \varphi_{2,5}^{I'} = \sqrt{\frac{7}{10}}\psi_{\pm\frac{7}{2}}^{\frac{7}{2}} \mp \sqrt{\frac{3}{10}}\psi_{\mp\frac{3}{2}}^{\frac{7}{2}} .$$

3.4. KOSTER-STATZ PERTURBATION OPERATOR MATRICES.

The matrices $M_\gamma(\alpha\times\beta)$ of perturbation operator $\hat{V}_k^\gamma$ represent a set of minimum number of nonzero ME of the type $V_{ikj}^{\alpha\gamma\beta} = \int(\varphi_i^\alpha)^* \hat{V}_k^\gamma \varphi_j^\beta d\tau$, which are allowed by symmetry reasons. In the ME all states i,j of IR α,β respectively are taken into account, $\hat{V}_k^\gamma$ being the k-component of the operator $\hat{V}^\gamma$ which transform in accordance with IR γ. The matrices can be obtained from the solution of the system of equations [16]:

$$V_{ikj}^{\alpha\beta\gamma} = \frac{1}{N}\sum_{i'k'j'} V_{i'k'j'}^{\alpha\beta\gamma} \sum_G \left(G_{i'i}^\alpha\right)^* G_{k'k}^\gamma G_{j'j}^\beta = \left[\int\left(\hat{\theta}\varphi_i^\alpha\right)^* \hat{\theta}\hat{V}_k^\gamma \hat{\theta}^{-1}\varphi_j^\beta d\tau\right]^* \qquad (1)$$

G_{lm}^δ -are the ME of the group operators, N-the number of the group elements, $\hat{\theta}$ - the time inversion operator, the first summation in (1) is realized through all group elements. We have solved these systems for the most often used operators - electric ($\vec{d}$) and magnetic ($\vec{\mu}$) dipol moment operators for all possible pairs of IR of IG. We present below as examples two matrices of Zeeman operator $-\vec{\mu}\vec{H}$ ($\vec{H}$ -the magnetic field) [1]:

$$M(F_2 \times G) = m_1 \begin{pmatrix} \dfrac{T}{2} & \dfrac{T^*}{2} & 0 & -H_z \\ 0 & \dfrac{T}{q} & \dfrac{T^*}{q} & 0 \\ H_z & 0 & \dfrac{-T}{2} & \dfrac{-T^*}{2} \end{pmatrix}; \quad M(G \times G) = m_2 \begin{pmatrix} -H_z & 0 & -T & 0 \\ 0 & -H_z & 0 & T \\ -T^* & 0 & H_z & 0 \\ 0 & T^* & 0 & H_z \end{pmatrix};$$

$T = H_x + iH_y$, $m_1 = -\int(\varphi_3^{F_2})^* \hat{\mu}_z \varphi_1^G d\tau$, $m_2 = -\int(\varphi_4^G)^* \hat{\mu}_z \varphi_4^G d\tau$. The Koster-Statz matrices are useful in determining the energy structure, resonance and other properties of local centers.

3.5. SPIN-HAMILTONIANS(SH)

There are a number of methods for obtaining the SH [16]. Using one of them we have:

$$\hat{W} = \hat{W}_C + \hat{W}_H, \qquad\qquad \hat{W}_H = \hat{W}_H^{(1)} + \hat{W}_H^{(2)} + \hat{W}_H^{(3)}, \qquad\qquad (2)$$

W_C and W_H are the crystal-field and Zeeman operators,

$$\hat{W}_C = d\left[\hat{\psi}_-^{6,5} - \sqrt{11/7}\hat{\psi}_0^6\right], \qquad \hat{W}_H^{(1)} = g_1\left(\vec{J}\cdot\vec{H}\right),$$

$$\hat{W}_H^{(2)} = g_2\left\{H_z\left(\hat{\psi}_0^5 - \sqrt{7/6}\hat{\psi}_{-1}^{5,5}\right) + \sqrt{5/24}\left[H_x\left(\hat{\psi}_{-1}^{5,1} + \sqrt{7/3}\hat{\psi}_+^{5,4}\right) + iH_y\left(\hat{\psi}_+^{5,1} - \sqrt{7/3}\hat{\psi}_-^{5,4}\right)\right]\right\},$$

$$\hat{W}_H^{(3)} = g_3\left\{H_z\left(\hat{\psi}_0^7 - 2\sqrt{6/77}\hat{\psi}_-^{7,5}\right) + \sqrt{2/7}\left[H_x\left(\sqrt{39/22}\hat{\psi}_+^{7,6} - \sqrt{3/44}\hat{\psi}_+^{7,4} - \hat{\psi}_-^{7,1}\right) + iH_y\left(\sqrt{39/22}\hat{\psi}_-^{7,6} + \sqrt{3/44}\hat{\psi}_-^{7,4} - \hat{\psi}_+^{7,1}\right)\right]\right\}$$

$\hat{\psi}_{\pm}^{L,m} = \hat{\psi}_m^L \pm \hat{\psi}_{-m}^L$, $m>0$. $\hat{\psi}_m^L (m = -L,...,+L)$ are the functions-operators, which consist of the linear combination of products of components $\hat{J}_i$ of angular momentum operator $\hat{J}$ [16]. $\hat{\psi}_m^L$ transforms as a basis of the IR D_L of rotation group. Constants d and g_i are the parameters of the SH. The number of terms contained in (2) depends on the value J(L≤2J). By J≤2 the nonzero term is only $\hat{W}_H^{(1)}$, by J≤5/2 - $\hat{W}_H^{(1)}$ and $\hat{W}_H^{(2)}$, by J≤3 - $\hat{W}_H^{(1)}$, $\hat{W}_H^{(2)}$ and $\hat{W}_C$, while by J≥7/2 all terms in (2) remain. Other terms of SH (hyperfine, nuclear quadrupole, nuclear zeeman interactions) can be easy obtained from (2) if we take into account, that the transformation properties of $\vec{J}, \vec{H}$ and nuclear spin $\vec{I}$ are the same. Therefore it is enough to make suitable substitutions of the type J↔I, J↔H, I↔H.

3.6. THE ELECTRON PARAMAGNETIC RESONANCE (EPR) SPECTRUM.

In the case of strong magnetic field approximation we have for the resonance value of magnetic field H_r (M↔M−1):

$$H_r = H_r^{(0)} + 30\Phi(\theta,\varphi)\left[\sqrt{3}dn_C(J,M) - g_2H_r^{(0)}n_H(J,M)/\sqrt{14}\right]/g_1 \qquad (3)$$

$$H_r^{(0)} = h\nu/g_1, \quad n_H = 7\left[3M^4 - 6M^3 + (9-2A)M^2 - 2M(3-A) + B/7\right]/12,$$

$$n_C = (M - 1/2)\left(11M^4 - 22M^3 + (49 - 10A)M^2 + 2M(5A - 9) + 5B/3\right)/20,$$

$$\Phi(\theta,\varphi) = \left(231x^6 - 315x^4 + 105x^2 - 5 + 42x\sin^5\theta\cdot\sin5\varphi\right)/16,$$

ν - the transition frequency, h - Plank constant, A=J(J+1), B=A^2−8A+12, x=Cosθ, θ and φ - the polar and azimuthal angles of vector $\vec{H}$. In the formula (3) the term $\hat{W}_H^{(3)}$ from (2) is not taken into account because of its small contribution into H_r. In the case d≠0(g$_2$=0) the EPR spectrum at any pair of angles θ and φ consist of 5 lines of different intensity I' (Fig.3). Each line has the angle dependence in accordance with the factor $\Phi(\theta,\varphi)$.

4. Noncentral Impurity Ions.

4.1. ENERGY STRUCTURE.

The establishment of an equivalent equilibrium position of noncentral ions (NI) and the rules of their transformation one into the other under the action of the symmetry operations of group Y_h allowed us to obtain a set of 11 RR. The dimensions of the latter coincide with the numbers of faces, apices and edges. We introduce the following denotations for them (table 5). We use capital letters A, F and E for RR representations corresponding to equilibrium position displacements along apices, faces and edges.

Table5. Energy structure of noncentral ions in icosahedral environment.

RR	IR									
	A_g	A_u	F_{1g}	F_{1u}	F_{2g}	F_{2u}	G_g	G_u	H_g	H_u
$A^i(12);F^d(12);F^f_5(12)$	1	0	0	1	0	1	0	0	1	0
$F^i(20);A^d(20);F^f_6(20)$	1	0	0	1	0	1	1	1	1	0
$E^i(30);E^d(30);E^f_{6-6}(30)$	1	0	0	1	0	1	1	1	2	1
$E^f_{5-6}(60);A^i(60)$	1	0	1	2	1	2	2	2	3	2

The upper indices d,i and f correspond to dodecahedron, icosahedron and fullerene, respectively. The dimension of RR is given in brackets. In the case of fullerene we introduce in addition inferior indices to distinguish the nonequivalent faces and edges. The energy structure corresponding to the particular RR is determined by the expansion of the latter into IR. The figures in table 5 show how many times IR designated in the caption of the column are contained in RR designated in the caption of the line.

4.2. PARAELECTRIC RESONANCE (PR).

When considering the resonance transitions within a set of energy levels shown in table 5, i.e. the field-free PR, it is sufficient, in general, to take into account each of IR A_g, F_{1g}, F_{2g} once, while each of IR F_{1u}, F_{2u}, G_g, G_u, H_g, H_u twice. In PR the transitions are realized under the action of an alternating electric field (EF) whose vector transforms according to IR F_{1u}. The analysis of direct products of the IR (tabl.4) shows that transitions are allowed between the following terms: $A_g \leftrightarrow F_{1u}$; $F_{1g} \leftrightarrow F_{1u}$, H_u; $F_{2g} \leftrightarrow G_u$, H_u; $F_{1u} \leftrightarrow H_g$; $F_{2u} \leftrightarrow G_g$, H_g; $G_g \leftrightarrow G_u$, H_u; $G_u \leftrightarrow H_g$; $H_g \leftrightarrow H_u$.

PR is often realized by passing the external static EF via resonance [17]. In this case EF for convenience is oriented along symmetry directions, so that certain elements of the point group are preserved. Table 6 shows symmetry group arising for such direction of the external EF. The presence of two

Table6. The point groups of figures in the presence of an EF

figure	direction from the centre toward		
	face	apex	adge
icosahedron	C_{3v}	C_{5v}	C_{2v}
dodecahedron	C_{5v}	C_{3v}	C_{2v}
fullerene	C_{5v}/C_{3v}	C_s	C_s/C_{2v}

point groups in a cell for fullerene is connected with two types of faces and edges. The upper symbols are attributed to pentagon and the edge - connecting penta - and hexagons. Table 7 presents data on the character of splitting of levels under the action of the external EF. The figures show how many times IR given together with the

Table 7. The splitting of tunnel levels under the action of EF applied along symmetry directions[1]

IR of gr. Y_h	Subgroups of group Y_h and their IR												
	C_{5v}				C_{3v}			C_{2v}				C_s	
	A_1	A_2	E_1	E_2	A_1	A_2	E	A_1	A_2	A_3	A_4	A_1	A_2
	z		x,y		z		x,y	z	y		x	x,y	z
A_g	1	0	0	0	1	0	0	1	0	0	0	1	0
F_{1g}	0	1	1	0	0	1	1	0	1	1	1	1	2
F_{1u}	1	0	1	0	1	0	1	1	1	0	1	2	1
F_{2g}	0	1	0	1	0	1	1	0	1	1	1	1	2
F_{2u}	1	0	0	1	1	0	1	1	1	0	1	2	1
G_g	0	0	1	1	1	1	1	1	1	1	1	2	2
G_u	0	0	1	1	1	1	1	1	1	1	1	2	2
H_g	1	0	1	1	1	0	2	2	1	1	1	3	2
H_u	0	1	1	1	0	1	2	1	1	2	1	2	3

[1] A - one-dimensional, E - two-dimensional IR. Letters x,y,z mean that these components are transformed according to such IR.

corresponding subgroup in the caption of the column is contained in the IR of group Y_h given in the caption of the line. IR A_u is omitted as, according to table 5, it is not actual.

5. Conclusion.

The results can be compared with the behavior of noncentral ions in an environment of other symmetries. The comparison shows that in the case of IS due to a large number of equivalent equilibrium positions the energy and resonance spectra are richer than in any other point groups. In the case of central ions the situation is opposite: the lower the symmetry, the richer is the energy and resonance spectra

6. References.

1. Lyubarskii G.Ya. (1957) Group Theory and Its Application in Physics, Tekh.-Teor. Lit., Moscow (in Russian).
2. Hamermesh N. (1964) Group Theory and Its Application to Physical Problems, Addison-Wesley Publ. Co., Reading (Mass.).
3. Landau L.D. and Lifshits E.M. (1989) Quantum Mechanics, Nauka, Moscow (in Russian).
4. Roitsin A.B. (1993) Icosahedral Symmetry, Priroda 8, 10-20.
5. Morokhov I.D., Petinov V.I., Trusov L.I. at al (1981) Structure and properties of small metalic particles, Uspekhi fiz. Nauk 133,653-692.

6. Shechtman D., Blech I., Gratias D. at al (1984) Metallic Phase with Long-Range Orientational Order and Translational Symmetry, Phys. Rev. Letters 53, 1951-1953.

7. Harter W.G. and Weeks D.E. (1989) Rotation-Vibration specta of icosahedral molecules, J. Chem. Phys. 90, 4727-4771.

8. Roitsin A.B., Maevskii V.M. (1989) Electron Paramagnetic Resonance of Solid State Surface, Uspekhi fiz. Nauk 159, 297-333.

9. Stathis J.H. and Pantelides S.T. (1988) Quantitative analysis of EPR and ENDOR spectra of D-centres in amorphous silicon: dangling versus floating bonds, Phys. Rev. (B).37, 6579-6582.

10. Krechmer V. (1992) The new carbon states, Priroda 1, 30-33.

11. Loktev V.M. (1992) Doped fullerite-first three-dimensional organic superconductor, Fiz.Nizk.Temp. 18, 217-237.

12. Bethune D.S., Johnson R.D., Salem J.R. at al (1993) Atoms in carbon cages: the structure and properties of endohedral fullerenes, Monthly nature 1, 67-72.

13. Roitsin A.B. (1993) Energy structure of Icosahedral Symmetry Centres, Fiz. Tverd. Tela 35,2548-2554.

14. Roitsin A.B. (1993) Energy Structure and Resonance Properties of Noncentral Ions in Icosahedral Environment, Phys. Stat. Sol. (b) 178, 275-280.

15. Roitsin A.B., Klimov A.A., Artamonov L.V. (1996) The Icosahedral Groups Representations, Fiz. Tverd. Tela 38.

16. Roitsin A.B. (1973) Some application of group theory to the Radiospectroscopy Problems, Naukova Dumka, Kiev.

17. Kopvillem V.Kh. and Saburova R.V. (1982) Paraelectric Resonance, Nauka, Moscow.

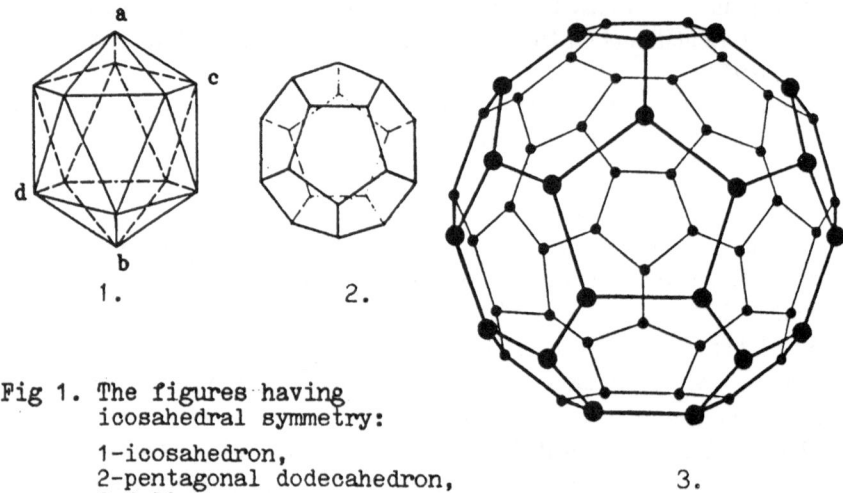

Fig 1. The figures having
icosahedral symmetry:

1-icosahedron,
2-pentagonal dodecahedron,
3-fullerene

3.

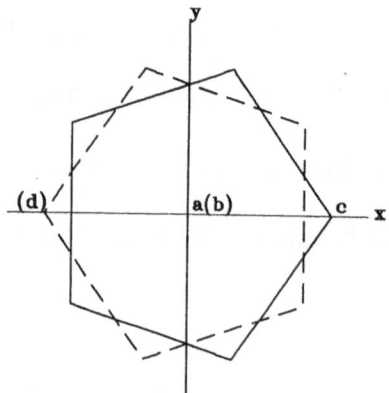

Fig.2. The projection of the icosahedron on the
xy - plane, passed across the centre of
icosahedron perpendicularly to the z -
axis, directed to the reader. Solid
(dotted) line and letters without (in)
brackets mean the part of the figure
situated over (under) the plane.

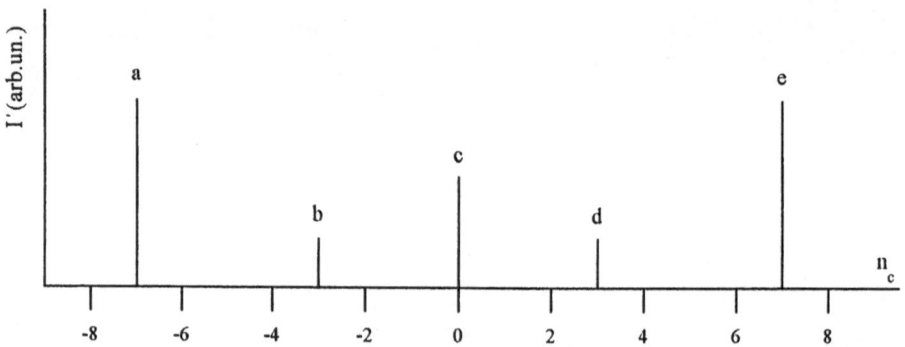

Fig.3. EPR - Spectrum of paramagnetic centre; J=7/2, g_2=0, d≠0.
a: 5/2↔3/2 and -1/2↔-3/2; b: -5/2↔-7/2; c: 1/2↔-1/2;
d: 7/2↔5/2; e: 3/2↔1/2 and -3/2↔-5/2.

MODELLING STRUCTURE AND DEFECTS IN ZEOLITES

A.A. SOKOL, C.R.A. CATLOW*
The Royal Institution of Great Britain
21 Albemarle Street, London W1X 4BS, UK

Abstract

Semi-classical lattice energy minimization and quantum chemical modifield INDO techniques have been applied to study the crystal structures of siliceous mordenite and ferrierite, and the relative stabilities to Al substitution of different T-sites in mordenite. The study evinced common features in the electronic structure of siliceous zeolites. Structural and electronic properties of Brønsted acid site, hydroxyl nest defect and hydroperoxy defects were examined using the periodic and Mott-Littleton models.

1. Introduction

Zeolites are an important class of microporous crystals widely used as solid catalysts and molecular sieves, properties in which are controlled by their microporous architecture [1-3]. The nature of these applications makes it particularly important to study zeolites at an atomistic level of theory.

However, problems arise due to the complexity of these systems - especially due to the large number of atoms in their unit cells (for example 36 atoms in siliceous sodalite and chabazite and 576 atoms in siliceous faujasite), and also because of structural disorder in real samples. But this complexity is based on an underlying simplicity. In fact, zeolites are aluminosilicates or silica polymorphs which are formed by interlinked tetrahedra: the centre of each tetrahedron (T-site) is occupied by a tetravalent silicon, or by the trivalent substituent, aluminium; in turn, the corners, shared between adjacent tetrahedra, are occupied by divalent oxygen anions. Tetrahedral coordination of Al in a zeolite framework necessitates charge compensation that can be accomplished through either extraframework cations or a proton attached to a bridging oxygen (the Brønsted acid site). So, although the medium-range order in zeolites can be highly complex, the short order is simple; and we note that short range structural features are a major factor in determining the electronic and defect properties of materials.

Rapid advances in computer technology in recent years have rendered it possible to start extended computational studies on zeolites and related materials such as zeotypes and clays. So far, most work has been done, on the one hand, with atomistic

R. C. Tennyson and A. E. Kiv (eds.),
Computer Modelling of Electronic and Atomic Processes in Solids, 125–135.
© 1997 *Kluwer Academic Publishers.*

simulation techniques based on ineratomic potentials using the Periodic and the Mott-Littleton cluster models [4, 5], and with *ab initio* techniques (both the Hartree-Fock and the Density Functional based methods) in the Molecular Cluster models [6. 7], on the other. From the early 1990s there have also appeared pioneering *ab initio* works using the Periodic and the Embedded Cluster models [8-11]. This type of calculation should eventually be the preferred choice but current computer facilities do not allow them to become common. For example, a single-point Hartree-Fock calculation with a minimal basis set on silicalite (288 atoms in a unit cell) was reported to take 18 hours on a CRAY-2S [12]. Hence, we need to separate the problems that can be successfully treated by applying relatively cheap, simple but reliable methods.

The atomistic simulation methods, mentioned above, allow us to solve a great many structural problems both for crystals and for defects within them. The method works well unless the defect or crystal itself, during the simulation, undergoes a transformation which leads to a radically different structure. In this case the interatomic potentials used are not necessarily suitable and the results are questionable. The simplest methods yielding the electronic structure of a crystal and its defects are semi-empirical, varying from the Extended Huckel to the NDDO type of schemes. They are slower than the atomistic simulations but give us additional information in a far shorter time than for the *ab initio* methods.

In fact, semi-empirical and simulation methods are complementary; and we have used both in our studies, although we plan in future work to investigate critical configurations using *ab initio* techniques. All results in this paper have been obtained using the Energy Minimisation method with interatomic potentials as implemented in the code GULP [13] and the semi-empirical, Modified Intermediate Neglect of Differential Overlap (INDO) method as implemented in the code SYMSYM [14].

Results of crystal structure simulations on siliceous mordenite and siliceous ferrierite are presented in Section 2. In Section 3 we discuss results of the electronic structure calculations on siliceous zeolites in the Large Unit Cell model [15]. Section 4 comprises examples of hydrogen containing defects in zeolites and their structural properties. In conclusion, we discuss some possible directions for further studies on defects in zeolites.

2. Crystal structure simulation

Real zeolites are often even more complex than their idealised crystallographic structures. All sorts of defects (including point and planar defects, and intergrowths) are present. Moreover, crystal structure determination is hindered by the fact that often only powder samples are available. And there is often controversy on matters related to space group assignment and T-sites distribution. Here we present our results for two such cases where our computational approach has been applied to consider problems presently not solved by experiment.

2.1. MORDENITE: STRUCTURE AND AL DISTRIBUTION

Attempts at crystal structure determination for mordenite go back to the 1930s and a whole set of different zeolites including ptilolite were in the end identified as having the same structure. However, the first reliable data were presented by W.M.Meier [16] who proposed the space group Cmcm. Since then there have appeared new data obtained on purer synthetic and natural samples; the latest studies still confirm the space group assignment (see, for example, [17]). A number of atomistic simulations were therefore undertaken on siliceous [18] and Al substituted [19] structures, using interatomic potentials especially derived for zeolites [4]. In all the calculations the space group Cmcm was assumed.

In our simulations of Al substitution in siliceous mordenite (which used the interatomic potentials reported in [4,20]) we unexpectedly encountered instabilities in the Cmcm crystal structure which made us return to the modelling of siliceous mordenite, for which we found a stable structure of lower symmetry with the $P2_12_12_1$ space group. Calculated X-ray diffraction patterns for the different space groups differ only slightly and the differences should be obscured by the defects present in real structures. The energetic difference between the new and the old structures is not large - only 0.02 eV per SiO_2 formula unit. But they allow us to account for 12 different T-sites as candidates for Al substitution and to open more T-sites for such a substitution without breaking the crystal symmetry. A crystallographic unit cell of mordenite consists of 48 formula units and natural samples contain 8 Al sites per unit cell. Thus in the case of the Cmcm structure only one crystallographically distinguishable T-site can be occupied by Al (Wyckoff special position g of multiplicity 8), and simulations place it inside the 8-ring channel. The new model, on the contrary, has two low energy positions inside the bigger 12-ring channel with each of them having multiplicity of only 4. The T-site locations in the mordenite framework along with the relative stability of Al substituted T-sites are shown below in Fig. 1.

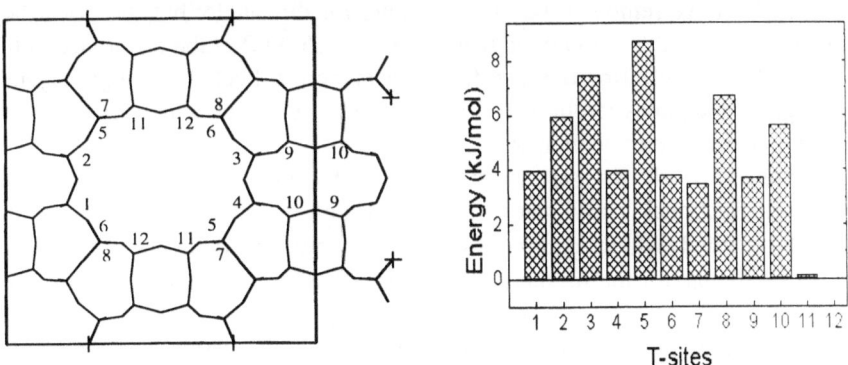

Figure 1. T-sites assignment in the $P2_12_12_1$ structure of mordenite and plot of relative energies for substitution of Al. All energies are relative to the value for T12 - the most stable, Al substituted T-site.

We note that all results on the relative stability of Al at different T-sites have been obtained using a Mott-Littleton Cluster model. Charge compensation has been achieved through the attachment of a proton to the oxygen atom nearest to an Al site, with results averaged over the first oxygen coordination shell. In real mordenite, interactions between the Al subtitutionals will have a significant effect. In particular, to avoid violation of Løwenstein rule, T11 and T12 sites should not be substituted by Al simultaneously. Nevertheless, with the constraints of the $P2_12_12_1$ space group there still remains enough freedom to place one of two inequivalent Al atoms at T11 or T12 sites and another at T9 or T10 (that are preferential within the Cmcm structure). Even if an Al distribution in real mordenite breaks the crystal symmetry, making the problem of the space group assignment undetermined, we still have here two contradictory predictions on energetically preferential sites for Al substitution (and, so, for potential chemical active sites). Which is true is an important question which should be solved by experiment.

2.2. FERRIERITE

Ferrierite is a zeolite closely related to the mordenite family. The main channels go along the [001] direction; 5-ring channels serve as an interface between the neighbouring bigger (10-ring) and smaller (6-ring) channels. Until recently two crystal modifications have been known: one, orthorhombic Immm, the other, monoclinic $P2_1/n$. However, after a new method of single crystal synthesis was reported [21], two groups [22, 23] independently proposed a new space group, Pmnn, for the orthorhombic variety of ferrierite.

Both orthorhombic structures were examined using the simulation techniques in [22]. The newly reported structure was found to be only slightly more stable than the older one. The key issue is the position of a bridging oxygen in a 5-ring channel, connecting two 10-ring channels. In the Immm structure it is directly at the inversion centre with fractional coordinates (¼,¼,¼). In the Pmnn structure the oxygen is moved from that symmetrical position by 0.3 Å. The Si-O-Si bond angle is 180° in the first case and about 170° in the second case. Thus, having introduced the lower symmetry, the authors of [22] have removed the unusual value for this angle, but the price they paid was a large scatter in the Si-O bond lengths (1.56-1.65 Å). Yet the difference in the value of the angle is small. Moreover, so far, all data were collected on single crystals containing organic templates (from 1 to 2 pyridine molecules per unit cell) or, again, using powders obtained after calcination.

In the present work we carried out a geometry optimisation of the lattice parameters and fractional coordinates on the orthorhombic structures using the modified INDO method (which required respectively ~15 and ~200 hours of CPU time on SG Power Challenge for the Immm and Pnnm structures). The results are illustrated in Fig. 2.

The bond lengths in our parameterisation of zeolites on average are reproduced with an accuracy better than 1%; at the same time the bond angles are exaggerated by about 3-5%. The calculated value of 173° for the critical Si-O-Si angle essentially confirms the earlier simulation result [22]. Our conclusions on the relative stability of the two structures are also in accordance with these calculations. In addition we found that the scatter of the Si-O bond length is significantly lower than that predicted by experiment. In this case, both structures are obviously plausible and there is no reason, why one of the structures should be considered as more realistic (at least, in a low temperature region).

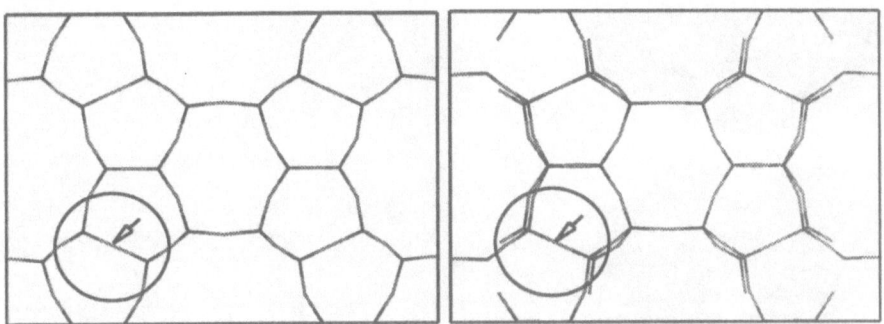

Immm: E=36.445 a.u. per SiO_2 unit   Pmnn: E=36.445 a.u. per SiO_2 unit
$\angle Si-O-Si=180°$ $\angle Si-O-Si=173°$

Figure 2. Siliceous ferrierite crystal structure. Semi-empirical INDO optimised models. Arrows in circles show the angle of specific interest. Structures are topologically equivalent and only slightly differ from one another.

In general, our calculations suggest that many zeolites may have several closely related, topologically equivalent , but crytallographically distinct structures. At a local level, these differences in the long range structures may be of little significance.

3. Electronic structure of siliceous zeolites

Our calculations on the electronic structure of perfect crystalline zeolites pursued two main goals. First, we wished to investigate the differences in the electronic structure induced by the long and medium order in different zeolitic systems. Second, the results obtained will be used in future cluster studies on defects in corresponding zeolites, providing us with "reference" systems.

In the present study the modified INDO parameterisation for alumino-silicates has been used to obtain one-electron Hartree-Fock densities of states, effective charges and total energies of four siliceous zeolites: sodalite, chabazite, faujasite, and mordenite. (For the latter, the calculation was done within the Cmcm space group.) The large unit cells for all four materials included at least one crystallographic unit cell; in the case of sodalite and chabazite, 2x2x2 supercells of the basic unit cell were chosen. The lattice parameters for all materials except mordenite have been optimised. However, the fractional coordinates were fixed at their experimental values (because even the semi-empirical periodic calculations on zeolites are still very expensive). The calculated spectra and relative stabilities are shown in Fig. 3.

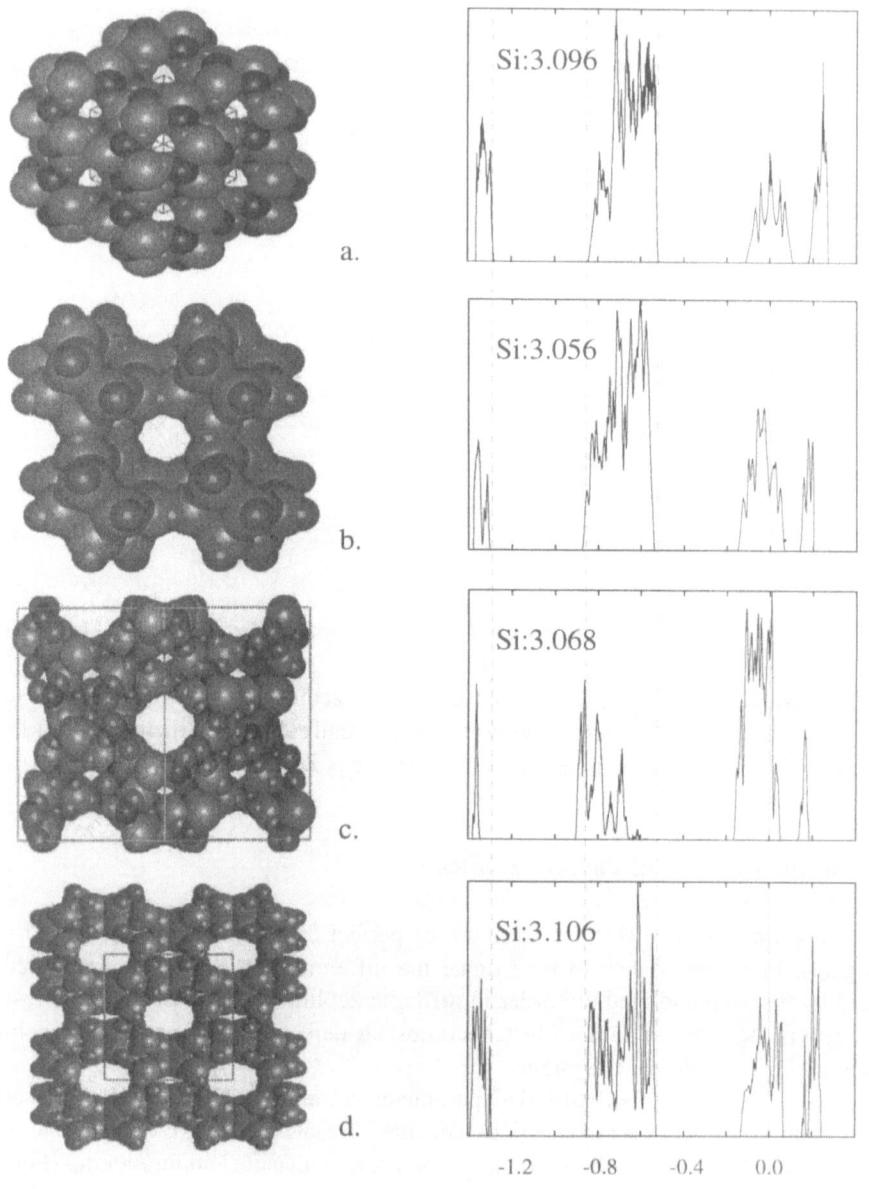

Figure 3. Comparison of electronic structures and lattice energies of siliceous zeolites. Densities of states in arbitrary units. Lattice parameters as found from simulations and INDO (second value).
a. Sodalite E=128.57 eV, a=8.77 and 8.82;
b. Chabazite E=128.55 eV, a=9.22 and 9.21, alpha=94.9 and 94.1;
c. Faujasite E=128.52 eV, a=24.31 and 24.35;
d. Mordenite E=128.59 eV, a=18.04, b=20.05, c=7.41 (not optimised with INDO).

We see that lattice parameters in our semi-empirical studies are reproduced reasonably well compared to the corresponding interatomic potentials based simulations. Total energies vary insignificantly from zeolite to zeolite. The electronic spectra of all four zeolites show important similarities both to each other and to that of α-quartz. There are two clear valence sub-bands, formed in turn by O-2s and O-2p states with approximately equal widths of 0.11 and 0.33 a.u, respectively (compared with *ab initio* HF results on α-quartz and silico-chabazite 0.36-0.44 a.u for the upper sub-band [8, 24]). The gap between the upper valence and the lower conduction band is about 0.4 a.u. Crystallographically inequivalent Si and O sites bear only slightly different effective charges. The same is true of the Madelung potentials: 0.85 on O, and -1.3 on Si. The largest variations were observed on mordenite and could be attributed to the fact that its lattice parameters have not been optimised. The largest noticeable distinction between the zeolites is in the structure of their upper valence band. Therefore we would expect that their most important optical characteristics should also be very similar.

It is generally found, even in complex solids that atoms of the same chemical type and valence are at sites with similar Madelung potentials and have similar charges. In the present case we suggest that a lattice rearrangement (such as the formation of very large pore structures) might result in substantial changes in Madelung potentials and consequent instability.

4. Hydrogen containing defects in zeolites

Zeolite growth both in natural conditions and laboratory synthesis inevitably involves the formation of various defects. Hydrogen, or rather hydroxyl containing defects arouse special interest due to their active rôle in chemical processes, in particular, catalysis. Synthesis and processing of zeolites often includes steaming and dealumination which can produce, in particular, the three types of defects described next. (Possible mechanisms for relevant reactions discussed in detail in [25].) All three defects are neutral centres and therefore are localised. Moreover, their neutrality makes them directly suitable for studying using the Periodic model.

4.1. BRØNSTED ACID SITE

In ion exchange reactions a univalent cation, balancing the charge on a zeolite framework due to an Al substitution, can be exchanged by a proton creating the usual Brønsted acid site. The defect is similar in different zeolitic materials and on different T-sites.

In Fig. 4 we present the calculated equilibrium electronic and ionic configuration of a Brønsted acid site in natural sodalite ($Cl_2Na_8Si_6Al_6O_{24}$). The result was obtained using a periodic supercell model (2x2x2) with the modified INDO method. The one-electron spectrum of perfect natural sodalite in comparison with its counterpart for a siliceous material has an additional feature: its upper valence band consists now of two

132

sub-bands - the lower one due to O-2p (as before) and the upper one due to Cl-3p states. The Cl-3p states are separated from the O-2p states by the gap of about 2.1 eV. The bottom of the conduction band is about 10.3 eV above the top of the valence band and is formed by Si-3sp hybridised states with a small contribution from Al-3sp states.

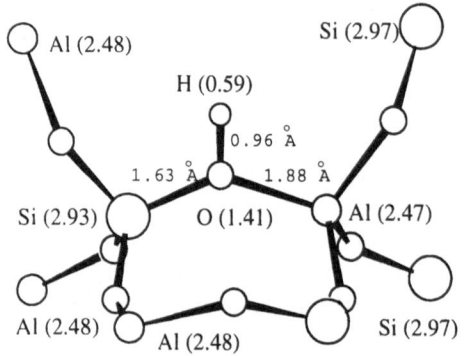

The Brønsted acid site introduces a new, highest occupied level (a very narrow sub-band in the present model) at 0.4 eV over the top of the Cl-3p sub-band, and a new, lowest unoccupied level at 1.6 eV below the Si-3sp sub-band. As a result we have the reduction of the gap by ~2 eV and our CIS calculation of optical transitions in the system shows approximately the same reduction in the optical gap from 9.6 eV to 7.8 eV.

Figure 4. Brønsted acid site in natural sodalite. Na and Cl ions are omitted from the picture for clarity. Numbers in parentheses indicate effective charges. Only the most apparent changes in local geometry are highlighted by means of distances. (Compare with regular bond lengths in a model: Si-O: 1.58 Å, and Al-O: 1.75Å.)

The geometry optimisation has given the same results as ab initio calculations on small molecules, embedded clusters and simulations (see, e.g. [26-31]). Structural perturbations are concentrated in the immediate vicinity of the hydroxyl group; the Al-O bond lengthens most and the O-H bond distance is practically the same as in a water molecule.

4.2. HYDROXYL NEST DEFECT

Zeolite dealumination in hydrophobic conditions, using hydrochloric acid, may result in a hydroxyl nest defect that is a vacancy on a T-site with the four adjacent oxygens saturated by protons.

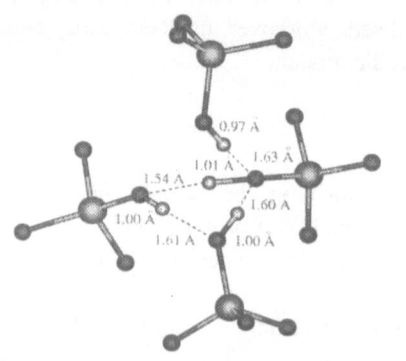

The importance of this defect has been proposed for a number of silica and silicate systems and was studied in these materials by both *ab initio* and simulation methods. We should note here the work of Purton et al. on α-quartz who used a local density approach [32], and Wright et al. [33,34] on olivine and grossular using the interatomic potential based simulations. A few zeolitic structures have also been considered by Lewis [35]. In the present study we applied the combined techniques of lattice simulation and INDO to obtain the electronic and ionic structures of the

Figure 5. Hydroxyl nest defect in siliceous ferrierite. Larger circles show Si; smaller dark circles are O, and yet smaller and light circles are H.

defect (as shown in Fig.5). A special feature of our result is the shortening of the hydrogen bonds compared to the usual range 1.8-2.1 Å (apparently, due to structural restraints imposed by the lattice). This shortening is accompanied by the lengthening of hydroxyl bonds. The effective charges on the hydrogen ions vary between 1.52 and 1.58, and on oxygen in the hydroxyl groups between 1.28 and 1.36. The effective charge on Si ions, adjacent the defect, drops by only 0.02; and further away from the Si vacancy the impact of the defect on the lattice becomes negligible.

In contrast to the Brønsted acid sites, the hydroxyl nest defect introduces no new levels into the gap, but it gives a noticeable contribution to the spectrum at the bottom of all sub-bands owing to mixing with corresponding hybridised Si-3sp states. Thus a hydroxyl nest defect is an "ideal" substitution of Si in a T-site from the point of view of electronic as well as ionic properties.

4.3. HYDROPEROXY BRIDGE

Attack at an oxygen bridge by a water molecule, especially if the framework fragment contains an electronic defect (such as a self-localised hole or exciton), could lead to a hydroperoxy defect formation. We first studied this defect with a simulated cluster model comprising two silanol groups on neighbouring T-sites in ferrierite. Geometry optimisation carried out from different initial configurations always generated the equilibrium structure of the defect shown in Fig. 6.

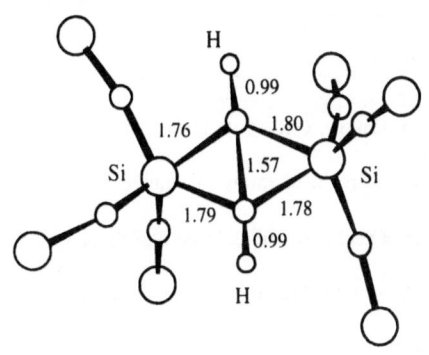

Figure 6. Hydroperoxy defect in the ferrierite framework. Larger circles, Si ions, smaller, O ions. All distances in Å.

As can be seen from the figure the interaction of the two silanol groups leads to a hydroperoxy bridge formation which was confirmed in our INDO calculations. However, there is an important difference between the result predicted by the two methods. INDO calculations also show that the two hydrogens form full chemical bonds with two oxygens, as is the case with typical extra-framework cations, e.g. Na+, K+. This double coordination cannot be reproduced by the interatomic based simulation method because of the model used.

Ab initio calculations are now clearly needed to resolve the uncertainty as to the structure of this most interesting defect. We should emphasise that such a study necessarily should include solid state effects since the double coordination of H is probably a consequence of the lattice restraints. We note that the existence of such a species can account for the luminescence observed in zeolites as in dense silicas, where it has been attributed to peroxy bridges [36].

Conclusion

Our calculations have shown that a number of problems concerning the structural and electronic properties of zeolites can be successfully solved using approximate and relatively simple methods such as interatomic-potential based simulations and semi-empirical INDO techniques. The computational results demonstrated that it is important at all stages of the modelling to account in a consistent manner for solid state effects which lead to significantly different results compared to studies on small molecular clusters. For some problems such as the hydroperoxy bridge structure it is necessary to use more accurate *ab initio* techniques. Such studies will be reported in the near future.

Acknowledgement

We are grateful to Dow Chemicals for supporting this work and to J.M.Garcés, A.Kuperman, J.M.Ruiz, and M.McAdon of Dow Chemical Research Centre, A.L.Shluger of UCL, and D.W.Lewis and P.E.Sinclair of the RI for many worthwhile discussions and advice concerning our studies. We would like to thank J.D.Gale, L.N.Kantorovich and A.I.Livshits whose computer codes helped us to perform our calculations.

References

1. Thomas, J.M. (1992) Solid acid catalysts, Sci.Amer. no.4, 82-88.
2. Modelling of structure and reactivity in zeolites (1992) Ed. Catlow, C.R.A., Academic Press. London - San Diego - New York - Boston - Sydney - Tokyo - Toronto.
3. Dag, Ö., Kuperman, A., and Ozin, G.A. (1995) Nanoclusters: new forms of luminescent silicon. Adv.Mater. 7, 72-78.
4. Jackson, R.A. and Catlow, C.R.A. (1988) Computer studies of zeolite structure. Molecular Simulations 1, 207-224.
5. Lewis, D.W., Catlow, C.R.A., Sankar, G., and Carr, S.W. (1995) Structure of iron-substituted ZSM. J.Phys.Chem. 99, 2377-2383.
6. Mortier, W.J., Sauer J., Lercher, J.A., and Noller, H. (1984) Bridging and terminal hydroxyls. A structural chemical and quantum chemical discussion, J.Phys.Chem. 88, 905-912.
7. Kramer, G.J. and Santen, van R.A. 91993) Theoretical determination of proton affinity differences in zeolites, J.Am.Chem.Soc. 115, 2887-2897.
8. Aprá, E., Dovesi, R., Freyria-Fava, C., et al. (1993) *Ab initio* Hartree-Fock modelling of zeolites: application to silico-chabazite, Modelling Simul.Mater.Sci.Eng. 1, 297-306.
9. Teunissen, E.H., Roetti, C., Pisani, C., et al. (1994) Proton transfer in zeolites: a comparison between cluster and crystal calculations, Modelling Simul.Mater.Sci.Eng. 2, 921-932.
10. Shah, R., Payne, M.C., Lee, M.-H., and Gale, J.D. (1996) Understanding the catalytic behaviour of zeolites: a first-principles study of the adsorption of methanol, Science 271, 1395-1397.
11. Allavena, M., Seiti, K., Kassab, E., et al. (1990) Quantum-chemical model calculations on the acidic site of zeolites including Madelung-potential effects, Chem.Phys.Lett. 168, 461-467.
12. White, J.C. and Hess, A.C. (1993) An examination of the electrostatic potential of silicalite using periodic Hartree-Fock theory, J.Phys.Chem. 97, 8703-8706.

13. Gale, J.D. (1991-1996) General utility lattice program (GULP), The Royal Institution of GB and Imperial College, London, UK.

14. Kantorovich L.N. and Livshits A.I. (1993) Symmetry adapted modified INDO code (SYMSYM), Chemical Physics Institute, Latvian University, Riga, Latvia.

15. Stefanovich, E., Shidlovskaya, E., Shluger, A., and Zakharov, M. (1990) Modification of the INDO calculation scheme and parameterisation for ionic crystals, phys.stat.sol. (b) **160**, 529-540.

16. Meier, W.M. (1961) The crystal structure of mordenite (ptilolite), Z.Kristall. **115**, 439-450.

17. Rudolf, P.R. and Garcés, J.M. (1994) Rietveld refinement of several structural models for mordenite that account for differences in the X-ray powder pattern, Zeolites **14**, 137-146.

18. Bell, R.G. Private communication.

19. Ouden, den C.J.J., Jackson, R.A., Catlow, C.R.A., and Post, M.F.M. (1990) Location of Ni^{2+} ions in siliceous mordenite: a computational approach, J.Phys.Chem. **94**, 5286-5290.

20. Schröder, K.-P., Sauer, J., Leslie, M., et al. (1992) Bridging hydroxyl groups in zeolite catalysts: a computer simulation of their structure, vibrational properties and acidity in protonated faujasites (H-Y zeolites), Chem.Phys.Lett. **188**, 320-325.

21. Kuperman, A., Nadimi, S., Oliver, S., et al. (1993) Non-aqueous synthesis of giant crystals of zeolites and molecular sieves, Nature **365**, 239-242.

22. Morris, R.E., Weigel, S.J., Henson, N.J., et al. (1994) A synchrotron X-ray diffraction, neutron diffraction, ^{29}Si MAS-NMR, and computational study of the siliceous form of zeolite ferrierite, J.Am.Chem.Soc. **116**, 11849-11855.

23. Lewis, J.E., Freyhardt, C.C., and Davis, M.E. (1996) Location of pyridine guest molecules in an electroneutral ((3) infinity) [SiO4/2] host framework - single-crystal structures of the as-synthesized and calcined forms of high-silica ferrierite, J.Phys.Chem. **100**, 5039-5049.

24. Nada, R., Catlow, C.R.A., Dovesi, R., and Pisani, C. (1990) An *ab initio* Hartree-Fock study of α-quartz and stishovite, Phys.Chem.Minerals **17**, 353-362.

25. R.M.Barrer. Hydrothermal Chemistry of Zeolites. 1982, Academic Press, London, New York.

26. Ahlrichs, R., Bär, M., Häser, M., et al. (1989) Nonempirical direct SCF calculations on sodalite and double six-ring models of SiO_2, and $AlPO_4$ minerals: $H_{24}Si_{24}O_{60}$, $H_{12}Al_6P_6O_{30}$. Chem.Phys.Lett. **164**, 199-204.

27. Kassab, E., Seiti, K., and Allavena, M. (1991) Theoretical determination of relative acidity in zeolite (faujasite), J.Phys.Chem. **95**, 9425-9431.

28. Brand, H.V., Curtiss, L.A., and Iton, L.E. (1992) Computational studies of acid sites in ZSM-5: dependence on cluster size, J.Phys.Chem. **96**, 7725-7732.

29. Datka, J., Broclawik, E., and Gil, B. (1994) IR spectroscopic studies and quantum chemical calculations concerning the O-H dissociation energies in zeolites NaHX and NaHY, J.Phys.Chem. **98**, 5622-5626.

30. Bleiber, A., Sauer, J. (1995) The vibrational frequency of the donor OH group in the H-bonded dimers of water, methanol and silanol. *Ab initio* calculations including anharmonicities, Chem.Phys.Lett. **238**, 243-252.

31. Zygmundt, S.A., Curtiss, L.A., Iton, L.E., and Erhardt, M.K. (1996) Computational studies of water adsorption in the zeolite H-ZSM-5, J.Phys.Chem. **100**, 6663-6671.

32. Purton, J., Jones, R., Heggie, M., et al. (1992) LDF pseudopotential calculations of the α-quartz structure and hydrogarnet defect, Phys.Chem.Minerals **18**, 389-392.

33. Wright, K., Freer, R., and Catlow, C.R.A. (1994) The energetics and structure of the hydrogarnet defect in grossular: a computer simulation study, Phys.Chem.Minerals **20**, 500-503.

34. Wright, K. and Catlow, C.R.A. (1994) A computer simulation study of (OH) defects in olivine, Phys.Chem.Minerals **20**, 515-518.

35. Lewis, D.W., Private communication.

36. Guzzi, M., Martini, M., Mattaini, M., and Spinolo, G. (1987) Luminescence of fused silica: observation of the O_2^- emission band, Phys.Rev. B **35**, 9407-9409.

THE NATURE OF A PHOTOINDUCED METASTABLE STATE IN α-SI:H

F.T.UMAROVA AND Z.M. KHAKIMOV

Institute of Nuclear Physics of Uzbekistan Academy of Sciences
Ulughbek, 702132, Tashkent, Uzbekistan

1. Introduction

Investigation of the metastable-defect generation phenomenon in hydrogenated amorphous silicon ($a - Si : H$) by excitation of its electronic subsystem is of considerable interest from both fundamental and practical viewpoints. This phenomenon induced by light in $a - Si : H$ was first discovered by Staebler and Wronski in 1977 [1] and has drawn considerable attention in subsequent years.

Now it is commonly admitted that light-induced defects (the D centers) are Si dangling bonds (DB) and H is involved in this phenomenon. In addition, the similarity of light-induced metastability observed recently in hydrogen passivated polycrystalline silicon ($poly - Si : H$) [2] with that in $a - Si : H$, and the results of an investigation of metastable-defect generation, depending on the amount of microstructure (microvoids and etc.) [3] in $a - Si : H$, indicated that DBs are presumably created on surfaces of grain boundaries and microvoids. Here both normal $Si - H$ and strained $Si - Si$ bonds are expected to play a significant role. Meanwhile, among different models proposed to explain how DBs can be created by light with the energy ~ 1.4 eV, the "weak" $Si - Si$ bond-breaking model (see [4]) is now commonly accepted, although no distinct role is ascribed to the $Si - H$ bond in either this or other models. The reason is that the rupture of this bond normally requires an energy $2 - 3$ eV or this bond is expected to be stable with respect to deviation of H from the direction along Si sp^3-hybrid orbitals. However, localization of carriers near this bond may significantly alter this situation, making a normal $Si - H$ bond unfavourable as compared to its reconstructed configuration, which may be either DB-like itself, or an intermediate one susceptible to consequently

137

R. C. Tennyson and A. E. Kiv (eds.),
Computer Modelling of Electronic and Atomic Processes in Solids, 137–141.
© *1997 Kluwer Academic Publishers.*

138

creating a true DB. In this paper, the model describing this possibility is proposed and supported by quantum chemistry calculations.

2. Model and Calculation method

Let us consider a fragment of $a - Si : H$ with the monohydride $Si - H$ bond (Fig.1). It is possessed of the C_{3v} symmetry, so that all the molecular orbitals of this system are divided into the onefold and twofold ones. Depending on the degeneracy of the "high occupied" (HOMO) and "low unoccupied" (LUMO) molecular orbitals, the capture of an electron or hole can make the system unstable with respect to the Jahn-Teller distortion which, in this case, consists of bending of the $Si - H$ bond. There are four cases to be distinguished. As seen from Fig.1, only in one case (i) removing an electron from HOMO or adding it to LUMO does not cause the system to lower the symmetry, but it does in the other three cases: (ii) - the hole capture, (iii) - the electron capture, and (iv) - both electron and hole capture. In the presence of excess electrons and holes supplied by illumination (electron irradiation etc.) the system with the monohydride $Si - H$ bond is rather subject to a distortion, removing the degeneration and lowering the energy.

To examine this, the simplest cluster was considered, which was similar to the fragment of $a - Si : H$ shown in Fig.1, where the unseen neigbors of 2nd, 3rd and 4th Si atoms were replaced by H atoms. The minimum energy configurations of this cluster were obtained for its different charge states by molecular dynamics (MD) simulation based on a new tight-binding total energy calculation method ([5], see also our related paper in this volume). The MD simulation was performed at an initial temperature of 0 K as follows. The equation of motion $\vec{F} = m(d^2\vec{r}/dt^2)$ was solved for each atom with time step $\Delta t = 0.1 \times 10^{-15}s$, where m is the atomic mass and $\vec{r}$ is the atomic positions at time t. Whenever the total kinetic energy achieved its maximum (ie., began to reduce), quenching force was applied to all atoms reducing their velocities to zero. In the next step, all atoms were allowed to accelerate. This procedure was repeated until the total kinetic energy and velocities reached a specified tolerance at this maximum. The simulation was carried out for different initial positions of H.

3. Results

While excess carriers are absent, H, as was expected, prefers the normal undistorted $Si - H$ bond. The equilibrium bond distances were 2.35 Å for $Si - Si$ bond and 1.54 Å for the $Si - H$ bond. The structure of HOMO and LUMO corresponded to the case of (ii) in Fig.1, so that one might expect a symmetry lowering distortion for the hole capturing case (ii). The

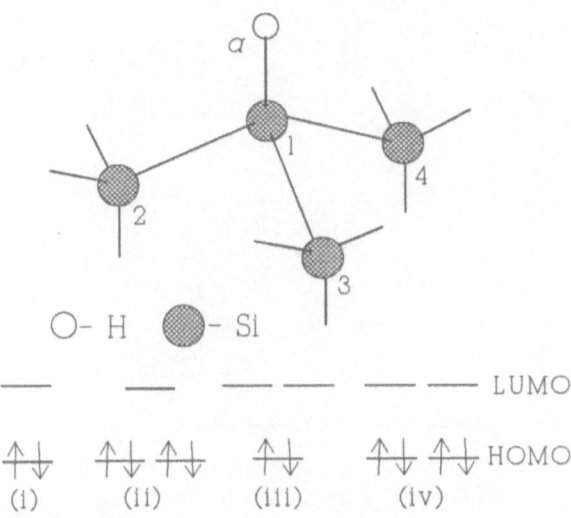

Figure 1. The perfect fragment of $a - Si : H$ with the monohydride $Si - H$ bond. Below the possible variants of the HOMO and LUMO degeneracy of this system are shown.

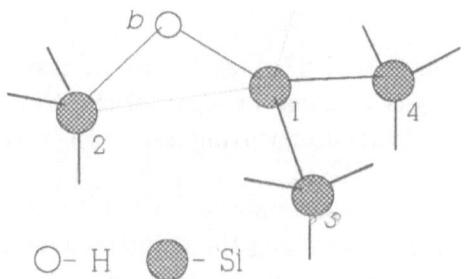

Figure 2. The distorted configuration of H after hole cupture.

distorted configuration of the $Si - H$ bond schematically shown in Fig.2, was indeed found, where $Si(1) - H$ and $Si(2) - H$ distances were 1.61 Å and 1.85 Å, respectively, and H bound with Si(2) and Si(1) atoms with comparable energies. Additionally, $Si(3) - H$ and $Si(4) - H$ distances were equal, so that there are three equivalent such configurations.

The simulation was repeated for the cluster containing one tensile bond $Si(1) - Si(2)$ with initial inter-atomic distance 2.69 Å. In this case the hole capturing led to such a configuration of H which deviated from the normal $Si - H$ bond more than in the absence of tensile bond. Here H formed a bend-bridge bond with Si(1) and Si(2) atoms, where $Si(1) - H$

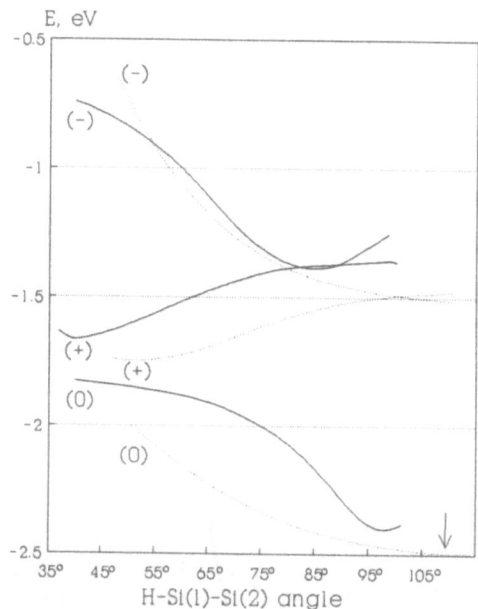

Figure 3. The configuration coordinate diagram for $H - Si(1) - Si(2)$ angle (the arrow indicates the tetrahedral angle) for different charge states of the cluster with one tensile bond $Si(1) - Si(2)$ (solid lines) and without it (dotted lines). The energies are measured relative to the dangling bond energies.

and $Si(2) - H$ bonds were 1.67 Å and 1.76 Å, respectively. Existence of a tensile bond also leads to some bending of the $Si - H$ bond toward the former in the case of electron capturing and even in the case of absence of excess carriers.

The results obtained are summarized in Fig.3 as the configuration coordinate diagram for H moving along the trajectory involving the positions a (Fig.1) and b (Fig.2). As is seen from Fig.3, there are two channels leading to the EPR-visible distorted configuration of H consisting of a bend-bridge bond H and an adjacent DB-like defect (Fig.2): hole capturing

$$e^+ + (Si - H) \rightleftharpoons (Si - H)^+ \tag{1}$$

and electron capturing with consequent simultaneous two electron emission

$$e^- + (Si - H) \rightleftharpoons (Si - H)^+ + 2e^-. \tag{2}$$

At first sight, the reaction (2) seems complicated and, consequently, less probable, but it may not be, because it can be considered as the single process of scattering of a "hot" electron on another electron near the $Si - H$ bond with sharing emitting energy.

However, for the Fermi energy position around the middle of the band gap which is characteristic of $a - Si$, the only stable charge state is the neutral one. Consequently, the DB-like defect exists while illumination or charge injection is taking place. Nevertheless, this configuration may play a significant role in light-induced degradation of both $a - Si : H$ and $poly - Si : H$. During prolonged illumination, a dynamic equilibrium should be established between H atoms in $Si(1) - H$ and $Si(1) - H - Si(2)$ bonds, in which the concentration of H in the latter may reach a level which will probably be no less than that of observed for DBs. These H atoms may dissipate, leaving behind true DBs, because H here are bound weakly and the barrier for migration may be decreased as low as ~ 0.4 eV.

In summary, a model of metastable-defect creation by carrier capture in $a - Si : H$, in which the distortion of the monohydride $Si - H$ bond plays a dominant role, is proposed and supported by the tight-binding molecular dynamics simulation. The distorted configuration obtained consists of the bend-bridge bond of H with the two nearest-neighbour Si atoms and an adjacent DB-like defect, which may then be readily converted into true DB.

References

1. Staebler D.L. and Wronski C.R. (1977) Reversible conductivity changes in discharge-produced amorphous silicon, *Appl. Phys.Lett.*, **Vol. no. 31**, pp. 292-300.
2. Nickel N.H., Jackson W.B., and Johnson N.M. (1994) Defect metastability in hydrogen passivated polycrystalline silicon, *Modern Phys.Lett.*, **Vol. no. 8**, pp. 1627-1642.
3. Bhattacharya and Mahan A.H. (1988) Microstructure and light-induced metastability in hydrogenated amorphous silicon, *Appl.Phys.Lett.*, **Vol. no. 52**, pp. 1587-1589.
4. Stutzmann M., Jackson W.B., and Tsai C.C. (1985) Light-induced metastable defects in hydrogenated amorphous silicon: a systematic study, *Phys.Rev.B.*, **Vol. no. 32**, pp. 23-47.
5. Khakimov Z.M. (1994) A new semiempirical electronic structure and total energy calculation method for solids and large molecules, *Comput.Mater.Sci.*, **Vol. no. 3**, pp. 95-108.

THE MOLECULAR DYNAMICS SIMULATION OF CONTACT MELTING: FOUR-COMPONENT IONIC SYSTEMS

V.S. ZNAMENSKI
KABARDINO-BALKARIAN STATE UNIVERSITY
P.O.BOX 46, NALCHIK-04, KBR
360004, RUSSIA

P.F. ZILBERMAN and I.N. PAVLENKO
KABARDINO-BALKARIAN STATE AGRICULTURE ACADEMY
NALCHIK, KBR
360004, RUSSIA

Abstract

The results of several molecular dynamics simulations are reported where an NaCl crystal is in contact with a KBr crystal. Molecular Dynamics (MD) computer simulations have been performed for a system of 216 ions interacting by means of the Born-Maier-Huggins' or Poling's inter-ionic interaction potentials under contact melting conditions over some range of densities.

1. Introduction

The fusion of two solids in a zone of contact when the temperature is lower than the fusion points of each crystal is known as contact melting (CM) [1-3]. Contact melting is widely used in various engineering processes and in physical chemistry analysis. It is successfully used to produce permanent joints, construction materials, alloys and chemical compounds. Contact melting serves in constructing fusibility diagrams and in simplifying investigations of diffusion processes. In particular, it allows one to easily determine inter-diffusion and partial diffusion coefficients.

For some time the problem of contact melting research on a nanometer-size scale was interest to scientists are engaged in studying this phenomenon. The basic questions are: what is occurring in the contact region and how do ions or atoms move there? Understanding the atomic processes occurring at the interface of two solids brought together at high temperature, is central to many technological problems, including soldering, and coagulation. We have applied a method of numerical simulation to find the answers to these questions - a molecular dynamics method (MD). The computer simulation gave a series of unsurprising results, from a physics point of view. For example, there are temperatures of contact melting and movement trajectories of surface ions. Some of our results are clearly an advance, but on the

R. C. Tennyson and A. E. Kiv (eds.),
Computer Modelling of Electronic and Atomic Processes in Solids, 143–148.
© 1997 *Kluwer Academic Publishers.*

other hand, we can also evaluate CM parameters from first principles, describing interactions of ions in the melt at the interface. Simulations can show which ions interact more strongly, which ions interact more weakly, what happens, if diverse crystals are in contact, which ions will strengthen the interaction, and which will ease interactions. These results can deduce understanding of the nature of this phenomenon and also pose many new questions, thus opening up a new subject for research - the nanophysics of contact melting. Some of these questions are:
- whether contact melting can occur without mutual penetration of the diverse substances;
- whether a crystal lattices destroyed on contact before penetration;
- what is the difference of radial distribution functions at heterogeneous and homogeneous phases, as well as at CM;
- what is the influence of the proximity of solid-phase borders on melt properties;
- what is the necessary number of particles for phase transition.

2. Contact Melting Problem

We addressed the problem of the contact-melting phenomenon on the nanometer scale by an MD-method, and our software was constructed on that basis. In our case, the application of the general MD-method is decided by the following two aspects of the problem:
- the modification of the setting of a standard MD-method for contact modelling;
- revealing special research themes of the contact-melting phenomenon, which previously could not be investigated by other methods.

The first aspect includes the following points:
1. Division of a calculation cell in two halves, a simulation, and accounting for of any statistical characteristics of the system separately for the particles which were originally placed in the first and second halves, divided by a contact border;
2. Monitoring features of dynamics of particles, and stipulated availability of a contact border. Comparison of the characteristics of a system when a contact border exists and after its disappearance (owing to diffusion of hashing of particles);
3. Variation of the mutual disposition of crystal lattices until the different state exists from a contact border.

The second aspect includes the following problems:
1. Research of the initial contact melting stage;
2. Research of dependence of contact-melting parameters from the fundamental characteristics; for example, the binary potential of interparticle interactions, effective radius of ions, rigidity of the repulsing part of potentials.

Software was developed for realisation of an MD-method for contact melting of ionic crystals. Personal computers permit one to study temperature dependencies of the ionic crystal and melt characteristics in a boundary zone. Analysis has shown, that the MD-method permits one to evaluate the characteristics by use of a personal computer of an IBM PC AT 486 type with arithmetic coprocessor. The methods accuracy is increased with application of more high-speed computers by means of expansion of a calculation base for statistical processing.

In the program part we used the following model:
1. The initial location of particles in the calculation box is represented by a cubic lattice; 1-2 and 3-4 ions occupy opposite halves of the box so as to imitate the contact between two heterogeneous substances. Here 1,2,3,4 are the numbers of the ion types.
2. The simulation box is a cube of constant edge length.
3. Periodic boundary condition are used.
4. Constant-temperature conditions are maintained by multiplying velocities by a calculated scaling coefficient.
5. The forces of pair interactions are initially calculated, and tables of f /r are used in simulations. Here f is the force; r is the distance between the particles; and i=1,2,...,10 is the type number of an interacting ion PRDF. Our simulations involved a total of 10 types.

3. Molecular Dynamics Simulations

The MD simulations were performed using a simple molecular dynamics algorithm for the NVT-ensemble (at constant N, number of molecules; V, volume; and T, temperature). The system NaCl-KBr studied consisted of 216 ions in a cubic box with the NaCl in the left side and KBr in the right side. The length of the box was variable for different experiments. The temperatures for calculations were chosen over a wide range including the contact melting temperature. In previous works; the calculations were executed for ionic three-component systems with a common anion or cation: sodium chloride-potassium chloride, sodium chloride-rubidium chloride, sodium iodide-potassium iodide, sodium chloride-sodium iodide, potassium chloride-potassium iodide [3].

The mean-square displacements of ions were calculated in all numerical experiments and were hereinafter served for determination of the diffusion factors. It was found resulted that the diffusion coefficient of the sodion is more than the diffusion coefficient of the potassium ion and diffusion coefficient of the chlorion is more than the diffusion coefficient of the bromion. It is possible to explain this by the different size and mobility of these ions. It is shown, that the calculated diffusion factors are sensitive to the contact melting temperature and can be used for its valuation. The temperature dependence of the diffusion coefficients correspond to

$$D=A\cdot\exp(-E/RT).$$

In fig.1 the change in diffusion coefficients by contact melting temperatures is shown.

The calculated results were obtained using potential approximations for pair interactions based on simplified Poling's and Born-Maier-Huggins' potentials. The results are compared with experimental data [1-3].

4. Partial Radial Distribution Functions

Partial radial distribution functions (PRDF) for different temperatures were also obtained. For the 900 K melt temperature, the parameters of PRDF (location of first nonzero value, location of first maximum, values of first maximum) are presented in table 1.

Table 1. Parameters of PRDF

	Na-Na	Na-Br	Br-Br	Na-K	Br-K	K-K	Na-Cl	Br-Cl	K-Cl	Cl-Cl
Loc. of first nonzeros, 10^{-10}m	3.30	2.78	4.14	3.72	2.99	3.93	2.67	4.03	2.99	3.82
Loc. of first maxima, 10^{-10}m	11.23	3.22	6.17	6.49	3.71	6.49	3.11	5.75	3.74	5.96
First maxima	1.7	4.1	1.9	1.9	4.0	2.0	4.5	1.9	3.7	1.8

The location of the first nonzero value for ions of different types (anion-cation) is less then the location of the first nonzero value for ions of similar type (anion-anion, cation-cation). The values of the first maximum for ions of different types (anion-cation) is more, than the values of the first maximum for ions of similar types (anion-anion, cation-cation).

The values of the first maximum for ions of different types (anion-cation) for Na-Cl is more than for Na-Br, K-Br, K-Cl.

The values of the first maximum for ions of similar types (anion-anion, cation-cation) for K-K is more, than values of the first maximum for Na-K, Br-Cl, Br-Br, Na-Na.

The temperature dependencies obtained the PRDF parameters are used for the zone of contact. The increase of temperature is accompanied by a reduction in space to the first nonzero value of PRDF.

We shall note, that location accuracy of the maximum of the second co-ordination sphere is far below, in view of the difference of these maxima.

Modelling conducted for various temperatures specified a significant extension of maximas for the first and the second coordinate sphere. Intensity of these maximas was reduced. It is possible also to note, that the position of the first maximum in a crystal is higher than for melts, and with growth of temperature is observed an insignificant increase is observed.

The computer simulation of contact melting is one more check on adequacy of using pair interaction potentials. As was shown, we used simplified Poling's potential and a more complicated Born-Maier-Huggins' potential [4]. The first of them, for a NaCl-KBr system, has given results in better agreement with experiment.

5. Normalized Velocity Autocorrelation Functions

The normalized velocity autocorrelation functions are calculated for ionic systems in a boundary zone by a contact-melting mode. An oscillation part is typical for velocity autocorrelation functions of ionic systems, as well as for liquid metals. We used the following data as characteristic parameters for velocity autocorrelation functions: (1) first passage times of the first minimum; (2) value of the first minimum. For this system it was found: 1- Na 84 fs, Br 190 fs, K 135 fs, Cl 113 fs; 2 - Na 0.45, Br 0.30, K 0.43, Cl 0.45. Comparing molecular-dynamic experiments with one-component ionic systems by repeating some other known analyses, has shown consistency of our results with the results of other authors.

6. Conclusions

1. The initiation of contact-melting is characterised by increased mobility of ions, which is observed by extended trajectories of ions on the contact border;
2. The mobility of ions in contact depends on a mutual disposition of lattices;
3. The microscopic mechanism of contact melting includes a formation of liquid-phase clusters in micro-volumes (size of the order 2 nm; for time scales it is of the order of program facilities). Cluster behaviour is unstable, but it can initiate the beginning of a liquid phase.
4. The beginning of contact melting is characterised in the contact zone by increasing distance between the nearest anion and cation, and the intensity of interaction grows, for those anion-cation pairs, which have in a similar system, higher intensity of interaction; for other pairs, the intensity decreases

7. Acknowledgements

This work has been supported by the Russian Foundation of Basic Research under Grant No. 95-01-01567. The financial support of the Scientific Affairs Division of NATO is gratefully acknowledged.

8. References

1. Savintsev, P.A. and Avericheva, V.E. (1958) On the Melting Point of a Crystal Contact   Layer, *Dokl. Akad. Nauk SSSR*,119, 5, 936-937.
2. Zalkin V.M. (1987) *Nature of Evtectical Melts and Phenomenon of Contact Melting,* Metallurgiya, Moskow.
3. Znamenskii, V.S. Savincev, P.A., and Zil'berman, P.F. (1993) Molecular Dynamics Study of the Contact Melting - Method. *Russ. J. Phys. Chem.*, 67, 7, 1349-1352.
4. Sangster, M.J.L. and Dixon, M. (1976) Interionic Potentials and their Use in Simulations of the Molten Salts, *Advanced in Physics*, 25, 3, 247-342.

$D \times 10^5$, cm^2/s

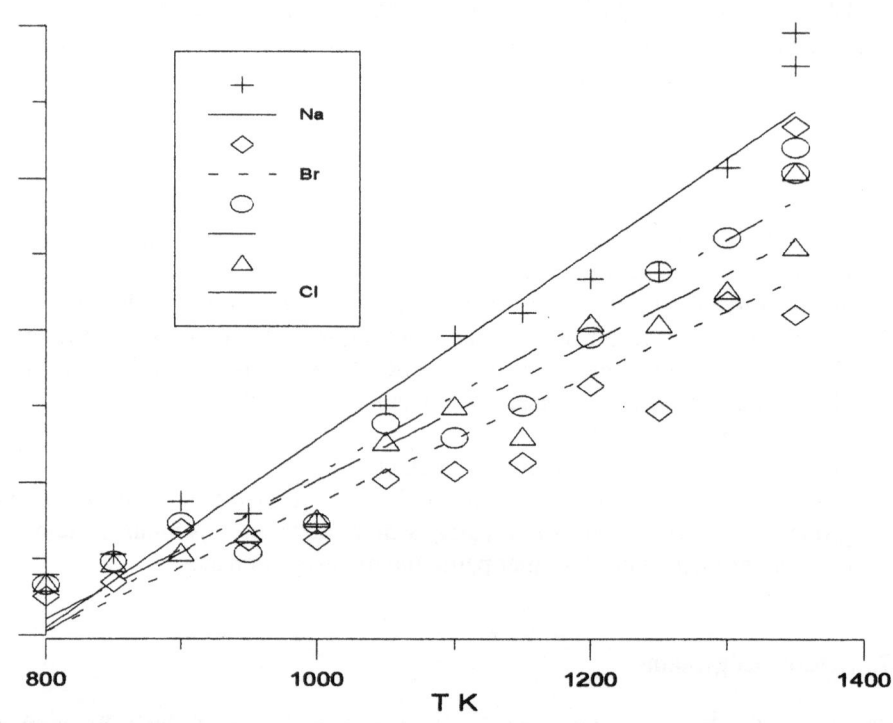

Figure 1. The temperature dependence of the diffusion coefficient $D \times 10^5$, cm^2/s. for Na$^+$, Br$^-$, K$^+$, Cl$^-$.

THE MOLECULAR DYNAMICS SIMULATION OF INTERACTIONS IN SHOCK-COMPRESSED SYSTEMS

V.S. ZNAMENSKI
Kabardino-Balkarian State University, Russia
P.F. ZILBERMAN and I.N. PAVLENKO
Kabardino-Balkarian State Agriculture, Russia

Abstract

A general study of some atomic aspects of shock and unloading phenomena is presented. Computer molecular-dynamics simulations based on the Born-Maier-Huggins' and Polings' potential models have been used to study the shock-compression phenomena.

1. Introduction

This work is an extension of our investigations on molecular dynamics simulation of contact melting [1-12]. Contact melting is a phenomenon of lowering of the melting point at the contiguity zone of two different crystals. Our molecular dynamics simulations showed that a system of two contact bodies has new physical properties. It happens when only masses and sizes of particles are different in the opposite sides of the boundary. We are interested in the question: what is the effect of differences in system parameters such as location of particles, group average velocities, etc. on the section border boundaries of two materials forming the contact? Thus we started the studies of shock-compressed systems [13].

The research on shock-compressed conditions is closely connected with work on the production of amorphous materials, explosive welding, making artificial diamonds, and other similar works. Molecular dynamics simulations have been broadly used to study microscopic features of substances. From the simulation, the time dependence of positions and velocities of particles are obtained by numerically integrating the equation of motion. Therefore, the MD method has been used to simulate a nonequilibrium phenomenon. The MD method can be used to simulate crystallization, glass formation [14] and melting processes for many systems [15-16].

Our studies were conducted for an ionic system. The alkali halides have been the preferred model system. Their physical and chemical properties have been widely investigated, and are well understood. There is a considerable interest in the

149

R. C. Tennyson and A. E. Kiv (eds.),
Computer Modelling of Electronic and Atomic Processes in Solids, 149–157.
© 1997 *Kluwer Academic Publishers.*

understanding of phenomena of extreme conditions, for example, pressure-induced solid-solid phase transition [17].

2. Some shock-compression physical phenomena

The feature of processes, which take place by shock-compression, is a structural modification of surface layers with essentially the absence of diffusion. In Figs. 1-2, one can see some simplified schemes of two aspects of shock-compression phenomena. The welding by explosion is shown in Fig. 1. The cumulative effect is shown in Fig. 2. The initial stages of these processes are alike. Initial location and velocity of surfaces are shown in Fig 3. Some combinations of the velocity V and the angle $\propto$ lead to welding or a cumulation. The cumulative effect occurs when surfaces have a specific angle. Shock-compression conditions in the region of the collision of two surfaces give these effects. It is known that welding by explosion can connect two substances, which cannot be connected by other ways. This has important technical significance due to the phenomenon of 'strengthening' by explosion, as shown in Fig. 4.

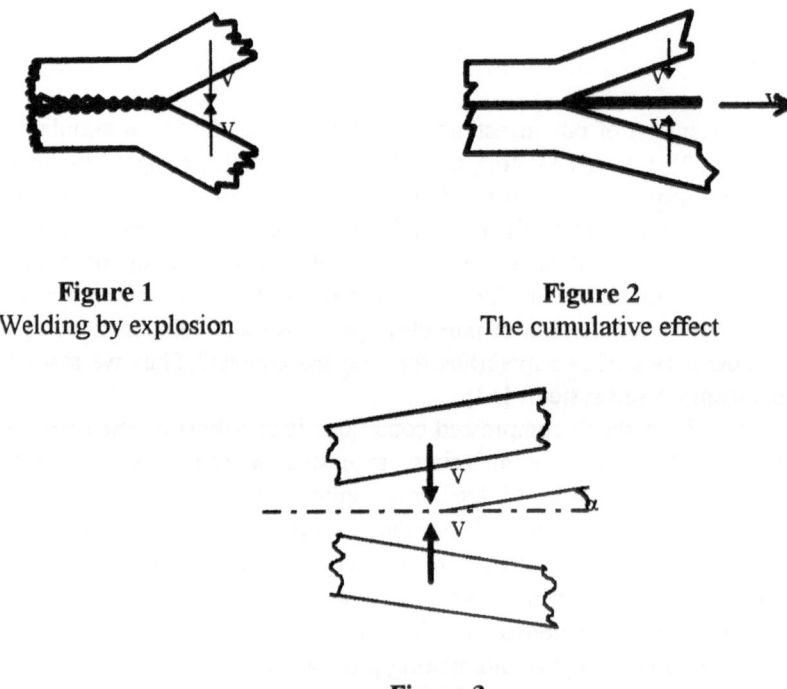

Figure 1
Welding by explosion

Figure 2
The cumulative effect

Figure 3
The initial configuration and movement
for the cumulative effect and welding by explosion.

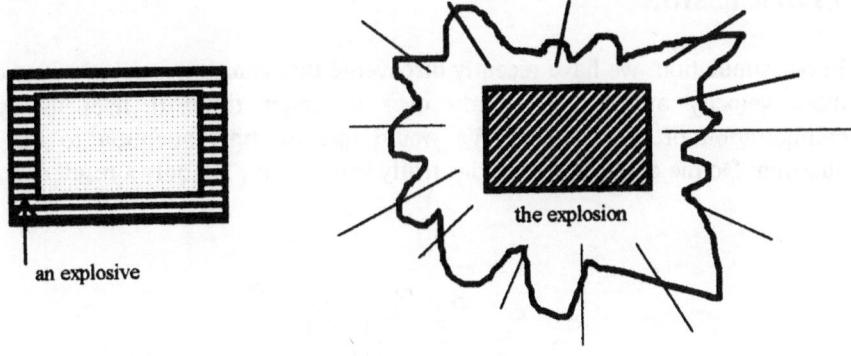

an explosive

the explosion

Figure 4
The phenomenon of the strengthening by explosion.

3. MD simulation of the cumulative effect

The use of simulation on an atomic scale is the best possible means of learning about the formation mechanisms under the actions of shock loads, and for revealing the processes which take place. The molecular dynamic method (MD) allows one to trace the movement of atoms or ions under the action of interparticle forces, and also to research the modification of its structures under the action of shock loads.

We made the simulation for ionic crystals with the BHM potential for various initial speeds. An estimate of structural characteristics was implemented by calculating radial distribution functions (RDF) and by visualization of particle trajectories. We note that the RDF do not allow one to find a great variety of structures which are formed. Therefore, it is possible to obtain the maximal information only by visualization of particle trajectories. We calculate the average of particle co-ordinates for the time-average in the fluctuation period, to the exception of the influence of thermal fluctuations on visualization. The MD-experiments had shown the existence of strong processes in the structural rebuilding and the transfer to the amorphous structure under shock interaction in the surface zone of several crystal layers. There were observed the cumulative effects on the atomic scale of size. In Fig. 5, one can see the first simplified scheme of the molecular dynamic experiment to study the shock-compression phenomena for welding by explosion and the cumulative effect. The initial coordinates and velocities were randomly varied to provide the best fit to the anticipated effect.

3.1 DISCUSSION

In our simulation, we have recently discovered that sometimes certain ions move with more velocity and more linearly over a longer time. It happens by random configuration of the particle. We would like to draw attention to the following question: Do the observed properties really exist or they are only a result of some

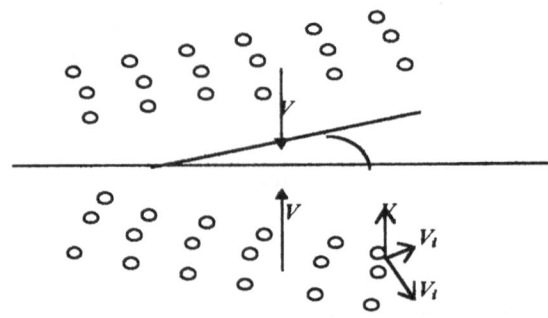

Figure 5

The scheme of molecular dynamic experiment to study shock-compression.
V is the group velocity. V_i is some velocity of ion or atom,
V_i^t is the thermal velocity. $V_i = V_i^t + V$

calculation divergence? We particularly examined situations when the calculation system had calculation divergence. This may occur, for example, when the initial position of two particles is at a very short, unrealistic distance. The system behaviour is absolutely different in this case. We think that the properties obtained are accurate.

One can say that there is some computer evidence for a commulation effect on the scale of a hundred particles. With our small number of computer experiments, it does not seem reasonable to draw any definite conclusions as to the mechanism of commutation. Our statistics are not rich enough to give conclusive results. Under the microscope, the shock-compression lattice becomes excited. We presuppose the stochastic shock influences are the ordinary events for normal conditions. The external shock action considerably increases the number of internal overshock events. These overshocks have a stronger specific effect. They are the basic reasons for such phenomena as welding by explosion and hardening by explosion. Atomic implications of this process leads to destruction of old structures and the formation of the new stable nodes. We should add, however, that MD simulation is very 'computer-time' consuming and our simulation result should be considered preliminary since our statistics are not rich enough to give an accurate confirmation of this ideas.

4. MD simulation of substance properties in the shock- or unloading-wave

A different way to study shock-compression is the simulation of a substance in (a) significantly increasing or (b) decreasing the inter-atomic space. It is the simulation of substance properties (a) in the shock-wave or (b) in the unloading-wave. Modelling of shock-compression was realized by changing of the size of the calculation box. For simulating these effects, we use the same models. The first model started the calculation of the ion movements from nodes of the cubical net. Temperature has a broad range of values. First, we constructed the initial coordinate and velocity of the ions. Then, by use of the numerical methods of temperature stabilization we made a run of MD simulations with different temperatures and values of size of the cubic cell. One initial structure for the particle system is a regular cubic lattice of ions as for NaCl, and variable velocities. A second initial structure is the same result of the preceding MD simulations. The system states have been examined by use of the mean square displacement $R^2_k(t)$ ions

$$R^2_k(t) = <r_i(t) - r_i(t) - r_i(t_0)\}^2>_k \qquad (1)$$

where $r_i(t)$ is the position vector of i-th ion at time t, $<>$ denotes the average, k denotes the type of ions.

We have developed a computer program which can be used for simulation of a many-body problem for classical systems with central interactions between particles, and can obtain an evolution in time for an ensemble of particles. The ensemble consists of 4 ion species, described by the corresponding ion positions and velocities. The program allows one to get the mean-square displacements of ions, diffusion coefficients, partial radial distribution functions, normalized velocity autocorrelation functions and other data. The simulation was realized using a traditional MD scheme with the BHM [18] and Pauling's potentials. The number of all ions is 216, and the number of each type of ion is 64. Molecular Dynamics Simulation 54 A^+, 54 B^-, 54 C^+, and 54 D^- were placed in the periodic cube. Our cubic simulation cell is divided initially by two equal parts corresponding to the AB and CD species respectively. The time step was 7-10 fs. At the initial stage of the MD simulation, the NTV ensemble was settled. The data analyses were done with the final 1000 steps. The general conditions for the numerical experiments are given in Table 1. The AB-CD notation denotes the following: chemical system AB is initially located in the first half of the cubic cell and system BC is initially located in the second half. The A^+, B^-, C^+, and D^- compound denotes a homogeneous ion mixture.

The diffusion coefficients $D_k(k =1,2,3,4)$ have been estimated by using the evolution parameters of the system and have been calculated by means of the relation

$$<r_i(t) - r_i(t_0)\}^2>_k = 6 D_k t + C_k, \ t_0 < t < t, \qquad (2)$$

The velocity autocorrelation function is defined as,

$$Z_k(t) = <V_i(0)V_i(t)>_k / <Vi^2>k \tag{3}$$

TABLE 1

The general statements of numerical experiments

System composition	Initial configuration of ions	Some details or/and purpose of experiment
NaCl-KBr	regular ideal cubic lattice	an effect of the ion exchange: Cl⁻ and Br⁻
Na⁺, Br⁻, K⁺, Cl⁻	amorphous state	
NaBr-KCl	regular ideal cubic lattice	
NaBr-KCl	regular ideal cubic lattice	locations of Br⁻ and Cl⁻ are fixed
Na⁺, Br⁻, K⁺, Cl⁻	amorphous state	locations of Br⁻ and Cl⁻ are fixed

where V_i is the velocity of ions, and k is a number of the ion type.

We have used here two potential approximations by Pauling [7] and Fumi-Tosi. The Pauling's potential between ions labelled "i" and "j" is given by

$$V_{ij}(r) = e^2/(4\pi\varepsilon_0 r)(q_i q_j r^{-1} + [(\sigma_i + \sigma_j)/r]^p/(p+1)) \tag{4}$$

where σ are parameters, $p = 8$ (the hardness parameter), ε_0 is the electric constant, q, e are the ionic charges, e is the electron charge.

The Fumi-Tosi potential is of the form

$$V_{ij}(r) = Z_i Z_j e^2/(4\pi\varepsilon_0 r) + (1 + Z_i/n_i + Z_j/n_j) \, b \, exp[(\sigma_i + \sigma_j - r)/\rho] - C_{ij}/r^6 - D_{ij}/r^8 \tag{5}$$

where Z is an ionic charge number, n the number of the electrons in an outer shell, b a repulsion parameter, σ a characteristic value of an ion size and ρ a softness parameter.

5. Results

We ran a number of numerical experiments, consecutively varying cell sizes and temperature, and calculating the diffusion coefficients. The beginning of sharp increases in the diffusion coefficients identifies the melting point. Herewith, sharp increases in the diffusion coefficients is observed simultaneously for all types of ions, the difference due to different values of the diffusion coefficients only. In Fig. 6, two areas of the phase diagram "solid -liquid" stand out. Comparison of the results for systems NaCl- KBr and NaBr-KCl show that for these systems, the "solid-liquid" phase diagrams are alike. If different systems are constituted of 'like ions,' the temperatures of transition "solid-liquid" are alike for these systems for given sizes of cells. However, values of the diffusion coefficients of ions are different in the fluid phase. For instance, diffusion coefficients of sodion in NaCl-KBr are smaller than in NaBr-KCl. Reduction of the rib length of the calculate cube from 2700 prior to 2550 pm raises the melting point from 800 prior to 1500 K for the NaCl-KBr system. Changing the pressure from 0 prior to 5 GPa corresponds to such a change in the melting point. Other results are observed for the unordered mixture. The amorphous state of such a system is unchanged when changing the cell size (for a given time of the experiment). Comparison shows that for the amorphous system, the diffusion coefficients are vastly higher when the accounting cell has a small size (Fig.7). Diffusion coefficient values for amorphous and regular systems become similar when increasing the sizes of cell prior to such, which are characteristic of the liquid state for the original regular system.

We looked for system factor effects on the ion diffusion. For instance, we stopped some ions, assigning to them very large masses, to estimate how they act on the diffusion of one ion moving other ions. When we had frozen all negative ions, the positive ions did not noticeably change their own diffusion coefficient values. This was characteristic for all system states examined.

Analysis of the radial distribution functions has shown the following. Increasing the temperature of the experiment brings about the reduction of the nearest distance between ions, which is reached due to heat moving all the ions for the original ranked system. An interesting dependency is observed when one varies the cell sizes. Increasing the size of a cell first brings about the growth in this feature, then its reduction. For instance, the least Na - Br distance increases from 2.7 prior to 2.9 A but then decreases again prior to 2.7. Such behaviour is explained by "solid-liquid" phase transition. In the solid phase, an increase in the average distance between ions is accompanied by an increase in the minimum attainable distance. In the liquid phase, a free volume appears by the cell size increasing, but the nearest ions again have a closer minimum distance to one another. Another result is observed for the unordered mixture of ions, that simulate the amorphous state. The initial increasing of the minimum distance between ions then changes to a constant. Similar dependencies are obtained for positions of first maxima of the radial distribution functions, characterizing average distances between nearby ions.

We have shown that the MD method gives some reasonably reliable way to determine the microscopic characteristics of a material for the shock-compression phenomenon. The simulation has shown that for the temperature of the experiment (T=700 - 1500 K), and for time of the experiment (t=1 ps), the original crystalline structure can be changed and formed into an amorphous or liquid state. Change of temperature and cell size is accompanied by changing of the time of the system transition into the amorphous state. Crystalline lattice is more stable under straining stresses at the wave of unloading, straining together with the system. The growing of

156

free volumes quickly becomes the main mechanism of expansion of the amorphous structure.

Acknowledgements

This work was supported by the Russian Foundation of Basic Research under Grant No. 95-01-01567. The financial support of the Scientific Affairs Division of NATO is gratefully acknowledged.

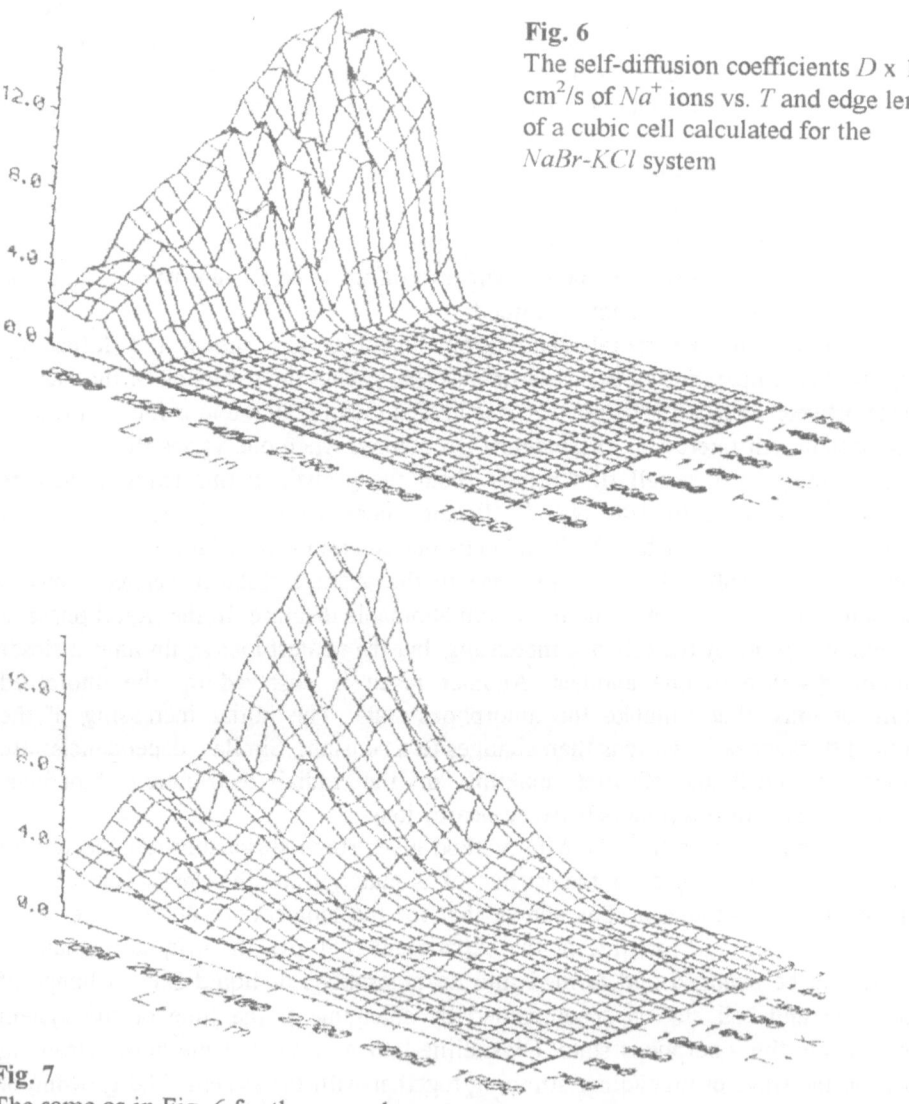

Fig. 6
The self-diffusion coefficients $D \times 10^5$, cm^2/s of Na^+ ions vs. T and edge length of a cubic cell calculated for the $NaBr$-KCl system

Fig. 7
The same as in Fig. 6 for the amorphous system

References

1. Savintsev, P.A., Znamenski, V.S., Zilberman, P.F. et al.(1995) Nanophysics of contact melting: simulation by molecular dynamics method, Bulletin of Kabardino-Balkarian State University, Series: Physics and Mathematics Sciences 1, 273-242.
2. Goncharenko, E.A., Znamenski, V.S., Zilberman, P.F. et al. (1995) The molecular- dynamics simulation of influence of external electromagnetic fields on contact melting of ionic crystals, The Physics and Chemistry of Material Processing (Russian) 3, 94-99.
3. Znamenski, V.S., Zilberman, P.F., Goncharenko, E.A. et al. (1995) The research by molecular-dynamics method into the contact melting and structure-dynamic properties of NaCl-RbCl system, Zhurnal Fizicheskoi Khimii 5, 845-848.
4. Znamenski, V.S. (1994) The statistical simulation of contact melting, Nonlinear Boundary-value Problems of Mathematical Physics and Their Applications. Collection of scientific works. Kiev. Mathematical Institute, 81-82.
5. Znamenski, V.S. et al. (1995) Parameters of contact melting in ionic systems: molecular dynamics simulations with Pauling and Fumi-Tosi potentials, Inorganic Materials 4, 483-485.
6. Znamenski, V.S. et al. (1993) Investigation of contact melting by methods of molecular dynamics, Zhurnal fizicheskoi khimii 7, 1504- 1507.
7. Znamenski, V.S. Savincev, P.A., and Zil'berman, P.F. (1993) Molecular dynamics study of the contact melting - Method. Russ. J. Phys. Chem. 7, 1349-1352.
8. Znamenski, V.S. et al. (1995) The some aspects of contact-melting research by molecular dynamics methods, The Russian Conference "The Physics of interphasic phenomena and interaction processes of energy streams with solids," Terscol, Russia. 3-6 October 1995, Thesis of Reports, Nalchik, K-BSU, 60-62.
9. Znamenski, V.S., Zilberman P.F. (1995) The problems of analysis of space configurations of particles, which is received at computer molecular-dynamics experiment, International Technological Seminar "Problems of Transmissions and Processing of Information in Information Networks", Moscow, Thesis of Reports, Scientific and Information Centre of Problems of Intelligent Property, 50-52.
10. Znamenski, V.S., Zilberman, P.F., Gelfand, T.V. at al. (1995) Computing techniques in physics of contact melting phenomenon, 10th Summer School GPS EPS on Computing Techniques in Physics "High Performance Computing in Science". Abstracts. Scalsky Dvur, Czech Republic. 5-14 September.
11. Znamensky, V.S. et al. (1994) Molecular dynamics simulation of the ionic crystals contact melting, Proceedings of the 6-th Joint EPS - APS International conference on physics computing. Lugano, Switzerland. 22-26 August, 645-647.
12. Savintsev, P.A., Zilberman, P.F., Znamenski, V.S. (1990) Molecular dynamics method in personal computer for analysis of contact melting for ionic crystals, Structures and properties for metal and slag melts. Scient. rep. of the VII All-Union conf. Cheliabinsk, USSR, 1,II, 207-208.
13. Znamenski, V.S., Zilberman, P.F. (1995) Numerical simulation of the surface-blow amorphization in explosive welding by the molecular dynamics method, Russian Seminar "Structural Heredity in Processes of Super- Rapid Hardening of Melts", Izhevsk, 26-28 September, 1995, Theses of Reports, 121-122.
14. Wataname, M.S. and Tsumuraya, K. (1987) Crystallisation and glass formation processes in liquid sodium: A molecular dynamics study, J. Chem.Phys. 8, 4891-4900.
15. Fincham, D (1994) Dynamic simulation of molecular and ionic materials, J.Mol.Graphics 12, 29-32.
16. Angel 1, C.A., Scamehorn, C.A., Phifer, C.C. at al. (1988) Ion Dynamics Studies of Liquid and Glassy Silicates, and Gas-in-Liquid Solutions, Phys Chem Minerals 15, 221-227.
17. Martin Pendas, A., Luana, V., Recio, J.M. et al. (1994) Pressure-induced B1-B2 phase transition in alkali halides: General aspects from first-principles calculations, Physical Review B 5, 3066-3074.
18. Sangster, M.J.L. and Dixon, M. (1976) Interionic Potentials and their Use in Simulations of the Molten Salts, Advanced in Physics 3, 247-3

Electronic Structure and Processes

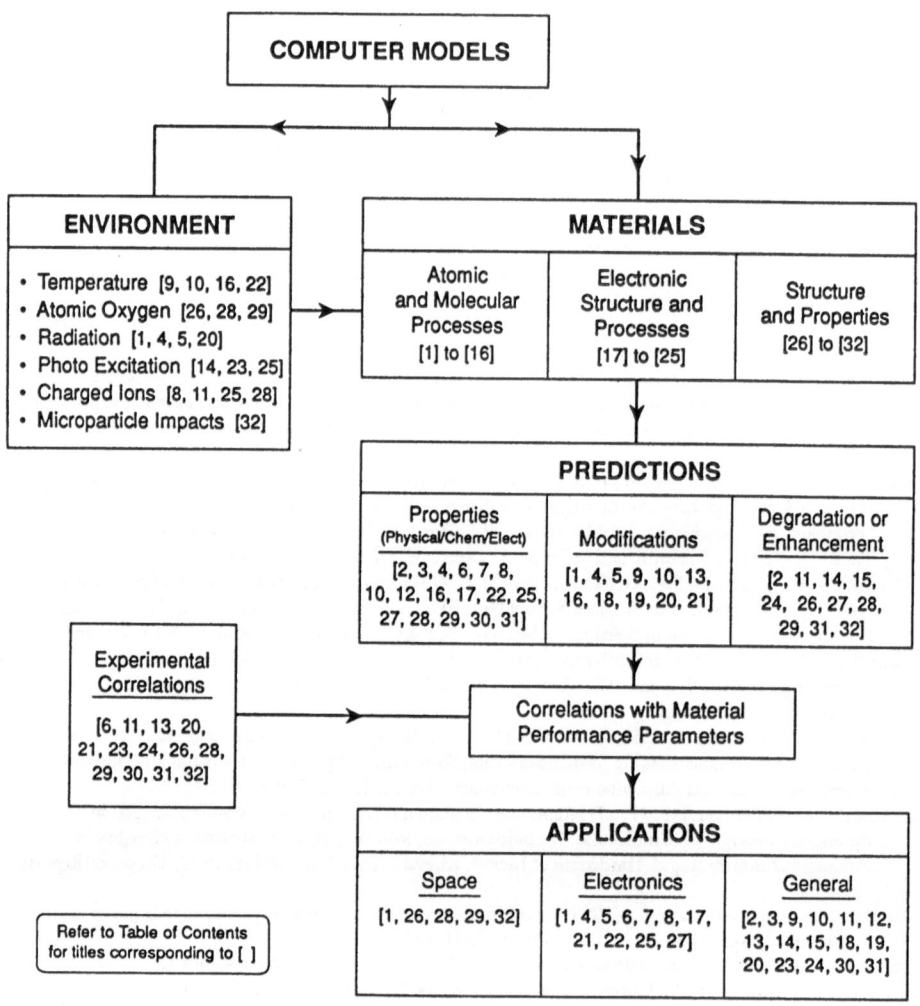

COMPUTER MODELS

ENVIRONMENT

- Temperature [9, 10, 16, 22]
- Atomic Oxygen [26, 28, 29]
- Radiation [1, 4, 5, 20]
- Photo Excitation [14, 23, 25]
- Charged Ions [8, 11, 25, 28]
- Microparticle Impacts [32]

MATERIALS

Atomic and Molecular Processes [1] to [16]	Electronic Structure and Processes [17] to [25]	Structure and Properties [26] to [32]

PREDICTIONS

Properties (Physical/Chem/Elect) [2, 3, 4, 6, 7, 8, 10, 12, 16, 17, 22, 25, 27, 28, 29, 30, 31]	Modifications [1, 4, 5, 9, 10, 13, 16, 18, 19, 20, 21]	Degradation or Enhancement [2, 11, 14, 15, 24, 26, 27, 28, 29, 31, 32]

Experimental Correlations

[6, 11, 13, 20, 21, 23, 24, 26, 28, 29, 30, 31, 32]

Correlations with Material Performance Parameters

APPLICATIONS

Space [1, 26, 28, 29, 32]	Electronics [1, 4, 5, 6, 7, 8, 17, 21, 22, 25, 27]	General [2, 3, 9, 10, 11, 12, 13, 14, 15, 18, 19, 20, 23, 24, 30, 31]

Refer to Table of Contents
for titles corresponding to [ ]

COMBINED DENSITY FUNCTIONAL AND CONFIGURATION INTERACTION METHOD FOR THE ELECTRONIC STRUCTURE OF SOLIDS WITH IMPURITIES

I. V. ABARENKOV
Physics Department
St.Petersburg State University
Russia

1. Introduction

This talk is based on the two papers [1,2] and describes the research done in collaboration between the Solid State Theory group of St.Petersburg University and the Theory of Condensed Matter group of Cambridge University.

One of the challenging problems of electronic structure calculations is the impurity atom(s) with open shells in the insulator or semiconductor host. The electronic structure of a solid can be obtained with the help of the density functional theory, e.g. in the local density approximation (LDA)[3,4]. However LDA fails to describe the "open-shells" atom with necessary precision and a method based on the many-determinant wave function should be employed instead, e.g. the many configuration self-consistent field (MCSCF) method [5,6]. At the same time, the MCSCF method cannot be applied to a solid as it stands because the necessary computer resources scale very badly with the system size. Still, if a solid contains only a few "open-shell" atoms, the MCSCF method can be applied not to the whole solid, but only to a small part of it. For these cases, a hybrid embedding scheme is discussed to incorporate the MCSCF method within solid state calculations done with LDA.

2. Outline of the method

Let us consider a solid with closed shell atoms, such as MgO, which can be adequately treated with the LDA method. Suppose an atom(s) with open

R. C. Tennyson and A. E. Kiv (eds.),
Computer Modelling of Electronic and Atomic Processes in Solids, 159–172.
© 1997 *Kluwer Academic Publishers.*

shell(s) is implanted in the bulk or on the surface of the solid. At present, we assume that the influence of the impurity atoms on the solid is a short-range one; that is, we suppose that the impurity atoms do not introduce an addition to the solid net charge or dipole moment. Moreover, we are interested only in those properties and processes which are confined to the impurity or its near vicinity and are influenced by the crystal environment. Accordingly, we propose to divide the whole system into three regions (subsystems) as follows.

One region, which will be referred to as A, contains all open-shells atoms and maybe some atoms of the solid whose electronic structure is strongly influenced by the impurity. We assume that all the properties and processes to be considered are localized mostly in the region A. Another region, which will be referred to as S, contains all atoms of that part of the solid where the influence of the impurity is negligible. The third region B contains those atoms of the solid neighboring the impurity which experience its influence, but this influence is comparatively small.

In calculating the electronic structure of the region A, one is forced to go beyond the LDA approximation. This region could be treated with the MCSCF method. The single region S could be adequately treated with the LDA approach as we assumed for the pure solid. The region B could be equally well treated with either the LDA or MCSCF methods. In principle, one could enlarge the A region to exclude the B and S regions completely and to employ the isolated cluster approximation. Unfortunately, the number of atoms which should be retained in this cluster to provide the desirable accuracy is far beyond the limits of the MCSCF method. In fact, the MCSCF method is so "resources demanding," that one is forced to employ the smallest possible region A.

In the method discussed, we are not satisfied with the usual divison of the system into two regions, A and S, and introduce in addition the third region, B. The reason for this is as follows. The electronic structure is attributed to the system as a whole. Calculating the electronic structure of one part of the system, it is necessary to first, take into account the other part, and, second, to match the results of calculations made for those different parts. It is this matching for which the third region B is proposed. Indeed, it is difficult, if at all possible, to match directly the MCSCF and LDA electronic structures on some surface dividing the two regions. It is much better to introduce the region for which the one-determinant approximation is good enough, and which could be therefore treated equally well with both the SCF and LDA methods. In this case, the matching could be done within the same method : B and S within LDA, A and B within MCSCF. So the overlap region B plays the role of the buffer to lessen the effect of using different methods on different parts of the whole system.

It is expedient to simulate the influence of one part of the system on the electronic structure of another part with the help of an effective potential in the manner of the pseudo-potential theory. In this aspect, the potential is usually referred to as the "embedding" potential. The main reason for choosing the potential for an agent is that the potential is a well-defined property in both the LDA and MCSCF methods. In general, the embedding potential should be non-local, but to make the potential least dependent on the particular method (LDA or MCSCF), its non-locality should be as small as possible.

Thus, the proposed combined MCSCF-LDA method of calculating the electronic structure can be expressed as the following sequence of steps.

- i B+S system is calculated with the LDA method.
- ii The embedding potential V_S, simulating the influence of the region S onto B, is extracted from calculations of the previous step.
- iii A+B system in the embedding potential V_S is calculated with the MCSCF method.

It could be that the accuracy of the results obtained in this stage will be already good enough. If not, the calculations should be continued.

- iv The embedding potential V_A, simulating the influence of the region A onto the region B, is extracted from calculations of the previous step.
- v B+S system in the embedding potential V_A is calculated with the LDA method.
- vi Return to the step ii and proceed until convergency.

The key point of this method is the construction of the embedding potentials V_S and V_A. We will return to it later in this paper.

3. Properties to calculate

In the present paper, the total electronic energy of the system, its dependence on the geometry of the system, and the excitation energy will be considered.

The total electronic energy of the system together with atomic core interactions govern the equilibrium positions of atoms in solids, vibration frequencies, the paths and barriers for an atom migration, etc. For systems with open-shell atoms, the one-electron scheme quite often fails to provide the correct descriptions of these properties, as for example, the population of different shells can be different for different atomic positions. Moreover, for the open-shell atoms with localized states, the electron-electron correlations could differ even qualitatively from that of the homogeneous electron gas model. All this shows that the many-determinant wave function should be employed in the total energy calculations in these cases.

The excitation energy manifests itself in the optical spectra. In the presence of the open-shells atoms, the use of the many-determinant wave function is crucial for this purpose, as the one-electron energy is an ill-defined property in this case. The distance between the energy levels of a one-electron scheme cannot represent the excitation energy in the open-shell case at all because the very positions of one-electron levels change as a result of excitation. They could even be reversed in the ground and excited states.

For the correct description of the observed absorption and emission spectra, the total energy calculations are also necessary, because the equilibrium positions of atoms in region A are different for the ground and the excited states, and for the ground state they differ from that of the ideal solid. For the absorption spectra, the electronic excitation energy should be calculated for the equilibrium position of the ground state, and the emission spectra - for the equilibrium positions of the excited state. Besides, vibrations make optical bands out of optical lines and to take account of them, the total energy should be known as a function of atomic displacements.

4. Prototype system

When developing a method, it is hardly expedient to test it on a realistic system, because it requires detailed computations, which are usually very computer resource demanding, so that every calculation is, in fact, unique. It is much better to employ a prototype system, which possesses the most important features of the real system and still is simple enough to make the test calculations practicable.

As described above, two different methods for the electronic structure calculations will be used : LDA and MCSCF. The most resource demanding of two is the MCSCF method, and it is the number of electrons that matters, especially the number of "active" electrons, whose distribution among the one-electron levels is different in different configurations. Therefore, in selecting the prototype system, we restricted ourselves to systems with the smallest possible number of electrons.

Next, it is not necessary for the prototype system to be a solid. This proposed method could equally well be applied to a cluster or to a molecule, where division into three regions can be meaningful. So it is expedient to take for the prototype system a cluster with the smallest possible number of atoms, every atom containing the smallest possible number of electrons.

For the building blocks of the prototype system, one-electron and two-electron atoms were chosen having a simple core potential of the form

$$V(r) = -Z\frac{1 - \exp^{-\alpha r}}{r} \tag{1}$$

Here Z is the number of electrons and α is an adjustable parameter. For the one-electron atom, α was adjusted so that the lowest energy coincides with the negative of the first ionization potential of the Li atom. The parameter α in the potential for the two-electron atom was adjusted so that the lowest energy of one electron in this potential coincides with the negative of the second ionization potential of the Mg atom. As the constructed one-electron and two-electron atoms only resemble Li and Mg atoms, respectively, and in many aspects differ from them, we will refer to these constructed atoms as L atom and M atom, respectively.

From these atoms, two molecules L_2 and M_2 were constructed. The electronic structure of the atom M and molecules M_2 and L_2 was calculated using the LDA approximation with the CASTEP program [7] and using the MCSCF approximation with the MOLCAS program [8], and with the numerical program for atoms and diatomic molecules [9]. The energy of the L_2 molecule calculated as a function of the internuclear separation R_{LL} shows that there is no equilibrium position in HF approximation and there is a shallow minimum near $R_{LL} = 6.5$ au with ≈ 0.1 eV binding energy in the MCSCF approximation. At the same time, there is no equilibrium position for the M_2 molecule in either the HF nor MCSCF approximations. As it is not necessary to use the equilibrium geometry in the method developed, the internuclear separation R_{MM} in the M_2 molecule was taken to be approximately double the atomic radius of the Mg atom, being slightly decreased to increase the M-M interaction. The particular values $R_{LL} = 6.548$ au and $R_{MM} = 4.804$ au were chosen to accommodate the grid requirements in the CASTEP program.

In the MCSCF method, the wave function of the system is a linear combination of many-electron functions of various configurations

$$\Psi(x_1, \ldots, x_n) = \sum_k C_k \Psi_k(x_1, \ldots, x_n) \tag{2}$$

From Table 1, where several of the largest weights of configurations $w_k = |C_k|^2$ are given, one can see that there is one configuration that dominates in the M atom and in the M_2 molecule, whereas at least two configurations have comparable weights in the L_2 molecule.

If the wave function (2) is known, the first order reduced density matrix can be calculated

$$\rho(x|x') = n \int \Psi(x_1, x_2, \ldots, x_n) \Psi^*(x_1', x_2, \ldots, x_n) dx_2 dx_3 \cdots dx_n \tag{3}$$

and its spatial part

$$\rho(\mathbf{r}|\mathbf{r}') = \int \rho(\mathbf{r}, \sigma | \mathbf{r}', \sigma) d\sigma \tag{4}$$

can be transformed to the diagonal form

$$\rho(\mathbf{r}|\mathbf{r}') = \sum_k \lambda_k \psi_k(\mathbf{r}) \psi_k^*(\mathbf{r}') \tag{5}$$

Here $\psi_k(\mathbf{r})$ are the natural orbitals and λ_k are the natural occupation numbers. From Table 2 where the biggest natural occupation numbers are given, it can be seen that two electrons of the atom M occupy mostly one 1s orbital (1.9 out of 2) and that four electrons of the molecule M_2 occupy mainly two orbitals $1\sigma_g$ and $1\sigma_u$ (3.8 out of 4). It is another indication that the wave functions of the M atom and M_2 molecule are close to the one-determinant function. Another situation is with the L_2 molecule where 0.3 out of 2 electrons occupy the second orbital. Therefore the one-determinant approximation for L_2 is a poor one. These considerations show that the LDA method could be well applied to the M atom and M_2 molecule, whereas the L_2 molecule should be calculated with the MCSCF method. The data shown in Table 3 confirm this conclusion.

The cluster consisting of the L_2 molecule and of two M atoms was chosen as the smallest possible cluster for the prototype system. Here, the L_2 molecule (L_I and L_{II} atoms) represents the A region, one M atom (M_I) closest to the L_2 represents the B region and another M atom (M_{II} farthest from the L_2 represents the S region. It was assumed that the centers of the L_2 molecule and of both M atoms are in one line. Two different geometries were considered. One geometry (L) is linear : all four atoms are in one line. Another geometry (T) is transversal : the axis of the L_2 molecule is perpendicular to the axis of the M_2 molecule.

L geometry   T geometry

Coordinates of atoms in both geometries are given in Table 4. The results of the MOLCAS calculations of the few lowest states of the prototype system are given in Table 5. In these calculations, approximately 10^5 configurations were used. One configuration dominates in the triplet states and two configurations dominate in the singlet states.

5. Embedding potential from LDA calculations of S+B regions

In the proposed method, the embedding potential $\widehat{V}_{embed}$ should be extracted from LDA calculations for the B+S region to make B+V_{embed} "chemically" similar to B+S from the B side. Hence, $\widehat{V}_{embed}$ should provide the correct density and the correct potential in the B region and its vicinity. It is evident that the localized orbitals should be employed here.

In the case of the prototype system, the B+S region is the M_2 molecule. It was calculated with the CASTEP program and the total energy, the electron density $\rho(\mathbf{r})$ and eigenfunctions and eigenvalues of the Kohn-Sham equation were obtained. The Kohn-Sham molecular orbitals were unitary transformed into the localized orbitals ϕ_I and ϕ_{II} to make the orbital ϕ_{II} localized on M_{II} as close to the orbital of the free atom M as possible. Then $\rho_{II}(\mathbf{r}) = |\phi_{II}(\mathbf{r})|^2$ was taken for the density of M_{II}, and the difference $\rho_I(\mathbf{r}) = \rho(\mathbf{r}) - \rho_{II}(\mathbf{r})$ was taken for the density of M_I. The reason for this choice is the following.

The density $\rho(\mathbf{r})$ could not be divided into two spherically symmetrical densities exactly. If in the electrically neutral system the electronic density is not spherically symmetrical (or the sum of the spherically symmetrical densities around each nucleus), the potential of the system will have a long range tail - dipole, quadrupole or whatever. Therefore, it is expedient to make the density of the atom M_{II}, which is to be removed, spherically symmetrical, because if the embedding potential reproduces the density ρ_I exactly, then there is no necessity to include the correction for the long-range part of the potential of the removed atom.

In the present work, the embedding potential was chosen local as an approximation to be removed later. The sum of two exponents localized on the M_{II} was used

$$V_{embed}(\mathbf{r}) = V_1 e^{-\alpha_1 r} + V_2 e^{-\alpha_2 r} \qquad (6)$$

The parameters α_1 and α_2 were chosen to be of different range and V_1 and V_2 were chosen to be of different sign to simulate the semilocality of the atomic pseudopotential. The particular values of parameters were found to approximate the density $\rho(\mathbf{r})$ in the M_I region as follows. For a given set of parameters, the Kohn-Sham equation for M_I in the field of the potential V_{embed} was solved and the orbital $\phi_v(\mathbf{r})$ was obtained. Then the density $\rho_v(\mathbf{r}) = |\phi_v(\mathbf{r})|^2$ was calculated and compared with $\rho_I(\mathbf{r})$. The integral

$$W = \int |\rho_v(\mathbf{r}) - \rho_I(\mathbf{r})|^2 \, \tau(\mathbf{r}) \, d\mathbf{r} \qquad (7)$$

was taken for the penalty function, the weight $\tau(\mathbf{r})$ having the maximum outside the M_1 atom from the L_2 side.

6. The prototype system in the isolated cluster and cluster in the embedding potential approximations

In the proposed method, the embedding potential V_{embed} obtained from LDA calculations for the B+S region is employed in the MCSCF calculations of the A+B region.

Two sets of calculations were performed for the prototype system. First the M_{II} atom was removed completely from the system and the molecule L_2M was calculated. This corresponds to the isolated cluster approximation. Next, the embedding potential found as described above from the LDA calculations of the M_2 molecule was applied to substitute the removed M_{II} atom. All calculations were performed for both L and T geometries. The calculated singlet-singlet and singlet-triplet excitation energies are given in Table 6 for the total system L_2M_2, the system in the isolated cluster approximation L_2M and in the cluster in the embedding potential approximation L_2M+V_{embed}. The difference of the ground states energies for the L and T geometries are also given in Table 6.

Inspection of the table shows that the local embedding potential, employed here, provides a good approximation for the energy and improves considerably the isolated cluster approximation. Still one can see that the particular embedding potential used here slightly overestimates the influence of the M_{II} atom. Besides, the local potential can only approximately simulate the s-p nonlocality. Therefore better, nonlocal approximations are desirable.

7. The separable embedding potential

The potential that simulates the unoccupied states of the B+S region can be constructed with the help of the pseudo-potential theory. In the problem considered, it is necessary to simulate as well the occupied states localized in the region B. This is a more difficult problem which was solved approximately in the previous sections. Below, it will be shown that by employing the separable operator for the potential, one can obtain the embedding potential that simulates the unoccupied states and *exactly* reproduces the occupied states localized in B.

The mathematical formulation of the problem is the following. Let the operator $\hat{H}_0$ have eigenfunctions ψ_k and eigenvalues ϵ_k

$$\hat{H}_0 \, \psi_k = \epsilon_k \, \psi_k$$
$$k = 1, 2, \ldots \tag{8}$$

Our aim is to construct the operator $\hat{H}$ which has

- i the given set of n orthonormal functions ϕ_p, $p = 1, \ldots, n$ as lowest eigenfunctions,
- ii the given values E_p, $p = 1, \ldots, n$ for the corresponding eigenvalues,
- iii all other eigenvalues equal to ϵ_k, $k = n + 1, \ldots$ from (8),
- iv the transformed functions ψ_k, $k = n + 1, \ldots$ from (8) as the corresponding eigenfunctions

$$
\begin{aligned}
\hat{H} \, \phi_p &= E_p \, \phi_p, & p &= 1, \ldots, n \\
\hat{H} \, \phi_k &= \epsilon_k \, \phi_k, & k &= n + 1, n + 2, \ldots
\end{aligned} \tag{9}
$$

For our problem, $\hat{H}_0 = \hat{H}_B + \hat{U}$ is the sum of the Fock operator $\hat{H}_B$ built up (not self-consistently) from occupied states ϕ_p, $p = 1, \ldots, n$ localized in the region B and of the pseudopotential $\hat{U}$ of the region S adjusted to reproduce the virtual states of B+S. The embedding potential we are looking for is $\hat{H} - \hat{H}_B$.

All functions ϕ_k must be orthonormal. Functions with $k = 1, \ldots, n$ are orthonormal by assumption, and those with $k > n$ should be constructed out of ψ_k. It is appropriate to employ here the orthogonalisation procedure introduced by Girardeau [10] adjusting it to our purpose.

Let χ_p be the functions

$$\chi_p = \phi_p - \psi_p \qquad p = 1, \ldots, n \tag{10}$$

and S be the $n \times n$ overlap matrix with elements

$$S_{pq} = \delta_{pq} - \langle \phi_p \mid \psi_q \rangle, \tag{11}$$

Let us make an (non-Hermitian) operator

$$\hat{R} = \sum_{p,q=1}^{n} | \chi_p \rangle T_{pq} \langle \phi_q | \tag{12}$$

where $T = S^{-1}$ is the inverse overlap matrix (we assume that χ_p are linearly independent, so the matrix S is non-singular). It is easy to show that

$$(1 - \hat{R}^+)\phi_p = 0, \qquad p = 1, \ldots, n \tag{13}$$

and

$$(1 - \widehat{R}^+)(1 - \widehat{R}) = 1 + \sum_{p,q=1}^{n} | \psi_p \rangle T_{pq} \langle \phi_q | + \sum_{p,q=1}^{n} | \phi_p \rangle T_{qp}^* \langle \psi_q | \quad (14)$$

Then functions

$$\phi_k = \left(1 - \widehat{R}\right)\psi_k, \qquad k > n \quad (15)$$

are orthogonal to ϕ_p, $p = 1,\ldots,n$ because of (13) and are orthonormal because of (14). Moreover, the infinite set of functions ϕ_k, $k = 1,\ldots$, makes the complete system and the operator, which has ϕ_k for eigenfunctions, can be written in the form

$$\widehat{H} = \sum_{p=1}^{n} | \phi_p \rangle E_p \langle \phi_p | + \sum_{k=n+1}^{\infty} | \phi_k \rangle \epsilon_k \langle \phi_k | \quad (16)$$

The eigenvalues of this operator are E_p and ϵ_k. Therefore this is an operator we were looking for.

The operator (16) can be written in closed form without any infinite sums. Indeed, because of (15)

$$\widehat{H} = \sum_{p=1}^{n} | \phi_p \rangle E_p \langle \phi_p | + (1 - \widehat{R}) \sum_{k=n+1}^{\infty} | \psi_k \rangle \epsilon_k \langle \psi_k | (1 - \widehat{R}^+) \quad (17)$$

By adding and subtracting $\sum_{p=1}^{n} | \psi_p \rangle \epsilon_p \langle \psi_p |$ to sum over k, one obtains

$$\widehat{H} = \sum_{p=1}^{n} | \phi_p \rangle E_p \langle \phi_p | + (1 - \widehat{R})\left(\widehat{H}_0 - \sum_{p=1}^{n} | \psi_p \rangle \epsilon_p \langle \psi_p |\right)(1 - \widehat{R}^+) \quad (18)$$

This operator can be written as

$$\widehat{H} = \widehat{H}_0 + \widehat{V} \quad (19)$$

where $\widehat{V}$ has the form of separable potential built up on $3n$ functions

$$\widehat{V} = \sum_{j,k=1}^{3n} | f_j \rangle V_{jk} \langle f_k | \quad (20)$$

and

$$
|f_k\rangle = \begin{cases} |\phi_k\rangle, & k=1,\ldots,n \\ |\psi_{k-n}\rangle, & k=n+1,\ldots,2n \\ \widehat{H}_0|\phi_{k-2n}\rangle, & k=2n+1,\ldots,3n \end{cases} \tag{21}
$$

The expression for the matrix element V_{jk} of the $3n \times 3n$ matrix V is cumbersome but easy.

Consequently the operator $\widehat{H}$ (9) is constructed and for the embedding potential we obtain $\widehat{U} + \widehat{V}$.

Acknowledgments

The author has the pleasure to thank Professor V. Heine for his warm hospitality in Cambridge, where a substantial part of this work has been done, and colleagues from the Theory of Condensed Matter group, Cavendish Laboratory, and from the Solid State Theory group, St.Petersburg University, for fruitful collaboration. The work was supported by the Royal Society Joint Project grant and by the Russian Foundation of Fundamental Investigations grant N 96-03-34074.

8. References

1. Abarenkov,I., Bulatov,V., Godby,R., Heine,V., Payne,M.C., Soushko, P., Titov, A., and I.Tupitsyn, (to be published) Electronic structure multiconfiguration calculation of a small cluster embedded in LDA host.
2. Abarenkov,I., and Tupitsyn, I., (to be published) A separable non-local embedding potential for ground and excited states.
3. Dreizler, R.M., and Gross, E.K.U., (1990) *Density functional theory* Springer-Verlag, Berlin
4. Parr, R.G., Yang, W. (1989) *Density-functional theory of atoms and molecules*, Oxford University Press, New York
5. Sheppard, R. (1987) The multiconfiguration self-consistent field method, in K.P.Lawley (ed.),*Ab-initio methods in quantum chemistry, Vol II*, John Wiley & Sons,pp 63-200
6. Roos, B.O., (1987) The complete active space self-consistent field method and its applications in electronic structure calculations, in K.P.Lawley (ed.),*Ab-initio methods in quantum chemistry, Vol II*, John Wiley & Sons, pp 399-444
7. Payne, M.C., Teter, M.P., Allan, D.C., Arias, T.A., and Joannopoulos, J.D., (1992) Iterative minimization technique for *ab initio* to-

tal energy calculations: molecular dynamics and conjugate gradients, *Rev.Mod.Phys.*, **64**, 1045-1097

8. Anderson, K., Blomberg, M.R.A., Fülsher, M.P., Kellö, V., Lindh, R., Malmqvist, P.-Å., Noga, J., Olsen, J., Roos, B.O., Sadlei, A.J., Siegban, P.E.M., Urban, M., and Widmark,P.-O., (1991) MOLCAS, Version2, University of Lund, Sweden

9. Kotochigova, S., and Tupitsyn, I.,(1995) Electronic structure of molecules by the numerical generalised-valence-bond wave functions, *Int.J.Quant.Chem.*, **29**, 307-312

10. Girardeau, M.D., (1971) Completely orthogonalised plane waves, *J.Math. Phys.*, **12**, 165-167

TABLE 1. Atom M and molecules M_2 and L_2 in the ground state; weights of the most essential configurations

n	M	M_2	L_2
1	0.951	0.901	0.840
2	0.040	0.024	0.142
3	0.009	0.012	0.012
4		0.011	0.006
5		0.009	
6		0.008	
7		0.007	
	1.000	0.972	1.000

TABLE 2. Atom M and molecules M_2 and L_2 in the ground state; natural orbitals occupation numbers

Orbital	M	Orbital	M_2	L_2
1s	1.902	$1\sigma_g$	1.910	1.681
2p	0.079	$1\sigma_u$	1.887	0.283
2s	0.019	$1\pi_u$	0.066	0.024
		$1\pi_g$	0.049	
		$2\sigma_g$	0.035	0.012
		$2\sigma_u$	0.027	
	2.000		3.974	2.000

TABLE 3. Energy data (in au) for atom M and molecules M_2 and L_2

Quantity	Method	M	M_2	L_2
	HF	-0.752	-1.389	-0.365
Ground state energy	LDA	-0.767	-1.433	-0.392
	MC-SCF	-0.783	-1.454	-0.400
Singlet-singlet excitation	MC-SCF	0.135	0.078	0.078
Singlet-triplet excitation	MC-SCF	0.107	0.062	0.016
Excitation energy	LDA	0.112	0.059	0.036

TABLE 4. Coordinates of atoms in prototype system (in au) in L and T geometries.

Atom	L z	T x	z
L_{II}	-6.548	-3.2745	0.000
L_I	0.000	3.2745	0.000
M_I	5.674	0.0000	5.674
M_{II}	10.478	0.0000	10.478

TABLE 5. Energies of the lowest states of prototype system (in au).

L geometry State	Energy	T geometry State	Energy
$^1\Sigma$	-1.826	$^1A_1$	-1.810
$^3\Sigma$	-1.804	$^3B_1$	-1.810
$^1\Sigma^*$	-1.755	$^1B_1$	-1.779

TABLE 6. Excitation energies (in au).

Geometry	Transition	L_2M_2	L_2M	L_2M+V_{embed}
L	$^1\Sigma \rightarrow 1\Sigma^*$	0.071	0.075	0.070
T	$^1A_1 \rightarrow {}^1B_1$	0.031	0.048	0.032
L	$^1\Sigma \rightarrow {}^3\Sigma$	0.022	0.019	0.023
T	$^1A_1 \rightarrow {}^3B_1$	0.000	0.005	-0.001
	L $(^1\Sigma) \rightarrow T(^1A_1)$	0.016	0.010	0.015

CONFIGURATIONS OF POINT DEFECTS IN SILICON UNDER CRITICAL CONCENTRATIONS

R. M. BALABAY, N.V. GRISHCHENKO
State Pedagogical Institute, Department of Computer Sciences
Krivoy Rog, Ukraine

Abstract

Ab initio pseudopotential calculations of the total energy electronic structure for a crystal model of silicon with large (more than 10^{20} cm^{-3}) concentrations of impurities (hydrogen, phosphorous and boron) have been performed. New configurations of localization of these impurities from the local-density-functional calculations have been found.

1. Introduction

Structural transformations in semiconductors relating to increasing concentrations of structural damage were studied in Ref's [1, 2]. Typical features of critical states in single-crystal silicon were discovered due to radiation effects, particularly, under high dose ion implantation. It is important to study critical states of semiconductor crystals with large dopant and defect concentrations theoretically. Originating from experimental data, the defect concentration which leads to structural transformations equals more than 10^{20} cm^{-3} [2] .

2. Model and Methods

The model formation idea consisted of the optimum combination of unit cell sizes (which determines possible calculation volumes) with the necessary defect concentrations in the crystal (which corresponds to the state that precedes structural transformations). In this unit cell, two defects account for six matrix atoms (Fig .1). The translation of that 8-atom unit cell in all directions allows examination of the infinite silicon crystal with 2:8 defect

173

R. C. Tennyson and A. E. Kiv (eds.),
Computer Modelling of Electronic and Atomic Processes in Solids, 173–180.
© 1997 *Kluwer Academic Publishers.*

174

concentrations. Features of that crystal are studied by the electron density functional method.

The use of accurate <u>first-principle pseudopotentials</u> [3] and a momentum space formalism [4] have permitted us to obtain precise total-energy-structure relations for solids with defects within <u>the local-density-functional approximation</u> (LDF) [5].

All our <u>calculations</u> of static structural properties via the evalution of the energy are based on the following assumptions: (1) that the electrons are in ground state with respect to the instantaneous position of the nuclei (the adiabatic Born-Oppenheimer approximation); (2) that the many-body effects are well accounted for within the local-density-functional formalism; and (3) what has been called the frozen-core approximation i.e.; pseudopotentials. Pseudopotential theory provides an excellent framework because the potentials are relatively weak, which makes it possible to use plane waves in the set of basis functions.

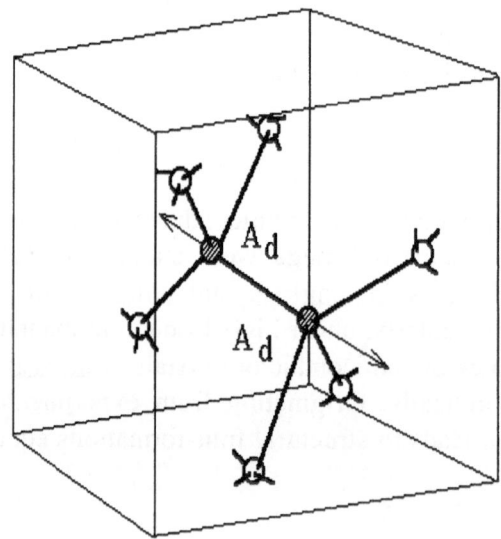

Figure 1. Atomic basis configuration of model crystal with dopant atoms A_d.

Owing to the translational symmetry of the bulk crystal, the total-energy expression can be simplified by formulating it in momentum space. The total energy per unit cell given by ($\hbar=m=e=1$)

$$E_{tot} = \frac{1}{N} \sum_{j, k} \omega(\mathbf{k}) \, \varepsilon_j(\mathbf{k}) + \int_n d^3 r \, \rho(\mathbf{r}) \, (\varepsilon_{xc} - v_{xc})$$

$$- 2\pi\Omega \sum_{G \neq 0} \frac{|\rho(G)|^2}{|G|^2} + \alpha \frac{Zn^2}{\Omega} - \gamma \frac{Z^2}{\alpha} \, . \tag{1}$$

Here $\varepsilon_j(\mathbf{k})$ are band energies, $\omega(\mathbf{k})$ is the number of elements in the star of $\mathbf{k}$, N is the number of elements in the normalization volume, and the summation extends over all occupied states in the irreducible segment of the Brillouin zone (BZ). The valence electron density is denoted by $\rho(\mathbf{r})$ and its Fourier coefficients by $\rho(G)$. The symbols Ω, α, Z, and n denote the cell volume, the cubic lattice constant, the valence charge, and the number of atoms per cell, respectively. The Ewald constant γ measures the strength of the ion-ion energy in a homogeneous negative background and α is the constant term in the power expansion of the local part $U(G)$ of the pseudopotential.

The coefficients of the electron density were obtained from

$$p(G) = \frac{1}{N\Omega} \sum_{j, k} \omega(\mathbf{k}) \, \frac{1}{g} \sum_{\alpha \in P} p(G) e^{-\alpha G \cdot t_\alpha} \, *$$

$$* \sum_{G'} \Psi_j(\mathbf{k} + \alpha G + G') \, \Psi_j(\mathbf{k} + G') \tag{2}$$

where $\Psi_j(\mathbf{k} + G)$ are the normalized coefficients of the plane-wave expansion from the band-structure calculation, and α are the symmetry operations in the point group P (g = number of elements in P) with associate nonprimitive translations t_a .

For the exchange and correlation energy per electron ε_{xc} and the related potential v_{xc} we used the results of Ceperley and Alder (CA) in the parameter forms defined by Perdew and Zunger[6].

The $\mathbf{k}$-space integrations were replaced by discrete sums over a grid in the irreducible segment of the BZ. Special $\mathbf{k}$ points were generated with the scheme in Ref. [7].

To obtain the plane-wave matrix elements of v_{xc} , the simplest procedure is to calculate the density

176

$$\rho(\mathbf{r}) = \sum_G \rho(G)e^{-t\,G\cdot t} \tag{3}$$

on a fine grid in the unit cell, to form $\nu_{xc}(\rho(\mathbf{r}))$ at each grid point, and to approximate the inverse transformation by a Fourier sum over the grid.

3. Results

Calculations were made for different <u>atom configurations</u>. The starting configuration was treated as the configuration in which all atoms positioned in ideal diamond lattice units, but dopant atoms located in neighboring planes and oriented along the {111} direction at the tetrahedron base. The relative displacement equaled 1/6 distance between atoms: (Fig.1). In all, 6 different configurations were considered and the <u>total energy</u> and <u>valence electron density maps</u> in the (110) plane were obtained.

Impurity atoms were placed in neighboring modes assuming this disposition is probable under such concentrations. Both the ion boron radius and the ion phosphorous radius differ little in the ion radius of matrix atoms. Mechanical stress does not lead to visible lattice deformation. The silicon atom frame we consider fixed. To search for stability, the impurity sites were displaced from nodes in the {111} direction. In Fig.2,

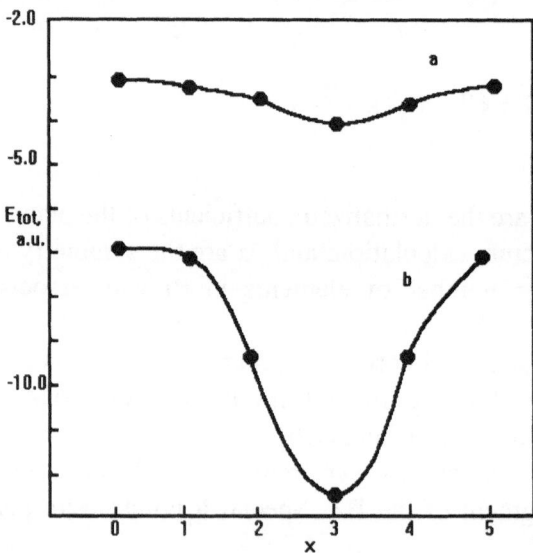

Figure 2. Potential relief for impurity atom displacement (phosphorous (a) and boron (b)) in case of its high concentrations. Along the x axis is displacement in the (1/6)•a units, where a is a between atom distance.

the total energy is plotted as a function intermediate positions of the impurity atoms. When this curve has a minimum, then we can confirm that the interstitial site for an impurity is favorable under such conditions. The $P_i V_2 P_i$ and $B_i V_2 B_i$ complexes will then be activation-free.

Consequently at large concentrations, both the boron impurity and the phosphorous impurity appear as a generator of divacancies and this process accelerates the crystal Si $\rightarrow$ amorphous Si transition. It was observed that the location of hydrogen atoms in neighboring vacant sites corresponds to the favorable energetic situation. (Fig.3) The displacement hydrogen atoms along the {111} direction of the tetrahedron basis is accompanied by a total energy increase. The results obtained confirm the conclusion of Ref. [8] on the formation of the two-layer configuration of atomic hydrogen in the {111} direction.

In addition, we calculated the electronic structure for selected systems using the method discussed earlier, but in a non self-consistent version. It was realized in the work [9] that in this case, it is possible to use qualitatively correctly, the electronic density and quantities which are received by its basis.

The charge distribution in a simulated crystal was estimated from the forming of chemical bonds between impurity atoms and silicon atom. Thus we find that the bonds between phosphorous and silicon are similar to the bonds between silicon and silicon, which are the defined by conditions for the covalent sp^3-configuration. For the case of boron in

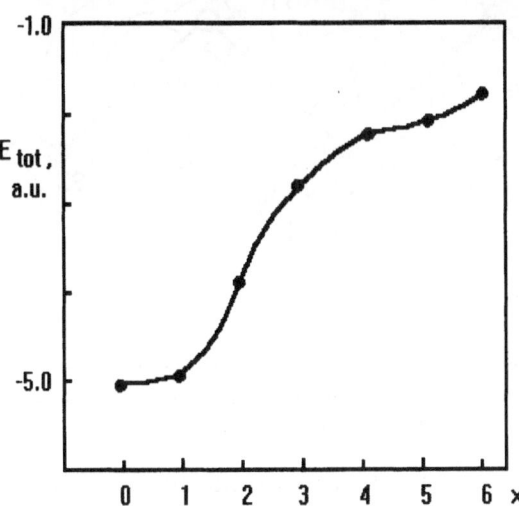

Figure 3. Adiabatic potential for the H₂ complex under movement of H atom in the {111} direction of one away the other. Along the x axis is displacement in the (1/6)·a units, where a is a between atom distance.

silicon, the charge distribution corresponding to the sp^3-valence electron configuration is violated (Fig.4 and Fig.5). A search of the local path of

178

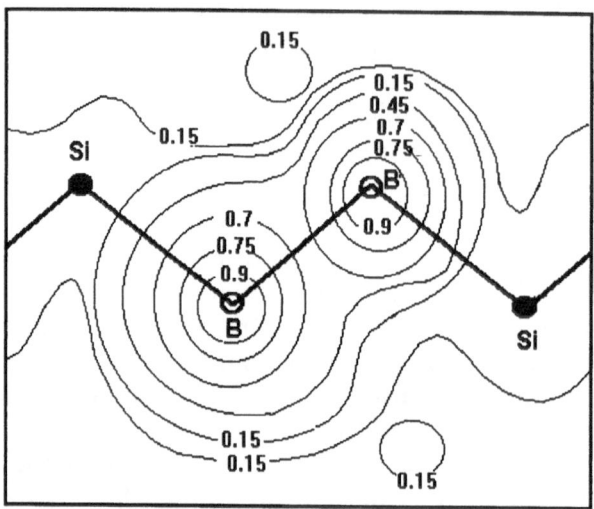

Figure 4. The distribution map of electronic density of valence electrons in the (110) plane. (The starting position of moving atoms.)

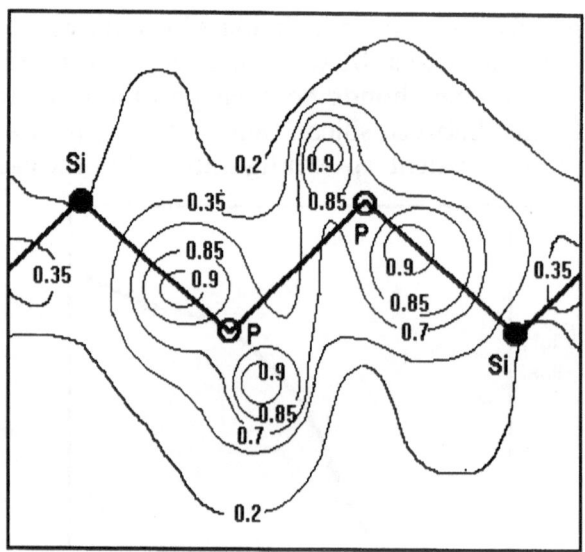

Figure 5. The distribution map of electronic density of valence electrons in the (110) plane. (The starting position of moving atoms.)

hydrogen in a silicon crystal was primarily stimulated by its ability to passivate broken bonds.

Inspection of the potential relief calculations for hydrogen migration discussed earlier shows that a two-atom hydrogen configuration is substantially motionless. This result corresponds to calculated maps of electron density distributions (Fig.6). It is obvious from the presence and type of closed-path contours of constant density on bond lines, there is interactions of the Si-H-H-Si system. This system is stable with respect to displacement of hydrogen atoms.

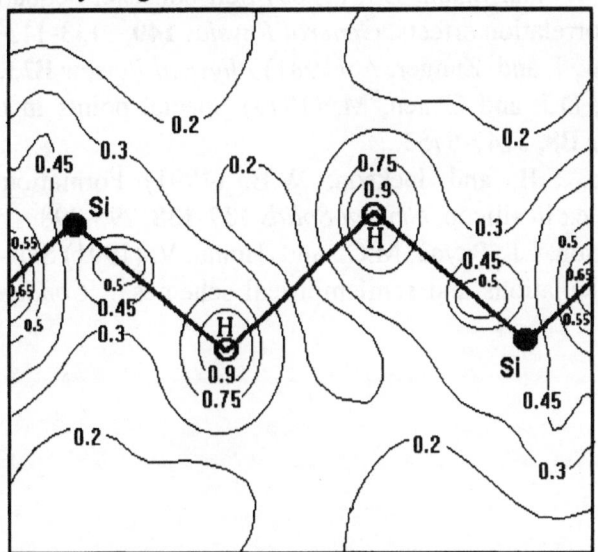

Figure 6. The distribution map of electronic density of valence electrons in the (110) plane. (The starting position of moving atoms.)

4. Conclusions

We have performed *ab initio* pseudopotential calculations of the total energy electronic structure for a crystal model of silicon with large (more than 10^{20} cm^{-3}) concentrations of impurities (hydrogen, phosphorous and boron). From the local-density-functional calculations, we found new configurations of localization of these impurities.

References

1. Smirnov, L.S. (1977) *Fizicheskie protsessy v obluchennyh poluprovodnikah*, Nauka, Novosibirsk.
2. Romanyuk, V.N. (1992) Protsessy ionno-stimulirovannogo massoperenosa i generirovaniya v poluprovodnikovyh planarnyh strukturah, Doktorskaya dissertatsiya.
3. Bachelet, G.B., Hamann, D.R., and Schluter, M. (1982) Pseudopotentials that work: From H to Pu, *Physical Review* **B62**, 4199-4228.
4. Ihm, J., Zunger, A., and Cohen, M. (1979) Momentum space formalism for the total energy of solids, *Solid State Physics* **12**, 4409-4422.

5. Hohenberg, P. and Kohn, W. (1964) Inhomogeneous electron gas, *Solid State* **2**, 864-871.

Kohn, W. and Sham, L.J. (1965) Self-consistent equations including exchange and correlation effects, *General Physics* **140**, 1133-1137.

6. Perdew, J. and Zunger, A. (1981) *Physical Review* **B23**, 5048.

7. Chadi, D.J. and Cohen, M. (1973) Special points in brillouin zone, *Physical Review* **B8**, 5747-5753.

8. Zhang, S.B. and Jackson, W.B. (1991) Formation of extended hydrogen complex in silicon, *Crystal Solids* **137-138**, 295-298.

9. Robertson, I.J., Payne, M.C., and Heine, V. (1991) Self-consistensy in total energy calculations and semiempirical schemes, *Condensed Matter* **3**, 8351-8667.

HYBRID QUANTUM-MECHANICAL AND POTENTIAL MODELS FOR STUDIES IN SOLIDS

A.H. HARKER
Centre for Materials Research
Department of Physics and Astronomy
University College
Gower Street, London, UK

Abstract. This paper provides an overview of the technique for computing defects and defect processes in ionic solids which embeds a central cluster, treated quantum mechanically, in a surrounding crystal, treated semi-classically. The justification of the method is discussed, as are the methods of joining together the two models. The paper also compares the hybrid technique with other embedding techniques, and reviews some of the applications of the hybrid scheme.

1. Introduction

Once the periodicity of a solid has been broken by a defect, it is necessary to reassess the usefulness of the computational methods which apply to perfect solids. In many cases the defect is in some sense well localized, and this localization can be exploited. For example, in a cluster calculation the defect's host is represented by only a small number of atoms, the remainder being represented by some long range potential, such as the Madelung potential, or even being ignored. In other cases the defects is not localized, and it is necessary to compute defect and band states simultaneously (although we are concerned here with electronic structure, the same considerations apply to vibrational properties). The aim of an embedding method is to achieve a blend of these two approaches, in which the cluster can be embedded in a representation of the host, and where the accuracy of this representation can be improved systematically.

Many successful calculations of defect energies and geometries in the solid state have been made using methods based on interatomic potentials,

181

R. C. Tennyson and A. E. Kiv (eds.),
Computer Modelling of Electronic and Atomic Processes in Solids, 181–192.
© *1997 Kluwer Academic Publishers.*

which can represent the elastic and dielectric response of the bulk. Problems which have been treated include vacancies, grain boundaries, and surfaces of ionic materials and, using many-body potentials, of metals. Many properties of defects, however, depend on changes of electronic structure relative to the perfect solid: electronic structure is even more important when defect processes or chemical reactions at surfaces are considered. Although it is clear that quantum mechanical methods are required in such cases, even with modern computational techniques the size of the region of crystal that can be treated is severely limited, and suitable boundary conditions must be applied to model the effect of the remainder of the crystal. By adopting a hybrid approach, in which the quantum chemical cluster is embedded in a potential model, it is possible to model a defect in an infinite solid.

In this paper we begin with a brief discussion of the computational methods available for perfect solids, and then describe the applications of cluster methods for defects. The hybrid method, as embodied in the ICECAP computer code, is then described, and give some examples of its applications.

2. Calculational Methods

The calculational techniques for solids are divided into potential models, involving semi-classical interatomic forces, and electronic structure calculations. Although valence-bond methods are used for solids[9, 43, 29], Hartree-Fock and density-functional techniques are more common.

2.1. POTENTIALS

The formal justification for an interatomic potential description of solids is related to the proof that, for a solid described by a product of N ion functions, the energy $U(R)$ may be written as[1]

$$U(R) = \sum_{i=1}^{N} E_i(R) + \sum_{i \neq j} V_{ij}(R) + \sum_{i \neq j \neq k} V_{ijk}(R) + \ldots. \qquad (1)$$

A subtle point is that the first term in this expansion is the energy required to create the ion in the electronic state that it has in the crystal, and this can not (especially in the case of 'sloppy' ions such as O^{2-}) be taken as a constant. However, it is possible[19] to absorb most of the effects of the internal rearrangement of ions into the short-range part of the pair potential (the second term). The polarizability of the ions is included using the shell model[12, 50], which links the electronic polarizability of ions to the forces exerted by the surrounding lattice, and is thus in a sense a many-body term. The shell model provides, for the solids we are concerned with here, a good description of the elastic and dielectric properties.

2.2. HARTREE-FOCK AND APPROXIMATIONS

The Hartree-Fock method is so well known[10] that it requires little comment. As in molecular applications, a key problem is the choice of basis set. Standard sets for molecules[21] may require improvement for solids[8, 34], with more flexible basis sets on anions being particularly useful[16]. Pseudopotentials are commonly used to reduce the size of the calculation: modern pseudopotentials are designed to avoid the problems of the 'orthogonality hole'[5, 47, 54, 24, 17]. Pseudopotentials designed for density functional theory appear to be suitable for Hartree-Fock calculations[53]. Correlation may be added to the Hartree-Fock model by many-body perturbation methods[52, 11]. Special care is needed to avoid the effects of basis set superposition in defect studies[16]. Approximations to Hartree-Fock theory, such as the semi-empirical methods, have been used with some success in defect studies (for example, the study[48] of the stabilization of cubic ZrO_2).

2.3. DENSITY FUNCTIONAL METHODS

Density functional methods rely on the Hohenberg-Kohn theorem[20] that the ground state energy of a many-electron system is a unique functional of the electron density. Most modern density functional calculations[36] use a basis set of plane waves, with pseudopotential cores to allow the plane wave energies to be cut off at a reasonably low value (typically about 500eV). Although described as 'ab initio' methods, the subtleties of developing adequate pseudopotentials and the choice of a functional for the exchange and correlation[38, 37], which may or may not depend on the gradient of the electron density, require some judgement by the users. The natural boundary conditions for the plane wave basis are periodic, and defect studies are therefore commonly carried out in repeated supercells (see below) of order 50 atoms in size. Despite the fact that long-range distortion and polarization must thus be excluded, the agreement between computed defect energies and those of the shell model is encouraging. As it is essentially a ground-state method, it is difficult to study defect processes which involve excited states by density functional methods. Recently, embedding procedures for density functional calculations have been proposed[49].

3. Cluster Methods

Starting from the idea that a relatively small cluster of atoms may be representative of the electronic structure of the bulk, calculations have been performed, both for Hartree-Fock (and approximations thereto) and density functional methods, on periodic clusters (supercells), simply terminated (bonds saturated with hydrogen atoms) clusters, and by formal embedding methods. Such methods have their strengths and weaknesses, but in all cases it is difficult to account for both the electronic response of the surroundings (high frequency dielectric response) and the lattice distortion (low frequency dielectric and elastic response). In ionic systems, both these effects can be accommodated by using a hybrid scheme, in which the response of the surrounding is handled by a polarizable potential model (the shell model) within a Mott-Littleton approach. We will not discuss periodic clusters here.

4. Embedding Methods

Embedding is the term applied to the formulation of a modified set of equations for the electronic structure of the cluster, embodying boundary conditions which represent the environment. In ionic materials, of course, the dominant effect is the Madelung energy (the site-to-site variation being particularly important), which may be included either by summation to infinity of the contributing terms[33], or by using a finite cluster of point ions chosen to mimic the field of the infinite crystal. The next interaction involves the propagation of electrons, from both bound states and resonances near the defect, into the bulk.

4.1. FORMAL EMBEDDING METHODS

The interactions which formal methods aim to include are those in which an electron leaves the cluster, propagates through the bulk and re-enters the cluster at a later time. Attempts to include these effects can be divided into two groups, which have been reviewed by Fisher[15]: one group modifies the Green function, the other group modifies the Hamiltonian.

4.1.1. *Corrective Operator Methods*

In the first group, the cluster Hamiltonian is diagonalized in the usual way, and the resulting Green function or density matrix is then modified[39, 40, 42]. The modifications are relatively easy to perform, but have the disadvantage that they are formally correct only in the limit where the defect perturbation goes to zero. This is because they rely on the calculation of the correction to the Green function in the perfect crystal; in the defective crystal, the perfect-crystal correction does not remove the spurious singularities introduced into the Green function by the cluster calculation[13, 14].

4.1.2. *Embedding Potential Methods*

The second class of corrections involves adding a new term, the *embedding potential*, to the Hamiltonian for the cluster region[23, 6, 13, 14]. The added term represents the quantum-mechanical amplitude for the propagation of electrons through the bulk and back into the cluster. It therefore includes retardation effects and is energy-dependent. The energy dependence greatly increases the difficulty of the calculations. These methods are formally equivalent to scattering theory, a 'perturbed-crystal' formulation which takes as its starting point the states of the perfect crystal. The embedding potential is a surface term, operating on states at the edge of the cluster, equivalent to a boundary condition constraining the wave-function to match its logarithmic derivative to the bulk.

Methods of this class are formally correct when the defect potential is non-zero, but rigorously confined to the cluster region. They are therefore particularly useful for neutral defects and for systems such as metals in which the screening length is short. When the defect potential contains a long-range component extending outside the cluster region the embedding potential, while still formally giving the correct solution to the problem, becomes very difficult to calculate. Accounting for long-range defect potentials is therefore very much an unsolved problem. The EMBED program[41] represents the state of the art of such codes.

4.2. HYBRID METHODS

Another way of dealing with the problem of joining the cluster to the surroundings is to make the cluster *more* like an isolated molecule by preventing the electrons from leaving it. This is possible because there is an arbitrariness inherent in the specification of the Hamiltonian if all that is required of it is that it should reproduce correctly the elements of the single-particle density matrix within the cluster. In particular, if the density matrix is

$$\hat{\rho} = \sum_{i \text{ occupied}} |\psi_i\rangle\langle\psi_i|, \tag{2}$$

then it acts as a projector onto the occupied eigenstates. Therefore the addition to the Hamiltonian (or, in a Hartree-Fock calculation, to the Fock operator) of any term of the form $\hat{\rho}\hat{W}\hat{\rho}$, where $\hat{W}$ is an arbitrary Hermitian operator, will simply mix the occupied levels and have no effect on $\hat{\rho}$. As it merely makes a unitary transformation among the occupied levels, it has no effect on the many-particle Slater determinant wave-function built up from them.

This arbitrariness in the Hamiltonian[3, 25] may be exploited[26, 27] by choosing $\hat{W}$ to cancel, as far as possible, the short-range components of the potential arising from outside the cluster (in particular the non-local exchange term of Hartree-Fock theory). The long-range Madelung contribution is then left as the only correction required to the cluster Hamiltonian and, since $\hat{W}$ is an attractive potential, the tendency of charge to spill out of the cluster is reduced. This approach assumes that the electronic structure of the ions outside the quantum cluster is not altered by the presence of the defect; this assumption is supported by some recent calculations of electronic structures of ions at surfaces[2], which may be regarded as a limiting case of a defective system. The distortion and polarization of these undistorted ions is then modelled by interatomic potentials, and may be summed to infinity by the Mott-Littleton method[32].

4.3. OTHER TERMINATION SCHEMES

One approach to cluster termination includes only the electrostatic effects of the bulk. Even here there is the problem of defining an effective charge. In many calculations an 'ionicity' less than unity is assumed, but this causes difficulties when a defect is formed by removing an ion which will have an integer charge. Another possibility, appropriate when there is a greater degree of covalency so that the electron states are spread more over bonds between neighbouring atoms, is to saturate the dangling bonds on the surface of the cluster with passivating states of some kind. The two most common possibilities are a hydrogen atom or a single hybrid one-electron state belonging to the neighbouring atom in the bulk. An alternative approach[22] is to use hydrogen atoms with specially modified atomic basis functions.

5. The ICECAP Approach

In many ways the embedding problem for an ionic crystal is simpler than for metals and semiconductors. This is because it is relatively straightforward to implement a model of an ionic system as an assembly of reasonably well localized ions. The difficult part of the problem (the overlapping charge densities giving rise to corrections to the Coulomb interaction and the exchange interaction) is therefore confined to a short-range correction; the long-range Coulomb problem, as mentioned above, can be treated by classical summation methods. Within the ICECAP program[18], however, it is recognized that accounting for electronic effects is only part of the problem: the distortion of the crystal around the defect is also important. The magnitude of the error which arises from neglecting the distortion is shown by some results on defects in Magnesium Oxide[16]. Formation energies of defects at a Magnesium site were computed with and without relaxation of the atomic positions. In each case the shells were allowed to move. To a reasonable approximation, the shell motion corresponds to the response of the electrons of the surrounding crystal (although it is dangerous to take the core-shell model too literally, this interpretation is supported by the fitting of shell model parameters to high and low frequency dielectric constants). The results obtained with empirical shell model potentials are shown in Table 5: the effect of distortion is very large.

TABLE 1. Effect of ionic relaxation on defect formation energies in MgO

Defect	Formation energy without relaxation of cores (eV)	Formation energy with relaxation of cores (eV)
Vacant Mg^{2+} site	40.704	24.295
Li^+ substituting for Mg^{2+}	17.323	13.650
Na^+ substituting for Mg^{2+}	22.957	18.002
Al^{3+} substituting for Mg^{2+}	-23.341	-29.163

The ICECAP[18] computer program adopts the hybrid approach to handle the distortion and polarization. Here the key problem is the different 'views' of the system which must be accommodated. An electron's view looking out from the embedded quantum cluster is dominated by electrostatic effects, but it will also see a short range potential caused by the electronic structure of neighbouring ions. From the point of view of ions outside the cluster, those short range interactions will be incorporated in the potential, loosely labelled as 'Pauli repulsion'. The localized orbital method, which is known[4] to give reliable charge densities, is used to minimize the inaccuracies here. An extra factor arises from the introduction of the potential model for the environment: it is necessary to check the consistency of the potential and the electronic structure model for perfect structures.

The separation of electrons into groups, which is central to the localized orbital method, has been used in other treatments of defects in solids[31, 30, 28], and may be generalized. Thus although the present implementation of the hybrid scheme in the ICECAP code is confined to ionic systems and extension to more covalent systems appears to be feasible. One effect which is not included in the present formulation, but which appears in more formally-based codes such as EMBED, is the transfer of electrons between one groups (i.e. between the central cluster and the surroundings). Additional work is necessary in order to quantify the importance of this effect, and the extent to which it may be accounted for by simpler models.

This distortion of the environment alters the potential seen by the embedded cluster, and a self-consistent treatment is required. The ICECAP scheme is as follows:

1. Estimate charge distribution in embedded cluster (e.g. ionic charges)
2. Compute distortion and polarization of environment
3.   – Perform quantum mechanical structure of embedded cluster in field of environment

       – Compute charge density in embedded cluster

       – If charge density has not converged, return to 2.
4.   – Relax atomic positions in central cluster

       – If total energy minimum not reached, return to 2.

There are thus *three* nested self-consistency loops. The innermost is that in the quantum mechanical calculation, performed by the Unrestricted Hartree-Fock method in ICECAP. The next is in point 2 above: this is referred to as the multipole consistency loop, as it ensures that the electrostatic multipole moments of the quantum mechanical embedded cluster are the same as those which determine the polarization and distortion of the embedding crystal. To remain within the spirit of the shell model, the program includes a distributed dipole model, which describes the multipole moments as dipole moments associated with the atoms of the embedded region (plus a correction term). The outermost loop of the program adjusts the positions of the atoms in the embedded cluster to minimize the total energy of the system.

6. Applications of ICECAP

Applications of the ICECAP method range from the determination of interatomic potentials for impurity systems through calculations on F-type centres to models of V_K centres and self-trapped excitons. The latter are particularly challenging, combining as they do a molecular bonding of two anionic species with a delocalized electron.

6.1. SUBSTITUTIONAL DEFECTS IN MGO

The ICECAP program has been used[35] to compute the distortions and the $p^6 \rightarrow p^5s$ excitation energies of O^{2-}, S^{2-} and Se^{2-} in magnesium oxide. The results, given in Table 6.1, show that an isolated cluster (lines labelled 'cluster') one anion and six cations is quite inadequate, that including the electrostatic potential of the environment ('cluster + point ions') is significantly better, but that the effects both of short-range interactions with the environment and of distortion ('ICECAP') are significant. These calculations also allowed the self-consistency of the quantum cluster and the potential to be checked: the distortion of the pure MgO was less than 1% whereas for the substitutional S and Se there were 6% and 8% outward distortions respectively.

TABLE 2. Energies and the $p^6 \rightarrow p^5s$ excitations of O^{2-}, S^{2-} and Se^{2-} in magnesium oxide.

| System | Method | Energy | | |
		Triplet	Singlet	expt
	cluster	9.65	11.32	
MgO	cluster + point ions	7.37	7.39	
	ICECAP	7.73	7.77	7.60
	cluster	13.24	13.89	
MgO:S	cluster + point ions	8.99	9.03	
	ICECAP	7.09	7.12	
	cluster	13.25	13.96	
MgO:Se	cluster + point ions	8.22	8.25	
	ICECAP	6.68	6.71	

6.2. THE V_K CENTRE

ICECAP has been applied to the V_K centre in NaCl[51], and the results compared with a "molecule in a crystal" model[7]. Although the bond length of the Cl^- molecular ion was similar in the two calculations, the displacements of the nearest neighbour sodium ions were quite different. Without self-consistent distortions[7] these ions are significantly displaced outwards, whereas the ICECAP calculations found them to be within 2% of their perfect lattice sites. There were correspondingly large differences in the energies of the excited states of the V_K centre, where the wavefunction extends over these neighbouring ions.

6.3. THE SELF-TRAPPED EXCITON

The problem of the self-trapped exciton has been the subject of some discussion, the question being whether the defect is on-centre (a V_K centre plus an electron) with D_{2h} symmetry or off-centre with C_{2v} symmetry. If it is off-centre, the precise nature of the triplet electronic state is unclear. Song and co-workers[46] used a simplified one-electron model and predicted very large off-centre displacements, with the defect resembling a coupled F centre and H centre binding the electron at the vacant site formed by the off-centre movement, whereas in KCl the semi-empirical INDO method was unable to show conclusively that an off-centre position is favourable. Recently Shluger and ço- workers[44] have applied the ICECAP model to the self-trapped exciton in LiCl. They followed Song in including a large number of floating Gaussian basis functions, centred on interstitial positions in the lattice, to allow for electron delocalization and polarization. They find a modest movement off-centre, by 0.07 Å, with a corresponding energy reduction of 0.15 eV: the electronic wavefunction is delocalized over the whole quantum mechanical cluster, but both the electron and hole components of the exciton are localized more on the Cl ion which is closest to the perfect lattice site and this is stabilized by the polarization of the surrounding lattice. The details of the electronic structure of the self-trapped exciton have not yet been unequivocally determined, and predictions of the electronic transition energies are not entirely satisfactory, but it is clear that a model which includes all the effects treated by ICECAP is required[45].

References

1. Abarenkov I V and Antonova I M (1970) Interatomic interactions in alkali halides *phys stat sol* **38** 783-797
2. Abarenkov I V and Frenkel T Yu (1991) The electronic structure of oxygen ions in the step on the (001) surface of MgO *J Phys Cond Mat* **3** 3471-3478
3. Adams W H (1961) On the solution of the Hartree-Fock equation in terms of localized orbitals *J Chem Phys* **34** 89-102
4. Asthalter T, Weyrich W, Harker A H, Kunz A B, Orlando R and Pisani C (1992) Comparison of quasi-Hartree-Fock wave-functions for lithium hydride *Sol St Commun* **83** 725-730
5. Bachelet G B, Hamann D R and Schluter M (1982) Pseudopotentials that work – from H to Pu *Phys Rev* **B26** 4199-4228
6. Baraff G A and Schlüter M (1986) The LCAO approach to the embedding problem *J Phys C* **19** 4383-4391
7. Cade P E, Stoneham A M and Tasker P W (1984) Self-trapped hole (V_K center) in NaCl-type alkali halides – lattice relaxation and optical properties for $MX - X_2$ systems *Phys Rev* **B30** 4621-4639
8. Causa M, Dovesi R, Pisani C and Roetti C (1986) Electronic structure and stability of different crystal phases of magnesium oxide *Phys Rev* **B33** 1308-1316
9. Cooper D L, Gerratt J and Raimondi M (1987) Modern valence bond theory, in *Ab initio methods in Quantum Chemistry (Advances in Chemical Physics Volume 69)* ed K P Lawley, New York; Wiley, 319-397
10. Daudel R, Leroy G, Peters D and Sana M (1983) *Quantum Chemistry* New York; Wiley
11. Davidson E R and Silver D W (1977) Size consistency in the dilute helium gas electronic structure *Chem Phys Lett* **52** 403-406
12. Dick B G and Overhauser A W (1964) Theory of the dielectric constants of alkali halide crystals *Phys Rev* **112** 90-103
13. Fisher A J (1988) Methods of embedding for defect and surface problems *J Phys* **C21** 3229-3249
14. Fisher A J (1989) *Theoretical studies of point defects* D Phil thesis, University of Oxford
15. Fisher A J (1991) How to tell a defect about the bulk; a survey of embedding

methods *Reviews of Solid State Science* **5**(2-3) 107-132, World Scientific Publishing

16. Grimes R W, Catlow C R A and Stoneham A M (1989) A comparison of defect energies in MgO using Mott-Littleton and quantum mechanical procedures *J Phys: Cond Matt* **1** 7367-7384

17. Hamann D R, Schluter M and Chiang C (1979) Norm-conserving pseudopotentials *Phys Rev Lett* **43** 1494-1497

18. Harding J H, Harker A H, Keegstra P B, Pandey R, Vail J M and Woodward C (1985) Hartree-Fock cluster computations of defect and perfect ionic crystal properties *Physica* **B131** 151-156

19. Harding, J.H. and Pyper, N.C. (1995) The meaning of the oxygen second-electron affinity and oxide potential models *Phil Mag Lett* **71** 113-122

20. Hohenberg P C and Kohn W (1964) Inhomogeneous electron gas *Phys Rev* **136** B864-871

21. Huzinaga S (1984) *Gaussian basis sets for molecular calculations* Amsterdam; Elsevier

22. Illas F, Roset L, Ricart J M and Rubio J (1993) Basis-modified hydrogen atoms as embedding atoms in ab initio chemisorption cluster model calculations on Si surfaces *J Comput Chem* **14** 1534-1544

23. Inglesfield J E (1981) A method of embedding *J Phys C* **14** 3795-3806

24. Kerker G (1980) Non-singular atomic pseudopotentials for solid state applications *J Phys C* **13** L189-194

25. Kunz A B (1969) Localized orbitals in polyatomic systems *phys stat sol* **36** 301-309

26. Kunz A B and Klein D L (1978) Unrestricted Hartree-Fock approach to cluster calculations. II. Interaction of cluster and environment *Phys Rev* **B17** 4614-4619

27. Kunz A B and Vail J M (1988) Quantum-mechanical cluster-lattice interaction in crystal simulations: Hartree-Fock method *Phys Rev* **B38** 1058-1063

28. LopezMoraza S, Pascual J L and Barandiarian Z (1995) Ab initio model potential embedded-cluster study of V^{2+}-doped fluoroperovskites – effects of different hosts on the local distortion and electronic structure of $^4T_{2g}$–$^4A_{2g}$ laser levels, *J Chem Phys* **103** 2117-2125

29. Lorda A, Illas F, Rubio J and Torrance J B (1993) Ab initio valence-bond cluster model for ionic solids: alkaline earth oxides, *Phys Rev* **B47** 6207-6215

30. Luana V, Florez M and Pueyo L (1993) Local geometry and resonant vibrations of Cu^+ : NaF – results of ab initio perturbed ion, cluster-in-the-lattice calculations involving clusters of 179 ions, *J Chem Phys* **99** 7970-7982

31. Mejias J A, Oviedo J and Sanz J F (1995) A method for including environment polarization effects in ab initio cluster embedded calculations – application to the water deprotonation over an ideal Al-MgO surface, *Chem Phys* **191** 133-139

32. Mott N F and Littleton M J (1938) Conduction in polar crystals. I. Electrolytic conduction in solid salts *Trans Farad Soc* **34** 485-499

33. Nijboer B R A and De Wette F W (1957) On the calculation of lattice sums *Physica* **23** 309-321

34. Pandey R and Vail J M (1989) F-type centres and hydrogen anions in MgO: Hartree-Fock ground states *J Phys Cond Mat* **1** 2801-2820

35. Pandey R, Zuo J and Kunz A B (1989) Excitonic states in pure and impurity-doped magnesium oxide *Phys Rev* **B39** 12565-12567

36. Payne M C, Teter M P, Allan D C, Arias T A and Joannopoulos J D (1992) Iterative minimization techniques for ab initio total-energy calculations: molecular dynamics and conjugate gradients *Rev Mod Phys* **64** 1045-1097

37. Perdew J P and Burke K (1996) Comparison shopping for a gradient-corrected density functional *Int J Quantum Chem* **57** 309-319

38. Perdew J P and Zunger A (1981) Self-interaction correction to density-functional approximations for many-electron systems *Phys Rev* **B23** 5048-5079

39. Pisani C (1978) Approach to the embedding problem in chemisorption in a self-consistent field molecular orbital formalism *Phys Rev* **B17** 3143-3153

40. Pisani C (1985) The embedding problem in ordered and disordered systems *Phil Mag* **B51** 89-99

41. Pisani C, Cora F, Nada R and Orlando R (1994) Hartree-Fock perturbed cluster treatment of local defects in crystals. I. The EMBED program – general features *Comp Phys Commun* **82** 139-156

42. Pisani C, Dovesi R, Nada R, and Kantorovich L N (1990) Ab initio Hartree-Fock perturbed-cluster treatment of local defects in crystals *J Chem Phys* **92** 7448-7460

43. Redondo A and Goddard W A (1982) Electronic correlation and the Si(100) surface: buckling versus nonbuckling *J Vac Sci Technol* **21** 344-350

44. Shluger A L, Grimes R W and Catlow C R A (1991) A new model for the self-trapped exciton in alkali halides *J Phys Cond Mat* **3** 3125-3138

45. Shluger A L, Harker A H, Puchin V E, Itoh N and Catlow C R A (1993) Simulation of defect processes; experiences with the self-trapped exciton *Modelling Simul Mater Sci Eng* **1** 673-692

46. Song K S and Leung C H (1990) The off-center self-trapped exciton in alkali halides *Reviews of Solid State Science* **4** (2-3) 357-381, World Scientific Publishing

47. Starkloff T and Joannopoulos J D (1977) Local pseudopotential theory for transition metals *Phys Rev* **B16** 5212-5215

48. Stefanovich E V, Shluger A L, and Catlow C R A (1994) Theoretical study of the stabilization of cubic-phase ZrO_2 by impurities *Phys Rev* **B49** 11560-11571

49. Stefanovich E V and Truong T N (1996) Embedded density functional approach for calculations of adsorption on ionic crystals *J Chem Phys* **104** 2946-2955

50. Stoneham A M, Harding J H and Harker A H (1996) The shell model and interatomic potentials for ceramics *MRS Bulletin* **21** 29-35

51. Testa A, Stoneham A M, Catlow C R A, Song K S, Harker A H and Harding J H (1991) The V_K centre in NaCl *Rad Eff and Defects in Solids* **119-121** 27-32

52. Thouless D J (1961) *The Quantum Mechanics of Many Body Systems* New York; Academic

53. Woodward C and Kunz A B (1987) Use of local-density pseudopotentials in Hartree-Fock calculations *Phys Rev* **B37** 2674-2677

54. Zunger A and Cohen M L (1979) First-principles nonlocal pseudopotential approach in the density functional formalism. II. Application to electronic and structural properties of solids *Phys Rev* **B20** 4082-4108

MECHANISMS OF DESTRUCTION OF SOLID SURFACES INDUCED BY ELECTRON EXCITATIONS

A.E. KIV
South-Ukrainian Pedagogical University,
26 Staroportofrankovskaya Str. 270020 Odessa Ukraine

1. Introduction

There are two ways of creating radiation defects in solids. The first is connected with atom displacements as a result of elastic collisions. Atoms can be displaced from their lattice sites, depending on the energy and the mass of the incident particles, as well as the mass of the target. The second is creation of defects induced by an excitation of the electron subsystem of solids. This case refers to subthreshold radiation effects.

The efficiency of subthreshold defect creation depends on the structure and lifetime of electron excitations, the peculiarities of electron-lattice interactions, the symmetry of the lattice site, etc. These conditions are the most favourable for atoms located near the surface. Thus, subthreshold radiation effects are most pronounced in the surface layers. That is why investigations of these processes in thin films and surface layers are of great interest. Below we shall consider the experimental situation in the field of subthreshold radiation influence on emission of atoms/ions from non-metallic solids and some results of quantum-chemical modelling of these processes.

2. Review of experimental results

Experimentally the ion and atom emission from electron- and phonon-irradiated solid surfaces has been studied quite extensively (see, e.g. the reviews [1, 2]). The situation has been most cardinally justified by Parks et. al. [3] who found that the threshold photon energy of the ejection of Na^+ and F^+ ions from NaF crystals coincides with that of the Na^+K^- shell ionization and concluded that the driving force for this injection was provided by neutral or positively - charged halogens arising as a result of the Auger cascade $Na^{2+} \rightarrow Na^+ + (F^+$ or $2F^0)$. There are many new facts that provide information on mechanisms of electron-induced emission (EIE) in numerous works of Ageev et. al. [2, 4—6]. Emission of absorbed ions/atoms Li, Na and Cs from the W

R. C. Tennyson and A. E. Kiv (eds.),
Computer Modelling of Electronic and Atomic Processes in Solids, 193–201.
© *1997 Kluwer Academic Publishers.*

194

surface initially covered by oxygen, silicon or tungsten silicide was investigated. The following primary results were obtained. Firstly, the thresholds of EIE were determined that were connected with inner shell ionization energy of metal or non-metal atoms/ions. Secondly, there is a strong dependence of EIE yields on the equilibrium distance between adsorbed particles. The yields of EIE for metal monolayers are very small. The energy distribution of ejected particles is wide (about several eV). Below we illustrate these conclusions by some concrete results [4 - 6]. Fig. 1 shows the dependencies of ion yields on the energy of incident electrons for EIE of Li (1) and Na (2) ions from the surface structure (W - Si). All threshold peculiarities can be clearly seen. The energy distributions of ejected Li$^+$ ions for two surface concentrations of ions are shown in Fig. 2 (2 10^{14} (1) and 6 10^{14} (2) sm^{-2}). Fig. 3 shows the difference of energy distributions of ejected Na$^+$ ions for energies of incident electrons (100 (1) and 300 (2) eV). As a result of electron energy increase from 100 to 300 eV, the second maximum in the energy distribution of Na$^+$ ejected ions appears. Another interesting experimental result is shown in Fig. 4. First, we see that the yield of Na$^+$ ions depends on the number of tungsten silicide layers and also on the concentration of adsorbed ions (the number of tungsten silicide layers increased from 1 to 5). The most important recent experiment was the experiment with alkali halide layers on W(110) [7]. Electron-bombardment-induced alkali- and halogen- ion desorption from NaCl, KCl, KBr, KI and CsI layers on W(110) was studied as a function of layer thickness from zero to several monolayers. Changes of the ion energy distributions with coverage of W surface by alkali halide layers were investigated in detail.

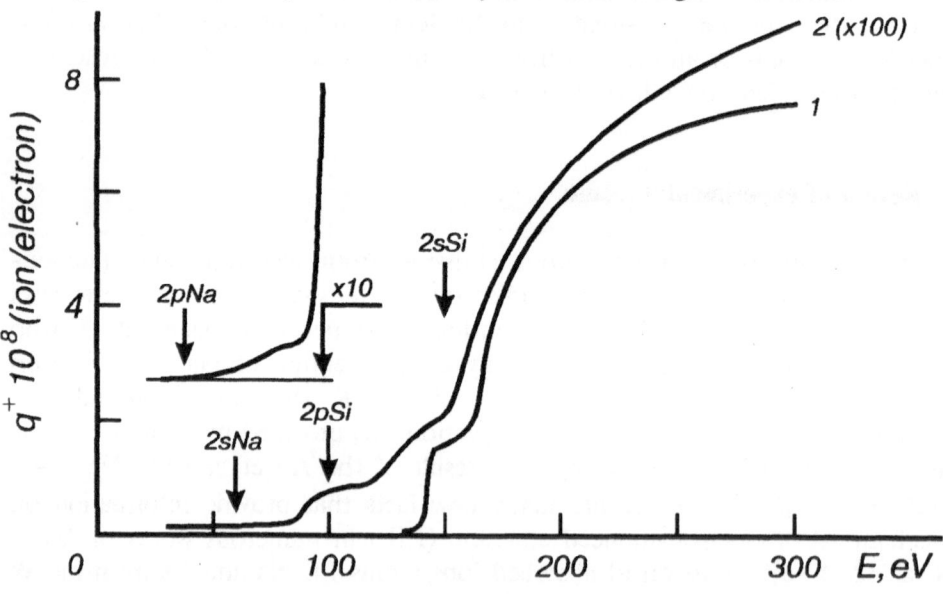

Figure 1.

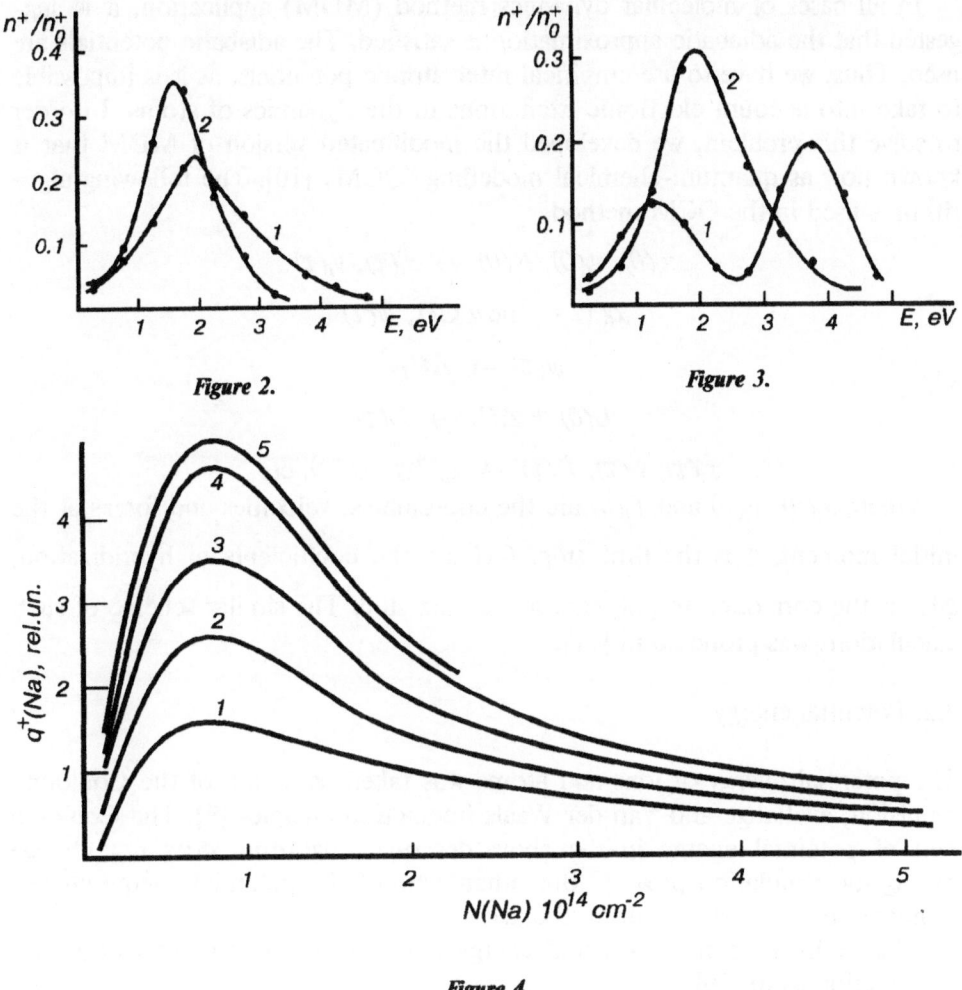

Figure 2.

Figure 3.

Figure 4.

Thus, all the results described above clearly show that EIE of ions/atoms is a consequence of inner shell ionization, the prerequisite of an Auger process, which produced the electronic excitation energy necessary for ion emission (see also [8]).

3. Model and physical simulation

3.1. Structure of algorithm

The cluster configuration was taken as a hemispherical fragment of an alkali halide crystal containing 500 point charges at the normal lattice positions of ions. The standard method of quantum-chemical modelling was used [9].

In all cases of molecular dynamics method (MDM) application, it is suggested that the adiabatic approximation is satisfied. The adiabatic potentials are used. Thus, we have to use empirical inter-atomic potentials, as it is impossible to take into account electronic excitations in the dynamics of atoms. In order to solve this problem, we developed the modificated version of MDM that is known now as quantum-chemical modelling (QCM) [10]. The following algorithm is used in the QCM method.

$$x_i(0),\ v_i(0),\ F_i(0) \rightarrow x_i(\tau),\ v_i(\tau)$$

$$x_i(\tau) \rightarrow \text{new CH},\ \psi(\tau)$$

$$\psi(\tau) \rightarrow \Delta U_1$$

$$U(0) + \Delta U_1 \rightarrow F_i(\tau)$$

$$x_i(\tau),\ v_i(\tau),\ F_i(\tau) \rightarrow x_i(2\tau),\ v_i(2\tau),\ \text{etc.}$$

Where $x_i(0)$, $v_i(0)$ and $F_i(0)$ are the coordinates, velocities and forces at the initial moment, τ is the time step, CH are the coefficients of hybridization, ΔU_1 is the correction to potential at the first step. The similar scheme of such calculations was proposed in [11].

3.2. Potential energy

The potential energy of ions and atoms was taken as a sum of the Coulomb, repulsion, exchange and Van der Waals interaction energies [9]. The Coulomb part of potential energy in the above-described algorithm does not change during the simulation process. But other parts of the potential energy change significantly as a result of atom displacement.

The exchange part of potential energy is calculated in pair interaction approximation as in [10]:

$$V_j^{ex} = \sum_{\mu=1}^{Z} \left\langle \psi_\mu^C * \left| \hat{H} \right| \psi_\mu^C \right\rangle$$

where $\hat{H}$ is the two-centre hamiltonian, Z is the lattice coordination number, μ lists the nearest neighbours of the given ion/atom, ψ_μ^c is the covalent part of the bond wave function $\psi_b = C' \psi_\mu^i + C \psi_\mu^c$, ψ_μ^i is the ionic part of the bond wave function, $\beta = C'/C$ is the ionicity parameter of the bond. ψ_μ^c was constructed as a Heitler-London function on the basis of Slater-type atomic function for corresponding ions and atoms.

In their angle-dependent parts, h_l $(l = a,c)$, the sp-hybridization of valence electrons was taken into account through the formula $h_l = \sum_n a_{nl} Y_n$, where

n corresponds to s- or p-symmetry of the orbital, a_{nl} are the hybridization coefficients, evaluated at each integration step demanding orthonormalization and a maximal overlap of hybrid orbitals of neighbouring atoms, Y_n is the angle-dependent part of the n-th electron function.

The Van der Waals energy was evaluated by the London formula [10]. The shell model was also used [12].

A dipole moment is induced in response to the electric field experienced by the ion from surrounding ions, and these dipoles in turn affect the interactions between ions. This is a many-body effect which cannot be represented within the simple pair-potential model for inter-ionic interactions. The shell model is a widely-used extension of the pair potential model which incorporates polarizability by a simple mechanical model. An ion is represented by a core and shell, both carrying a charge, and linked by an harmonic spring. The field polarises the ion by producing a separation between core and shell. The mass of the ion is assigned completely to the core, the shell being taken as mass-less. Thus, the shell is always found at a zero force position, and the internal degree of freedom of the ion can contain no kinetic energy. This reflects the idea that polarization of the ion is a result of a distortion of the electron cloud, but that according to the Born-Oppenheimer Principle the electrons always remain in their ground state. The non-Coulombic part of the inter-ionic pair potential, consisting primarily of a short-ranged repulsion, acts between the shells of the ions. This produces a necessary coupling between the short-range forces and ionic polarization which damps the self-induced polarization between a nearby pair of ions. Correction to the London formula was calculated with the help of the shell model procedure [12].

3.3. Cluster size effects

In our calculation including long-range Coulomb potential, it is extremely important that the results were insensitive to the size of simulation fragment. To be sure of this, we went through the following procedure. We simulated the motion of the central ion R_0, keeping the coordinates of all other ions fixed. After the stabilization of R_0, the ions of the first, second, etc. coordination spheres were subsequently released and the changes of all R_j checked. It turned out that these displacements approach zero if the number of ions inside the sphere approaches several tens. Thus, we confirmed that the choice $N = 500$ with excess guarantees the independence of results on the size of the model fragment.

4. The lifetime problem

It is established [10] that the subthreshold defect creation depends on the lifetime of electron excitation τ_e and, in particular, on the time of its localization τ_l ($\tau_l < \tau_e$). If the atom vibration frequency is denoted by ω_0, two versions of subthreshold mechanisms of atom displacement are possible: potential displacement ($\tau_l > \omega_0^{-1}$) and impact displacement ($\tau_l < \omega_0^{-1}$). Time-dependent forces F_e (Coulomb, Pauli, etc.) arise as a result of the electron subsystem excitation. The efficiency of subthreshold defect creation is determined by the momentum value:

$$P = \int_0^{\tau_l} F_e \, dt$$

So it is important to have information about τ_l corresponding to different types of electronic excitations and to different atom locations.

But exact information about τ_l is absent, as we can see in Fig. 4. Fig. 5 (from [5]) illustrates changes of lifetime τ_l connected with the tunnelling effect for EIE of Li^+ ions from W target covered by the Si monolayer. The maximum in the energy distribution of Li^+ ions depends linearly on the overlap integrals S for Li^+ ions on the surface. It is clear that $\tau_l \sim S^{-1}$ and we usually have a large uncertainty in values. So it is appropriate to consider some distribution of τ_l and to use the following scheme of calculation.

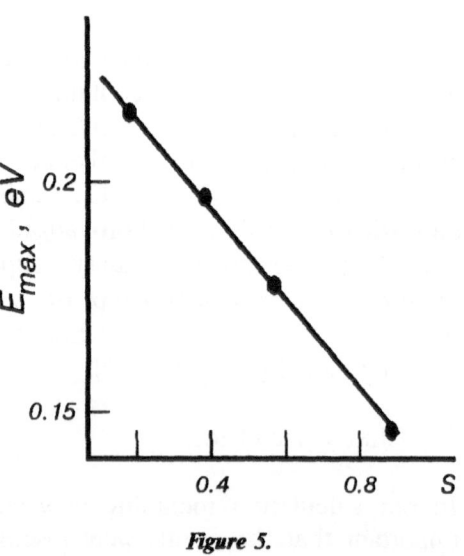

Figure 5.

Scheme 1

K-level				
$\tau_K^{(1)}$	$\tau_K^{(2)}$	$\tau_K^{(3)}$	...	$\tau_K^{(m)}$

$\downarrow$

L-level				
$\tau_L^{(1)}$	$\tau_L^{(2)}$	$\tau_L^{(3)}$	...	$\tau_L^{(m)}$

$\downarrow$

...

Examples of τ_l values for some energy levels of alkali and halide ions are given in [9]. The final algorithm of subthreshold radiation effects quantum modelling includes Scheme 1. The distribution of τ_l is suggested as random.

4. Results and discussion

The simulation procedure was applied for ions at the (100) surface of five alkali halide crystals - LiCl, LiBr, LiI, NaI and KI (see ref. [9]). In addition, the EIE from thin layers of alkali halides was investigated. New procedures of EIE processes simulation including Scheme 1 were used in order to get energy distributions of ejected ions/atoms and dependencies of yields of ejected particles on the concentration of adsorbed ions at the target surface. For each case of an ejection process investigation, 35-50 versions were calculated that corresponded to all possible combinations of lifetimes shown in Scheme 1. Such simulation allowed us to obtain wide energy distributions of ejected particles similar to those of Fig. 2 and 3 and also to those obtained in [7]. It was established that Pauli shock plays an important role in all processes of EIE. By considering the Pauli shock as one of the steps for the necessary energy accumulation by ejected particles, we can explain most of the EIE peculiarities. In all cases, it is important to take into consideration two types of Pauli shock mechanisms. The first one is initiated by the cation inner shell ionization. Realization of this mechanism is connected with two stages of charge relaxation, for instance: $M^{++} \rightarrow M^+$ and $M^+ \rightarrow M^0$. So a particle gets two impulses connected with Pauli forces.

The second type of Pauli shock is initiated by the anion ionization and its relaxation to the neutral state: $X^+ \rightarrow X^0$. The anion X^0 and the cation draw together under Van der Waals forces. The next step of relaxation $X^0 \rightarrow X^-$ leads to the arise of Pauli forces. Thus, we see that by taking into consideration the Pauli shock mechanism, we can explain ejection of neutral metal atoms and negative halide ions. The Pauli forces are larger in comparison with Coulomb ones. This is a reason for the large yields of neutral atoms. Lastly, the essential point in EIE theory is that the efficiency of processes considered depend strongly on the ionicity of the material β. There exist different dependencies of σ_C / σ_P (σ_C and σ_P are cross-sections of Coulomb and Pauli shocks) on β. In the case of a Pauli shock initiated by halide atom ionization we observe the decrease of σ_C / σ_P with β increase and in the case of metal atom ionization, the result is opposite.

5. Conclusion

Mechanisms of subthreshold atomic emission from solid surfaces have been analyzed. It is shown that elementary act of subthreshold displacement of surface atoms is determined by a wide spectra of electronic excitation lifetimes. A new procedure within the traditional method of quantum-chemical modelling is proposed that allows us to take into account the dispersion of electronic excitation lifetimes on each stage of calculations. The influence of boundary conditions on cluster parameters was also considered. By virtue of a modified scheme of quantum-chemical modelling, the processes of electron-ion/atoms emission from ionic surfaces were investigated. The Pauli shock mechanism of atom displacement participates in almost all processes of EIE.

6. References

1. Elango M. and Kiv A. (1986) Radiation induced Coulomb instability in non-metalic solids, *Crystal Lattice Defects Amorphous Mater*, **11**, 305-311.
2. Ageev V., Burmistrova O. and Kuznetsov Yu. (1989) Desorption induced by electron excitations, *Uspehi Fizicheskih Nauk*, **158**, 389-416.
3. Parks C., Hussain Z., Shirly D., Knotek M., Loubriel G. and Rozenberg R. (1983) Photo-stimulated desorption Na^+, F^+ from NaF surface, *Phys. Rev. B* **28**, 4793-4797.
4. Ageev V., Burmistrova O., Magomedov A. and Yakshinskii B. (1989) Primary moving of ions by the electron-stimulated desorption, *Pisma V JTF*, **15**, №13, 10-13.
5. Ageev V., Burmistrova O., Magomedov A. and Yakshinskii B. (1990) Electron-stimulated desorption of Li^+ and Na^+ ions from W surface covered by the monoatomic Si film, *Fizika Tverdogo Tela*, **32**, №3, 801-809.
6. Ageev V., Magomedov A. and Yakshinskii B. (1991) The dependence of electron-stimulated Li^+ and Na^+ ions desorption on conditions of silicon film formation on the W, *Fizika Tverdogo Tela*, **33**, №1, 158-165.
7. Stawinski U. and Bauer E. (1993) Alkali halide layers on W(110): Electron-stimulated desorption of ions, structure and composition, *Phys. Rev. B* **47**, №19, 12820-12832.
8. Uustare T., Aarik J. and Elango M. (1994) Oxygen depletion of the crystalline (anatase) TiO_2 initiated by ionization of the K shell, *Appl. Phys. Lett.* **65**, №20, 2551-2552.
9. Kiv A., Elango M., Britavskaya E. and Zakharchenko I. (1994) Mechanisms of subthreshold atomic emission from solid surface, *NIM B* **90**, 257-260.
10. Vavilov V.S., Kiv A.E. and Niyazova O.R. (1981) *Mechanisms Of Crea-*

tion And Migration Of Defects In Semiconductors, Nauka, Moscow.

11. Colombo L. (1994) Tight-binding molecular dynamics: present status and perspectives, *Proc. of the 6th joint EPS-APS International Conference on Physics Computing, Lugano, Switzerland*, 231-238.

12. Fincham D. and Gillan M. (1994) Shell model molecular dynamics simulation of ionic materials, *Proc. of the 6th Joint EPS-APS International Conference on Physics Computing, Lugano, Switzerland*, 231-238.

13. Parilis E.S. (1969) *Auger Effect*, FAN, Tashkent.

14. Kiv A., Kovalchuk V. and Elango M. (1991) Pauli shock in solids, *Izv. AN Latvii, Ser. Fiz. I Tehn. Nauk*, **32**, №11, 1181-1187.

NEW METHOD OF COMPUTER SIMULATION OF DEFECT CONFIGURATIONS IN SEMICONDUCTORS

Z.M. KHAKIMOV AND F.T.UMAROVA
Institute of Nuclear Physics of Uzbekistan Academy of Sciences
Ulughbek, 702132, Tashkent, Uzbekistan

1. Introduction

The (semi)empirical tight-binding method (TBM) [1] and extended Hükkel technique [2] are the simplest and most convenient methods for electronic and total energy calculations. These methods, however, have a number of inherent problems.

The primary difficulty is that in these methods, electron-electron interaction (EEI) energies are supposed to be implicitly involved in Hamiltonian matrix elements. Since the electronic energy, unlike the total, involves the EEI energy twice, the contribution of the EEI energy to the total remains unknown. A number of adaptions, like the models of Chadi [3] and Harrison [4] provide simple methods for calculating configurational energies of defects and surfaces. However, these problems have not been properly resolved until recently [5], which is obviously responsible for the absence of TBM, allowing the calculation of both the total energy and electronic structure (or configurational and spectroscopic energies) in the framework of the single computational scheme[1].

To this end, a new tight-binding method was proposed [5] based on a more accurate total energy expression consisting of well-shared and self-reduced terms that minimize errors of semi-empirical calculations. The first application of this method was successful for a number of defects in Si [7, 8, 9]. In this paper we describe this method in brief (Sec. 2) and the results of

---

[1]Methods based on neglecting differential overlap approximation [6] also cannot perform that function; the errors of explicit parametrization of EEI energy can be compensated by the errors of other terms only in either the electronic or total energies.

R. C. Tennyson and A. E. Kiv (eds.),
Computer Modelling of Electronic and Atomic Processes in Solids, 203–211.
© 1997 *Kluwer Academic Publishers*.

calculations of the vacancy, self-interstitial migration, and Watkins reaction in Si, and compare them with local density functional calculations (Sec.3).

2. The self-consistent tight-binding total energy and electronic structure calculation method

The proposed total (binding) energy expression is [5]

$$
\begin{aligned}
E_{tot} = & \sum_\mu \{ \sum_{\nu > \mu} [Z_\mu^{scr}(R_{\mu\nu}) Z_\nu^{scr}(R_{\mu\nu}) / R_{\mu\nu} + Q_\mu(R_{\mu\nu}) Q_\nu(R_{\mu\nu}) / R_{\mu\nu} \\
& + \sum_{i \in \mu} \sum_{j \in \nu} P_{\mu i, \nu j} H_{\mu i, \nu j}] + (W_\mu - W_\mu^0) \}.
\end{aligned}
\tag{1}
$$

Here W_μ^0 and W_μ are total energies of atoms in non-interacting and interacting systems, respectively. P is the bond order matrix, $Z_\mu^{scr}(R_{\mu\nu})$ and $Z_\nu^{scr}(R_{\mu\nu})$ are the screened nuclear (or core) charges

$$
Z_\mu^{scr}(R_{\mu\nu}) = \sum_{i \in \mu} N_{\mu i}^0 A_\mu \exp(-D_\mu R_{\mu\nu} / R_{\mu i}^0)
\tag{2}
$$

and $Q_\mu(R_{\mu\nu})$ and $Q_\nu(R_{\mu\nu})$ are the non-point ionic charges

$$
Q_\mu(R_{\mu\nu}) = \sum_{i \in \mu} (N_{\mu i}^0 - N_{\mu i}) \eta_{\mu i}(R_{\mu\nu}),
\tag{3}
$$

where $N_{\mu i}^0$ and $N_{\mu i} = P_{\mu i, \mu i}$ are occupancy numbers of atomic orbitals (AO) in non-interacting and interacting systems, respectively. $\eta_{\mu i}(R_{\mu\nu})$ are functions characterizing a degree of "non-pointness" of ionic charges [5].

The energies and wave-functions of electrons are obtained by self-consistent solving of the TBM secular equation with the following Hamiltonian matrix elements

$$
H_{\mu i, \mu j} = -(E_{\mu i} + U_{\mu i}^{mad}) \delta_{ij},
\tag{4}
$$

$$
H_{\mu i, \nu j} = -[h_{\mu i} h_{\nu j}]^{1/2}, \; \nu \neq \mu,
\tag{5}
$$

$$
h_{\mu i} = B_\mu E_{\mu i}^0 \exp(-D_\mu R_{\mu\nu} / \bar{R}_{\mu i}^0),
\tag{6}
$$

$$
U_{\mu i}^{mad} = \sum_{\nu \neq \mu} \eta_{\mu i}(R_{\mu\nu}) Q_\nu(R_{\mu\nu}) / R_{\mu\nu},
\tag{7}
$$

where $E_{\mu i}^0$ and $E_{\mu i}$ are energies of AOs in non-interacting and interacting systems, respectively, $\bar{R}_{\mu i}^0 = (n + 1/2) / \xi_{\mu i}^0$ and $R_{\mu i}^0 = n / \xi_{\mu i}^0$ (n is the principal quantum number and $\xi_{\mu i}^0$ is the Slater exponent) are the mean and the most probable distances between electron and nucleus, respectively,

f_{ij} is the function of mutual orientation of i-th and j-th AOs [1]. A_μ, B_μ and D_μ are fitting parameters. Details of parameterization procedures and calculations based on using H-saturated clusters can be found in [5, 7, 8].

3. Application to defects in silicon

3.1. VACANCY

For the silicon vacancy, the quantitative theory of deep-level defects in semi-conductors provided one of the most interesting results. It was predicted [10] as an "Anderson effective-negative-U" system and this prediction was then confirmed by experiment [11]. In order to clarify the difference between our calculation [5, 7] and those of [10] let us consider the latter in brief. The total energy calculations in [10] were done using the functional

$$E(N,Q) = \frac{k_{++}}{2}(Q - Q_{++})^2 + N[\epsilon(Q) + \frac{N-1}{2}U(Q) + E_F], \qquad (8)$$

where N is the number of electrons on the localized t_2-level of the vacancy in the gap of Si, k_{++} is the effective force constant for the doubly positively charged state of the vacancy V^{++} (N=0), $\epsilon(Q)$ is the energy required to add an electron to the vacancy level from the top of the valence band, $U(Q)$ is the electron-electron repulsion energy, E_F is the Fermi energy, Q is the displacement of the first-neighbors of the vacancy away from the minimum energy configuration Q_{++} of the state V^{++}.

The values of ϵ and U (0.32 eV and 0.25 eV, respectively) for the un-relaxed vacancy (at $Q = 0$) were calculated [10] correctly enough by the self-consistent Green's-function technique (GFT) in the framework of the Local Density Functional (LDF) approximation and using the transition state theory of Slater [12]. However, a number of additional approximations were used as well to avoid complicated *ab initio* calculations and difficulties of the GFT in considering lattice relaxations and defect charge states. (i) The force constant k was evaluated using a modified valence-force-field model of Keating [13]. (ii) Since this model does not distinguish charge states, the obtained value ($k_b = 7.5$ eV/Å² for the breathing mode displacements and $k_E = 14.8$ eV/Å² for the tetragonal distortions) for V^0 was used for V^+ and V^{++} as well. (iii) The dependence of ϵ on Q was considered to be the same as that of the one-electron level of the vacancy. Hence, the calculated deformation potentials were much more likely un-derestimated due to a well-known underestimation (about 50 %) of band gap of semiconductors in the framework of the LDF. This disadvantage of the LDF was obviously responsible for suppression of a quadratic term in $\epsilon(Q)$ in [10]. (iv) The (111) surface relaxation and breathing distortions around the vacancy are supposedly alike. However, this analogy is not complete enough to be useful. The interaction of dangling bonds of the vacancy which are all directed to one point must be actually different from that of the (111) surface dangling bonds which are parallel each to other. Consequently, equilibrium distortions for the vacancy should differ from those for the surface (111).

Thus, the calculation performed in [10] is in need of confirmation by independent means, done without the above approximations. We have performed such calculations using the hydrogen saturated cluster $X Si_4 Si_{12} H_{36}$ (where $X = Si$ or V).

The formation energy of silicon vacancy in neutral, single and double positive charge states was calculated taking account of displacements of the first-nearest-neighbors (1NN) of the vacancy. This energy is defined as the difference of the total energies of clusters with vacancy and without it. The results for the breathing mode displacements are presented in Fig.1. The points in Fig.1 correspond to calculated values, and curves correspond to parabolas fitted to these values

$$E(N, Q) = \frac{1}{2}(k + \Delta k)(Q - Q_N)^2 + N\Delta E. \qquad (9)$$

Here $N = 0, 1, 2$, $Q_0 = Q_{++} = -0.11$Å (negative sign corresponds to inward displacements), $Q_1 = Q_+ = 0.12$Å , $Q_2 = Q_o = 0.24$Å , $k = 7.33$ eV/Å^2, $\Delta k \simeq k/2$, $\Delta E = 0.21$ eV. Our calculations thus show strong dependence of the force constant on the charge state of the vacancy. Moreover, the charge dependence obtained of Q qualitatively differs from the estimations made in [10] using the above mentioned "analogy": $Q_{++} = 0.23$Å , $Q_+ = 0.17$Å and $Q_o = 0.10$Å .

Comparing (9) with (8) we get immediately

$$\epsilon(Q) \quad = \quad 0.25 - 2.10Q + 1.83Q^2, \qquad (10)$$
$$U(Q) \quad = \quad 0.31 - 0.13Q. \qquad (11)$$

The obtained values $\epsilon = 0.28$ eV (or 0.25 eV from (10)) and $U = 0.26$ eV (or 0.31 eV from (11)) for the unrelaxed vacancy ($Q = 0$) are in good agreement with those of [10]: $\epsilon = 0.32$ eV ($\epsilon = 0.24$ eV [14]) and $U = 0.25$ eV. However, we found a quadratic dependence in $\epsilon(Q)$ and a linear dependence in $U(Q)$ from our direct total energy calculations.

The regions of stability of V^0, V^+ and V^{++} before and after relaxation are shown in Fig.2. Before the relaxation, the state V^+ has a wide enough region of stability which is reduced to only a point $E_F = 0.21$ eV after relaxation. Activation energies are $E(+/++) = E(0/+) = 0.21$ eV. The parameter $\eta = (relaxation\ energy) - U$, positive values of which indicate negative-U character of vacancy, is $\eta \approx 0$. The corresponding best results

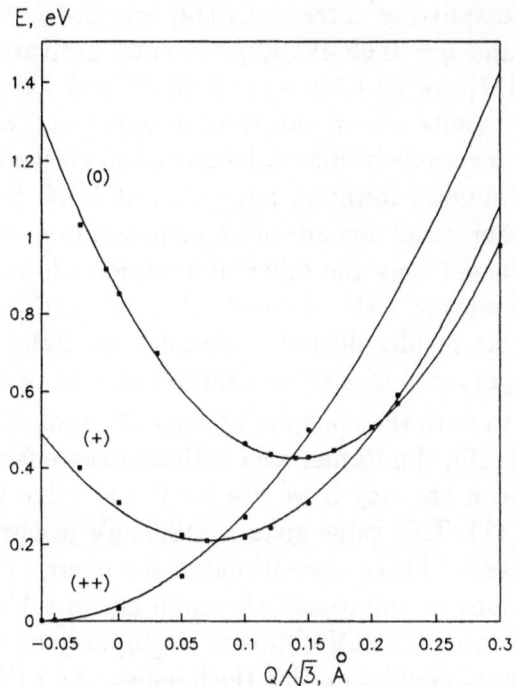

Figure 1. Configuration coordinate diagram for the energy of the vacancy in Si. Q is breathing mode displacement of the first-nearest-neighbors of the vacancy. Zero of Q corresponds to the unrelaxed vacancy; negative and positive signs of Q mean inward and outward displacements, respectively; zero of energy corresponds to the minimum energy of V^{++}, $E_{min} = 7.92eV$.

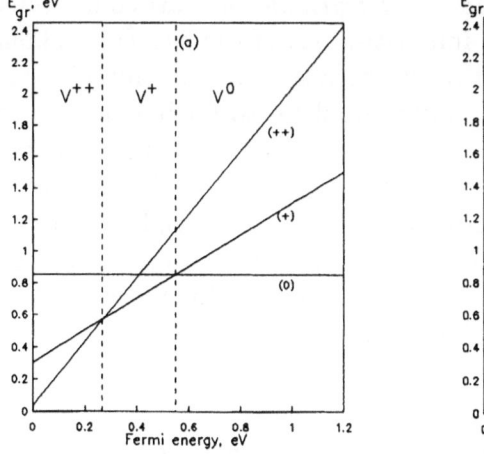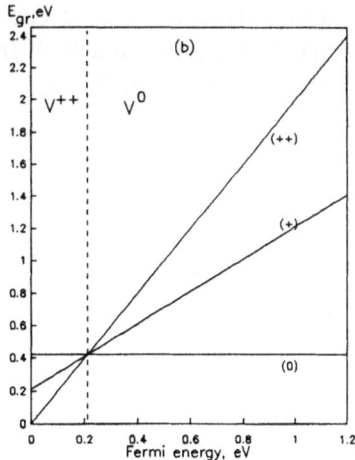

Figure 2. Groundstate energy $E_{gr} = E_{min}(N) - (N-2)E_F$ of the vacancy as a function of Fermi-level position in the band gap of Si for different charge states of the vacancy before (a) and after (b) the symmetric relaxation of its first-nearest-neighbors.

of [10] with multiple splitting corrections [14] are $E(+/++) = 0.07$ eV, $E(0/+) = 0.05$ eV and $\eta = 0.05$ eV. Experimental activation energies obtained by Watkins [11] are $E(+/++) = 0.13$ eV and $E(0/+) = 0.06$ eV. It is clear that our results are in substantial agreement with the experimental ones if we allow a relaxation of the second-nearest-neighbors (2NN) that facilitates a symmetry-lowering relaxation of 1NN. Since the cluster $VSi_4Si_{12}H_{36}$ is rather small for adequate consideration of relaxations of 2NN, we have considered only the different symmetry-lowering distortions (including pairing mode) of 1NN. However, these distortions were found to be unfavorable. These results do not contradict the Jahn-Teller theorem but indicate the importance of 2NN-relaxations in this case.

It is interesting to note that in spite of some discrepancies between our results and those of [10], the former also indicates the effective-negative-U character of the silicon vacancy if we use for U the value 0.25 eV instead of ~ 0.30 eV from (11). This value gives $\eta = 0.05$ eV in our case as well. It is reasonable because we likely overestimated the energy U, confining the vacancy wave functions in comparatively small clusters $VSi_4Si_{12}H_{36}$. On the other hand, the value 0.25 eV obtained in [10] may be rather less than the actual one because underestimating the band gap by LDF will in general result in underestimation of the degree of the localization of vacancy wave functions, consequently decreasing the electron-electron repulsive energy.

3.2. SELF-INTERSTITIAL

Self-interstitial atoms in Si are known to be highly mobile at low temperatures under irradiation and can easily reach various lattice imperfections, react with them and even kick impurities out of the site [15]. Though the mechanism of high mobility of the self-interstitial at extremely low temperatures (4.2 K [15]) is a long-standing problem, so far it is not completely understood. A detailed quantitative picture of the Watkins replacement reaction, to the best of our knowledge, is also absent. In this section the results of our studies of the self-interstitial migration and its reaction with substitutional Al are considered in brief. Calculations were performed using the same cluster as in the vacancy case . More detailed descriptions and discussion of these calculations will be presented elsewhere.

3.2.1. *Migration*

Rapid migration of the self-interstitial at 4.2 K requires either extremely low barriers to migration, or an electron-assisted transport, for instance, by alternating capture and emission of electrons [16, 17]. The latter case becomes possible if only the self-interstitial has (i) more than one charge state and (ii) the minimum and maximum energy positions of one charge

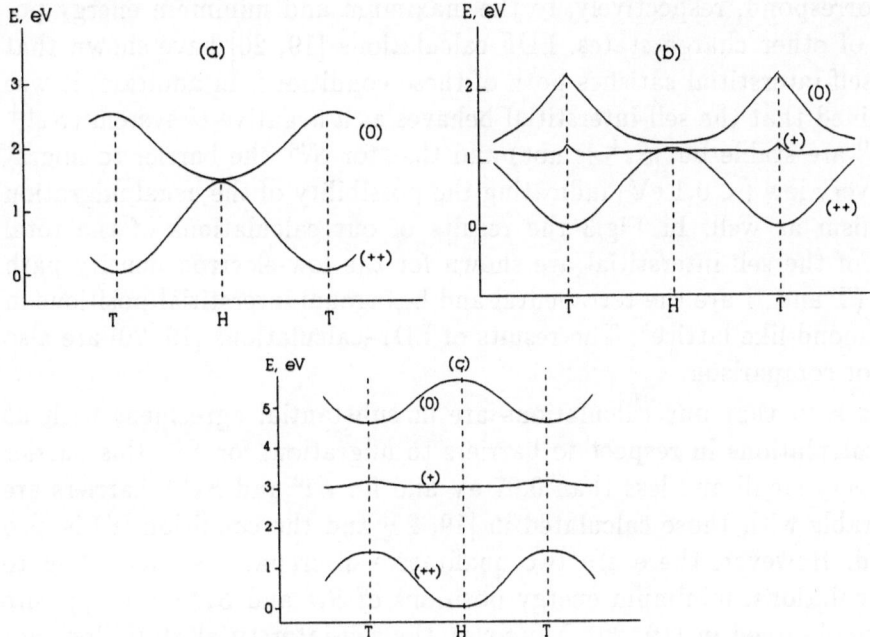

Figure 3. Configuration coordinate diagram for the total energy variation of the self-interstitial migrating along the $T - H - T$ channel; (a) and (b) correspond to the works [19] and [20], respectively, and (c) – to this work.

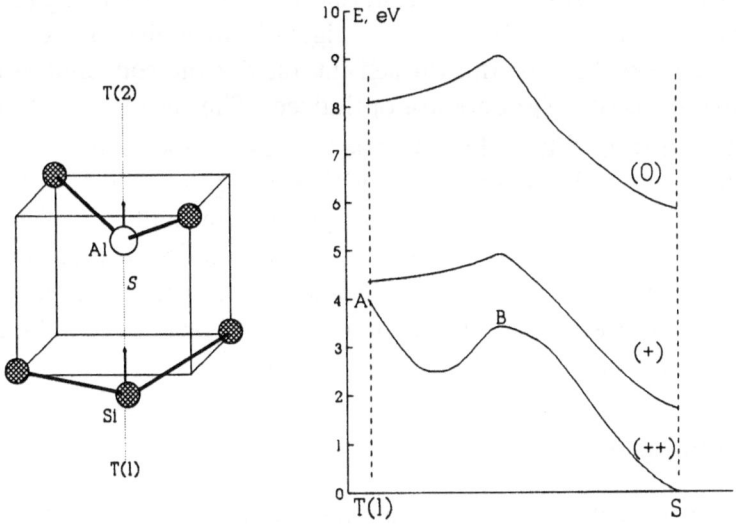

Figure 4. Configuration coordinate diagram for the total energy variation of the self-interstitial approaching to the substitional Al. S denotes the site occupied by Al.

state correspond, respectively, to the maximum and minimum energy positions of other charge states. LDF-calculations [19, 20] have shown that the Si self-interstitial satisfies both of these conditions. In addition, it was determined that the self-interstitial behaves as a negative-U system (Si^{++} and Si^0 are stable but Si^+ is not) and that for Si^+ the barrier to migration is very low (< 0.1 eV) indicating the possibility of the usual migration mechanism as well. In Fig.3 the results of our calculations of the total energy of the self-interstitial are shown for the low-electron density path T-H-T (T and H are the tetrahedral and hexagonal interstitial positions in the diamond-like lattice). The results of LDF-calculations [19, 20] are also given for comparison.

It is seen that our calculations are in substantial agreement with *ab initio* calculations in respect to barriers to migration: for Si^+ this barrier is also very small and less than 0.07 eV and for Si^0 and Si^{++} barriers are comparable with those calculated in [19, 20] and the condition (ii) is also satisfied. However, there are two qualitative discrepancies. According to our calculations, minimum energy positions of Si^0 and Si^{++} are opposite to those obtained in [19, 20]. Moreover, the self-interstitial atom does not exhibit a negative-U property and the only stable state is Si^{++}, which is, however, in agreement with another LDF-calculation [18].

3.2.2. *Watkins replacement reaction*

Watkins replacement reaction was calculated as follows. The self-interstitial atom was moved step by step from the tetrahedral site along (100)-direction toward the lattice site which was originally occupied by Al. For each distance between this site and the self-interstitial the coordinates of Al and its four first-nearest-neighbors are optimized. The results are shown in Fig.4.

As seen from Fig.4, the activation energy of ejection of Al from the site is lowered with the increase of positive charge of the $Al + Si$ system and for $(Al + Si)^{++}$ the ejection takes place athermally. In this case Si and Al can also form the metastable state (see the local minimum between A and B) at the distance of 2.2 Å between them. But in all cases Al and the self-interstitial atom can not remain together inside the volume formed by the four Si atoms around the site.

4. Conclusion

We have presented the results of the application of the new tight-binding total energy and electronic structure calculation method [5] to the silicon vacancy, the Si self-interstitial migration, and the Watkins replacement reaction. The results in whole are in good agreement with the LDF-calculations and available experimental data (see also [8, 9] and our related paper in this volume). They demonstrate this method [5] to be the most promising of semi-empirical methods developed so far with respect to its concurrent ability with *ab initio* methods in calculating defect properties of semiconductors.

References

1. Slater J.C. and Koster G.F. (1954) Simplified LCAO method for the periodic potential problem, *Phys.Rev.*, **Vol. no. 94**, pp. 1498-1524.
2. Hoffman R. (1963) An extended Hükkel theory, *J.Chem.Phys.*, **Vol. no 39**, pp. 1397-1412.
3. Chadi D.J. (1979) Surface atomic structure of covalent and ionic semiconductors, *Phys.Rev.B.*, **Vol. no. 19**, pp. 2074–2091.
4. Harrison W.A. (1981) Total energies in the tight-binding theory, *Phys.Rev.B.*, **Vol. no. 23**, pp. 5230–5245 .
5. Khakimov Z.M. (1994) A new semiempirical electronic structure and total energy calculation method for solids and large molecules, *Comput.Mater.Sci.*, **Vol. no. 3**, pp. 95–108.
6. Segal G.A.(ed.) (1977) *Semiempirical methods of electronic structure calculations*. Plenum Press, New York.
7. Khakimov Z.M., Mukhtarov A.P., and Levin A.A. (1994) About negative-U properties of the silicon vacancy, *Fizika i tekhnika poluprovodnikov*, **Vol. no. 28**, pp. 571-576.
8. Khakimov Z.M., Mukhtarov A.P., Umarova F.T., and Levin A.A. (1994) Equilibrium positions and migration paths of hydrogen in crystalline silicon, *Fizika i tekhnika poluprovodnikov*, **Vol. no. 28**, pp. 1727-1734.
9. Mukhtarov A.P., Pulatova D.S., Sulaymonov N.T., and Khakimov Z.M. (1994) About electrical levels of sulphur in silicon, *Fizika i tekhnika poluprovodnikov*, **Vol. no. 28**, pp. 1015-1019.
10. Baraff G.A., Kane E.O., and Schlüter M. (1980) Theory of silicon vacancy: an Anderson negative-U system, *Phys.Rev.B.*, **Vol. no. 21**, pp. 5662-5686.
11. Watkins G.D. and Troxell J.R. (1980) Negative-U properties for point defects in silicon, *Phys.Rev.Lett.*, **Vol. no. 44**, pp. 593-596.
12. Slater J.C. (1974) *The self-consistent field for molecules and solids*. McGraw-Hill, New-York.
13. Keating P.N. (1966) Effect of invariance requirements on the elastic strain energy of crystals with application to the diamond structure, *Phys.Rev.* **Vol. no. 145**, pp. 637-645.
14. Lannoo M., Baraff G.A., and Schlüter M. (1981) Multiplet splitting and Jahn-Teller energies for the vacancy in silicon, *Phys.Rev.B.*, **Vol. no. 24**, pp. 955-963.
15. Watkins G.D. (1967) Defects in irradiated silicon: EPR and ENDOR of the aluminium and vacancy pair, *Phys.Rev.*, **Vol. no. 155**, pp.802-805.
16. Oksengendler B.L. (1971) To the theory radiation-enhanced diffusion, In: *Method of radiation influence in study of structure and properties of solids*. Fan, Tashkent. pp. 16-32.
17. Bourgoin J. and Corbett J. (1972) A new mechanism for interstitial migration, *J.Phys.Lett.*, **Vol. no. 38A**, pp. 135.
18. Baraff G.A., Schlüter M., and Allan G. (1983) Theory of enhanced migration of interstitial aluminum in silicon, *Phys.Rev.Lett.*, **Vol. no. 50**, pp. 739-742.
19. Bar-Yam Y. and Joannopoulos J.D. (1984) Barrier to migration of the silicon self-interstitial, *Phys.Rev.Lett.*, **Vol. no. 52**, pp. 1129-1132.
20. Car R., Kelly P.J., Oshiyama A., and Pantelides S.T. (1984) Microscopic theory of atomic diffusion mechanisms in silicon, *Phys.Rev.Lett.*, **Vol. no. 52**, pp. 1814-1817.

COMPUTER MODELLING OF DIELECTRIC PROPERTIES OF COMPOSITE MATERIALS

V. V. NOVIKOV, O. P. POZNANSKY
Odessa Polytechnik University, Odessa, Ukraine

1. Introduction

Many publications have been devoted to the problem of the analytic definition of the properties of composite materials [1 - 5]. The complexity of the theoretical definition of physical properties of the stochastic microheterogeneous materials (MHM) is connected with the correct description of the structure for which knowledge of the distribution function is demanded [2, 5]. In particular, the main problem is to describe the properties of MHM near the geometrical phase transition, where unconnected areas of one of the phases form the percolation cluster. This problem had not been solved in terms of a classical physical approach. Only the percolating theory gives the right description for systems where the ratio of properties $a = c_1/c_2$ approached zero [6]. The next step in the description of property-structure dependencies is to define scaling properties of systems [7, 8]. One of the prospective methods for describing structure and properties of MHM is computer modelling and using developed renorm group approaches which are used in the theory of phase transitions [7, 8].

2. The model of MHM structure

The modelling of a stochastic structure is carried out on lattices with randomly-distributed parameters [9]. The sites of such lattices define inhomogeneities (the phases of the system) in materials, and bonds between sites provide contacts with neighbors. The contact conditions affect the main macroproperties. That is why we will be considering the bond's problem.

The basic set of bonds M_n of a 2D lattice is analyzed with an iteration process. On the first step ($n = 0$), we consider the probability p_0 of black bonds and $(1-p_0)$ of white bonds. The bonds of the same color have the same propertes. On the next step ($n = 1, 2, ...m$) all bonds of the lattice are substituted with a lattice which is described in the previous step. The iteration process ends when the properties of the lattice do not depend on the iteration number n. It is possible to analyze a lattice of linear size L_n longer than the correlation length, and to form a self-similar fractal set [11, 12].

Let both black and white phases be randomly distributed at the square lattice LxL, according to the following partition density function:

$$P_0 (c) = p_0\delta (c - c_1) + (1 - p_0)\, \delta (c - c_2) \qquad (1)$$

213

R. C. Tennyson and A. E. Kiv (eds.),
Computer Modelling of Electronic and Atomic Processes in Solids, 213–218.
© 1997 *Kluwer Academic Publishers.*

where c_1 is the color (property) of the bond, p_0 is the probability of appearance of such color on this bond, c_2 is the other color (property) of the bond, $(1 - p_0)$ is the probability of appearance of such color on this bond,

$$\delta (t) = \begin{cases} 1, \text{ if } t = 0, \\ 0, \text{ if } t \neq 0. \end{cases} \tag{2}$$

This function (keeping p_0 = const) may be realized by different ways with many possible bond configurations on a given lattice. The new partition density function $P_n (c)$ arises on every iteration step. It will connect opposite sides of lattice configurations of black bonds among all possible ones. The connecting set (CS) $M_n(l_0 p_0)$ is characterized by a geometrical correlation length and fractal dimension d_f which depends on l_0 and p_0. At $p > p_c$ and $L_n > \xi$, CS becomes homogeneous with constant density, where p_c is the percolation threshold [9].

Let the probability of appearance of CS be $R(l_0 p_0)$ which is defined as the ratio of the number CS to the number of all possible configurations (keeping p_0 const and l_0 const). The disordered nature of CS is preserved in the scale interval $1 \ll \xi \ll L$, where l and L are the smallest and largest size of CS, respectively. The evolution of the probability of appearance of CS after n steps of renormalization inside these intervals may be expressed as

$$p_n = R (p_{n-1}) \tag{3}$$

It should be clear that the iteration process is completed in the fixed points 0 or 1:

$$P_{00} = \begin{cases} 1, \text{ if } p_0 > p_c \\ 0, \text{ if } p_0 < p_c \end{cases} \tag{4}$$

An unsteady fixed point is defined from the curve of $R(l_0, p_0)$ as

$$p_c = R(p_c) \tag{5}$$

The fractal dimension d_f is caclulated at this point as

$$d_f = \ln(M_n(p_0, l_0)) / \ln(L_n) \tag{6}$$

See Table 1.

Table 1

Percolation thresholds p_c and fractal dimensions d_f for different original cells LxL

LxL	p_c	d_f
3 x 3	0.4267	1.623
6 x 6	0.4766	1.790
9 x 9	0.4883	1.843

3. Dielectric breakdown

The relevant model is the Z_d ($d = 2$) lattice with the S-structure as a conducting phase, and its supplement $Z_d \setminus S$ as a dielectric phase [10]. It is assumed that the S-structure grows by one unit in ech dielectric breakdown (DB) step until the opposite electrode is touched at the onset of the bonded configuration. Discretizing the Laplace equation, one obtains

$$\sum_{ij} \alpha_j \, (\phi - \phi_j) = 0 \qquad (7)$$

where (i, j) are the nearest neighbors in $Z_d \setminus S$ and S, respectively.

Before DB, the conductivity σ is assumed to be distributed over the square lattice according to the partition function (1), where p_0 is the concentration of resistors of conductivity σ_m and $(1 - p_0)$ is the concentration of the dielectric of conductivity σ_d.

The elementary acts of transition from one phase into another will be simulated by irreversible changes of conductivity. Assume that a constant voltage $V < V_c$ is applied to a resistor before DB (where V_c is a certain "critical" voltage); the DB at $V > V_c$ is accompanied by a conductivity increase from σ_d to σ_c, and remains thereafter unchanged at all values of V.

The system of linear equations (7) may be solved numerically using the following algorithms:

    i. calculate the conductivity σ with partition function (1) for each bond of the square lattice so as to ensure the onset of a disconnecting cluster between opposite electrodes;

    ii. apply the boundary condition U to opposite electrodes;

    iii. solve eq. (7);

    iv. calculate the conductivity of a system before DB;

    v. check every resistor (excluding those in which DB has already occurred) and replace σ_d by σ_c in the condition ensuring DB, $V > V_c$;

vi. stop the process if the electrodes are connected by a cluster of "damaged particles," otherwise, relax the boundary condition so as to ensure the DB condition at one of the resistors is available and return to step iii.

Let the limiting voltage corresponding to the onset of the first cluster of "damaged particles" be defined as the dielectric breakdown voltage of a given unit cell; the mean will be obtained by averaging over different configurations. The effective dielectric breakdown voltage is estimated in each step of the renorm group transformation. In the next step, both the effective conductivity and dielectric breakdown voltage V_c are assigned to each bond and the process of DB voltage renormalization with the number of iteration steps is continued until V_c eventually becomes independent of n.

As can be seen from the mathematics of this procedure, the DB voltage in the vicinity of the percolation threshold obeys the following asymptotic law

$$V_c = AL^{-b} \tag{8}$$

where b is the corresponding critical index. The values of the cited critical indices for different lattices LxL are listed in Table 2.

Table 2
Critical indices of scaling behaviour of dielectric breakdown voltage for different original cells LXL

L x L	b
3 x 3	1.19
6 x 6	1.08
9 x 9	1.03

4. Scaling behaviour of dielectric properties in metal-insulator composites

Composite materials have long been known to have electrical properties very different from those of their constituents. The differences are particularly marked near a percolating threshold, i.e., a point at which one of the two components of the composite first forms a closed connected path extending throughout the sample.

We summarize the scaling formalism for a composite of two metallic and dielectric components at low frequencies. The metallic component is assigned a real impedance, $Z_m = R$, characteristic of a pure resistor. The dielectric, on the other hand, is given a purely capacitive impedance, $Z_d = 1/(iwC)$, with C real. The impedances Z_m and Z_d are placed at random on the bonds of a two-dimensional rectangular network. We carry out the random placement according to the rules of bond percolation. That is, we denote the bonds on the network as "conducting" or "insulating" with probability p and 1 - p, respectively.

To use the scaling form at finite frequencies, we must consider the effective dielectric function of the composite. Hence, we must consider the effective "dielectric functions" of bonds, defined by

$$\varepsilon_m = 1 \, (Z_m iw) = 1/(iwR) \tag{9}$$

$$\varepsilon_d = 1/(Z_d iw) = C \tag{10}$$

In principle, a capacitance should also be added in parallel with the resistor across each metallic bond. But at sufficiently low frequencies, such a capacitance will have little effect, and we do not include it here.

We calculate the effective dielectric function of the network, denoted $\varepsilon_{ef}(w)$, by applying an AC voltage difference across a particular realization of the random network and solving Kirchhoff's equations to obtain the electric displacement across each bond. The complex displacement across bond (ij), denoted Dij(w), is related to the complex potential difference $V_i(w) - V_j(w)$ by

$$Dij(w) = \varepsilon_{ij}(w)[V_i(w) - V_j(w)] \tag{11}$$

where $\varepsilon_{ij} = \varepsilon_m$ or ε_d according to whether the bond (ij) is considered metallic or dielectric. The condition that no displacement current shall build up at the i-th node is equivalent to the equation

$$\sum_{ij} \varepsilon_{ij} (V_i - V_j) = 0, \tag{12}$$

where the sum runs over all sites j with a nonzero bond connected to site i. The total displacement current across the network is that of an equivalent uniform network in which each bond has a dielectric function $\varepsilon_{ef}(w)$. Eq. (12) is solved directly to compute the voltage at each site. The potential is fixed a V = 0 on the right side of the cell and $V = V_0$ on the left side of the cell. Many realizations of random impedance are required to calculate the average of the dielectric function $\varepsilon_{ef}(w)$.

The dielectric function $\varepsilon_{ef}(w)$ of the bonded and unbonded renormalization cluster is estimated on each n-th step. The step-by-step averaging for the dielectric function is carried out making use of Eq.(1) and Eq.(3). The dielectric function scales with L as

$$\varepsilon_{ef} = BL^{f} \tag{13}$$

where f is the corresponding exponent. The critical dielectric index for different original cells is listed in Table 3. They are in good agreement with that reported in [11].

Table 3
Critical indices of scaling behaviour of dielectric function
for different original cells LxL

L x L	f
3 x 3	1.28
6 x 6	1.04
9 x 9	0.94

5.   Conclusions

We have considered two problems of dielectric properties of metal-insulator composites. First, we have described a dielectric breakdown scaling relation using the renormalization group approach. Application of the real space renormalization group transformation technique to the problem of dielectric breakdown of percolating systems permitted estimations of the corresponding critical indices. Second, we have evaluated low-frequency scaling relations for dielectric functions. The renormalization group approach for evaluating dielectric functions gives good agreement with experimental literature data.

References

[1]   Dulnev G., Novikov V. (1991) *Transport process in heterogeneous media*, Energoatomizdat, Leningrad
[2]   Shermergor T. (1977) *Theory elasticity of heterogeneous media*, Nauka, Moscow
[3]   Cristensen R. (1982) *Introduction in composite mechanics*, Mir, Moscow
[4]   Privalko V., Novikov V., Yanovsky Yu. (1991) *Foundation of thermal physics of polymer materials*, Naukova Dumka, Kiyv
[5]   Hashin Z. (1983) Analysis of composite materials, *J. Appl. Mech.*, v.50, N2, p. 481-505
[6]   Shklovsky B., Efros A. (1976) *Electronic properties of semiconductors*, Nauka, Moscow
[7]   Stauffer D. (1985) *Introduction to percolation theory*, London Philadelphia
[8]   Mandelbrot B. (1982) *Fractal Geometry of Nature*, Freeman, San Francisco
[9]   Novikov V., Belov V. (1994) Reversible renormgroup transition in the problem of bond's percolation, *J. of theor. and exper. phys.*, v. 106, p. 780-789
[10]   Koss R., Stroud D. (1987) Scaling behavior and surface-plasmon modes in metal-insulator composites, *Phys. Rev.*, v. 35, n. 17, p. 9004 - 9013
[11]   Niemeyer L., Pietroniero L., Wiesmann H. (1984) The model of dielectric breakdown, *Phys Rev. Lett.*, v. 52, p. 1033 - 1038

SIMULATION OF RECOMBINATION PROCESSES IN POROUS SILICON

YA.O. ROIZIN, V.A. VOROBYEVA, A.B. KORLYAKOV,
Odessa State University, 270100 Odessa, Ukraine
E.RYSIAKIEWICZ-PASEK
Wroclaw Technical University, Wroclaw 50-370, Poland

Abstract

We have performed a direct Monte-Carlo simulation of radiative recombination processes in porous silicon (PS) that accounts for the energy transfer in the system of quantum domains. First we simulate the spatial structure of PS and then analyze spatial migration of nonequilibrium electrons and holes created by a UV laser prior to their radiative recombination. Both direct tunneling between quantum domains and hopping via localized states are considered in the corresponding model. The results obtained are in good agreement with time-resolved measurements of PS photoluminescence (PL) on different time scales.

1. Introduction

The luminescence of porous silicon fabricated by electrochemical anodization of crystalline silicon, and by a number of other techniques, has been extensively studied in the recent years. Nevertheless, to this date our understanding of light emission from PS is still incomplete. Two principal models that explain PL are generally recognized. The PL was attributed by Canham [1] to quantum confinement of exitons in Si clusters of nanometer sizes. Further research performed by different authors confirmed the quantum-confinement idea [2-5]. On the other hand, with a large surface-to-volume ratio in a highly porous structure, chemical compounds at the surfaces of silicon nanocrystallites can be responsible for observed luminescence [6-8]. Various kinds of studies have been made including infrared and Raman spectroscopy, electron and tunnel microscopy, various luminescence measurements, etc. [9-12] to distinguish between the quantum confinement and surface models in PS samples. The main argument for the quantum-confinement model is the dependence of PL spectra on the sizes of silicon clusters in PS. In particular, it is possible to explain the blue shift of PL in partialy oxidized specimens by the decrease of Si cluster sizes. The main arguments for the surface model are changes of PL efficiency after various treatments of PS in different mild chemicals. There has been much controversy over most aspects of both models and many authors introduced further models to explain one or another detail of the experimental behavior of PL. These models incorporate both nanostructures for quantum confinement and luminescent centers at the surfaces or between the silicon clusters ("hybrid" model) [13,14]. Luminescent centers, as in most surface models, are attributed to compounds of silicon with oxygen and hydrogen, amorphous silicon phase or chemisorbed water [15]. According to the "hybrid" model, the absorbed in PS radiation

219

R. C. Tennyson and A. E. Kiv (eds.),
Computer Modelling of Electronic and Atomic Processes in Solids, 219–230.
© 1997 *Kluwer Academic Publishers.*

creates electron-hole pairs in Si clusters but the nonequilibrium charge carriers can diffuse out and recombine through luminescent centers. All models noted have serious difficulties   explaining the wide range of time constants in radiative recombination processes in PS [9,16]. These time constants range from picoseconds to seconds. The typical bell-shaped form of PL spectra of porous silicon fabricated in various experimental conditions, and changes that they undergo after oxidation, are also puzzling. To explain spectral dependencies, most authors utilize peaked size distributions of quantum clusters, though these distributions are not always consistent with microstructure observations. Moreover, several PL mechanisms are generally taken into account to explain emission properties in the wide spectral range and for different lifetimes [17].

The principal question that remains open in all of the discussed models is whether the recombination process is geminate (electron recombines with the hole created in the same cluster) or electrons and holes created by excitation light intermix before recombination [18]. In the latter case, the dispersive transfer of charge carriers can take place before the radiative recombination of distant pairs in a manner similar to disordered semiconductors [18,19].

In this paper, we propose a new approach to the problem of PL in porous silicon. Our time-resolved PL measurements in a wide temperature range, as well as spectral and polarization measurements for both fresh and oxidized PS, strongly imply that energy transfer associated with the migration of electrons and holes between quantum-size clusters must be taken into account [16]. The energies of band edges of these clusters depend on their sizes, thus forming a staircase of energy states for downward hopping. To verify these qualitative ideas we have performed a direct Monte-Carlo simulation of PS structure and a subsequent simulation of distant-pair radiative recombination that takes place after the end of the excitation light pulse. The results obtained provide an internally coherent picture of PL in porous silicon. In particular, the results of our calculations are shown to be in good agreement with experimental recombination kinetics and light emission spectra. There is a strong evidence that energy migration processes dominate the recombination kinetics even at temperatures as low as 15 K.

2. Simulation of recombination processes in porous silicon

2.1. SPATIAL AND BAND STRUCTURE OF RADIATIVE PS.

We visualize luminescent PS as a gel-like material on the walls of micrometer-sized cavities. This was directly observed in our SEM investigations and was also in agreement with the results of other authors [20]. The X-ray microprobe analysis shows that even our initial (nonoxidized) PS incorporates a large amount of oxygen (up to 30 molecular percent). The luminescent PS thus consists of quantum-sized silicon crystallites surrounded by voids and SiO_x:H fragments. Porosity measurements (weighing of free-standing films) have shown that the integral porosity of our specimens exceeds 70-80 %. We assign this value also to the luminescent gel-like region of PS. This was justified by measurements of fractal dimensions in this material that were about 2.8 in the range of scale from 10^{-2} to 10 μm.

Porous silicon can be considered as a typical disordered system. Inhomogeneous potential relief is inherent to PS due to variations in the sizes of silicon domains that result in variations of effective band-gap energies E_r, where E_r is the band-gap of clusters having radius r. The shapes of quantum-size crystallites are rather complicated. A reasonable approximation that enables calculations, is that silicon domains have a spherical form. Spheres can contact or intersect to a certain extent, forming clusters consisting of these spheres. The radiative recombination in our model is supposed to occur in quantum-size domains. Their size is directly connected with the emission energy. Following [21], an implicit relation between the emission wavelength and the radius of the sphere r can be written:

$$\lambda = \frac{2\pi\hbar c}{E_{Si} + \Delta E_1 + \Delta E_2}, \qquad \Delta E_{1,2} = \frac{\hbar^2 k^2}{2m_{n,p}} \qquad (1)$$

where k is the solution of the equation

$$\sin(kr) = \pm kr\sqrt{\frac{\hbar^2}{2mr^2 U_0}}, \quad ctg(kr) < 0, \quad \text{(level with l=0),} \qquad (2)$$

m_n and m_p - effective masses of electrons and holes confined in silicon crystallite, respectively;

U_0 - potential barrier at the surface of the cluster (about 4-5 eV). The results of calculations are in agreement with $\lambda(r)$ dependencies obtained by other authors by cluster calculations of hydrogen-terminated crystallites [22].

We do not make certain assumptions concerning the structure of the media interconnecting Si clusters. We argue that it is a kind of complicated nonstoichiometric silicon-oxygen-hydrogen porous compound incorporating a large number of traps both for electrons and holes. These centers can obviously act as sites through which hopping of charge carriers can occur. Of course, the change of intercluster media structure and content results in variations in position of the PS mobility edges. In particular, charge transport between Si crystallites can be suppressed if the intercluster media is substituted for high-quality silicon dioxide.

We have performed a Monte-Carlo simulation of PS structure in the following way. A set of spheres was taken with the radii ranging from 10 to 15 Å with Gaussian size distribution $(\bar{r} < r_{min})$. The parameters of this distribution were the only fitting parameters in our simulation of the PS structure. Spheres of different sizes were randomly located in space to obtain the experimental value of PS porosity (~80 %). The side of the cube intended for realization of the radiation recombination scenario was about 500 Å. 150 various spatial configurations were generated to simulate the recombination processes. The simulation results were averaged over these configurations. A typical cross-section of the model PS structure is shown in Fig.1 (solid lines).

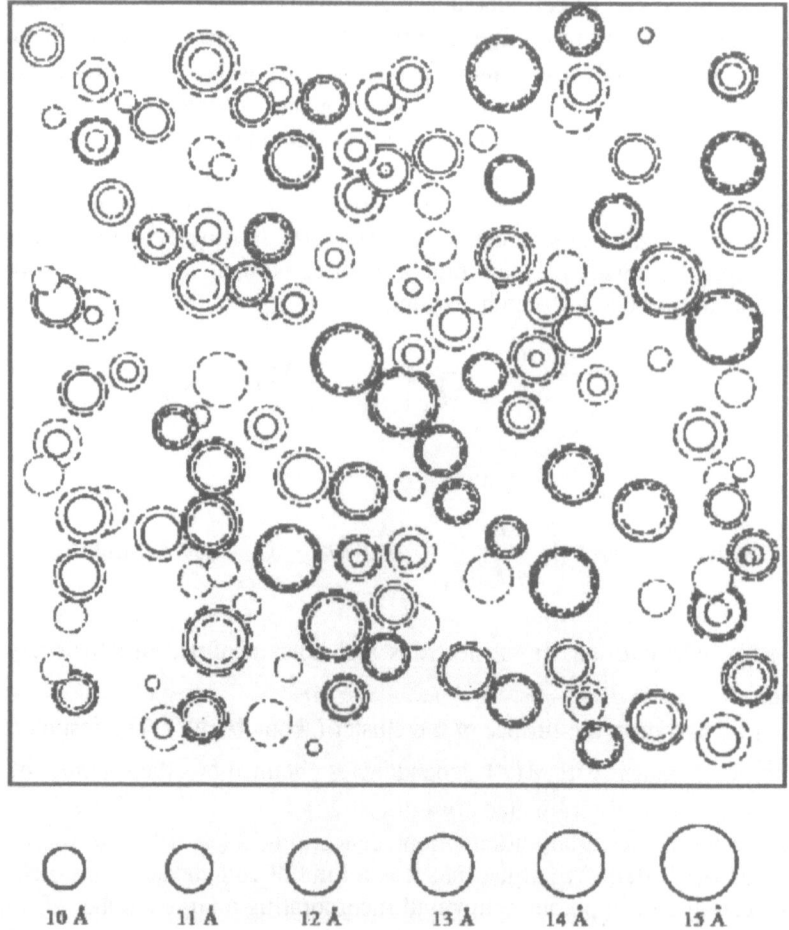

Fig.1. Cross-section of the model spatial structure

The dashed and dash-dotted lines show the cross-sections by planes parallel to the plane of the main cross-section at distances equal to the 5 Å. Fig.2 shows a schematic band diagram of the studied system. The regions of Si domains are separated by energy barriers consisting of microporous SiO_2 (nonstoichiometric and containing large amounts of hydrogen). Electrons and holes can tunnel through these barriers prior to recombination. The typical tunnel length for electrons $\alpha = \dfrac{\hbar}{(2m\Phi)^{1/2}}$ is of the order of 1Å ($\Phi = \Phi_0 - \Delta E$, $\Phi_0 \cong 3.1$ eV for the barrier at the crystalline Si-SiO_2 interface). It is clear that there may be some additional charge transfer paths between silicon crystallites. A probable candidate is the hopping path via localized states in strongly-disordered intercrystallite media. Parameters of this path should be extremely sensitive to preparation conditions, subsequent treatments, handling and aging.

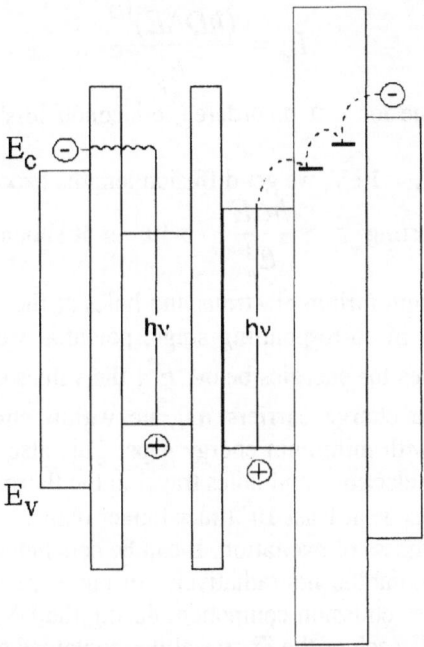

Fig.2. Schematic energy diagram of PS consisting of quantum crystallites that have various band-gaps. Arrows show hops of electrons and holes between crystallites prior to recombination. (solid lines - tunneling, dashed lines - hopping via localized states)

2.2. RECOMBINATION KINETICS

Little is known about the excited charge carrier thermalization process in porous silicon. In the experiments, a large density of electron-hole pairs was generated by a nanosecond laser pulse (10^{20} - 10^{21} cm^{-3}). The energy of excitation quanta in our experiments [16] was about E_e = 3.7 eV (nitrogen pulsed laser). We consider that the initial stage of the thermalization process is similar to that commonly accepted in amorphous silicon [18] . The charge carriers thermalize rapidly to the band edges of individual crystallites in PS. In this process, they can diffuse some distance apart which will depend upon the excitation quanta energy, the energy gap of the crystallite where they are excited, and the values of PS mobility edges (E_{me} for electrons). The mean energy that is dissipated before the carrier is at the mobility edge E^* of PS is given by $\Delta E \cong h\nu - \dfrac{E_{gi}}{2} - E^*$, where $h\nu$ is the energy of excitation quanta, E_{gi}- the bandgap of $r = r_i$ crystallites. We argue that the most efficient channel of energy relaxation is excitation of local phonons at the surfaces of silicon nanocrystallites. The characteristic phonon energy E_p in this case is about 0.1 eV. A simple estimate based on the indeterminacy relation gives for the minimum diffusion length L_d:

$$L_d = \frac{(\hbar D \Delta E)^{1/2}}{E_p} \qquad (3)$$

Assuming a typical value for a disordered semiconductors' diffusion coefficient of

$D \cong 1$ cm^2 /s and for $\Delta E = 1$ eV, we get diffusion lengths exceeding 25 Å in the rapid

thermalization process lasting $\tau_t > \frac{\hbar \Delta E}{E_p^2} \sim 5 \cdot 10^{-14}$ s. It should be noted that we pre-

sume supposed that nonequilibrium electrons and holes at the mobility edge lose their remaining excess energy in corresponding single potential wells (crystallites). If the diffusion process continues for energies below E^*, the values of L_d should be further increased. Moreover, the charge carriers migrate within clusters of Si crystallites moving to the domains with minimum energy gaps. This also leads to the significant increase of distances that electrons and holes travel in the thermalization process.

The excitation pulse is at least 10^4 times longer than τ_t. Thus, recombination is already efficient in the process of excitation. It can be concluded that at this stage most of electron-hole pairs recombine nonradiatively. In our experiments, we did not observe an additional intense emission component during the UV laser pulse, though for employed excitation levels each of the Si crystallites contained at least several electron-hole pairs. If recombination was radiative, this should lead to the existence of an intense PL component with time-constants on a time-scale below tens of nanoseconds. Most of the PL quanta were emitted after the end of the excitation pulse. For simplicity, we will consider below that all nonequilibrium electron-hole pairs that exist in PS after the end of the excitation pulse recombine radiatively. This corresponds to 10^{18} -10^{19} cm^{-3} electron - hole pairs in the case of experiments reported in Ref.16.

For the quantities of electron-hole pairs noted above some of the silicon crystallites are free of captured electrons or holes after the excitation pulse. Two principal scenarios of the recombination kinetics can be considered:

- If the energy barriers between the clusters are high enough ($\Delta E < 0$), electrons and holes are thermalized in the same or nearest potential wells. First, the geminate pairs thermalized in one crystallite (cluster of crystallites) recombine. In the next step, those geminate pairs recombine that correspond to nearest crystallites, while pairs with larger separations remain. We argue that this scenario is valid for the case of oxidized porous silicon when only a blue emission component with a lifetime of the order of 10 ns exists.

- The other limiting situation consistent with our experimental results [16] and that will be analyzed in detail below, is the distant-pair recombination. In this case, nonequilibrium electrons and holes travel a certain distance by hopping to crystallites of larger sizes before radiative recombination (Fig.2). The recombination rate due to tunneling between crystallites with surfaces at a distance x is given by

$$\nu_1(x) = \frac{1}{\tau_p} \exp(\frac{-2x}{a}), \qquad (4)$$

where $\tau_p = 10^{-8}$ s , $a \cong 1$-2 Å for the case of Si - SiO$_2$ barrier. ν_1^{-1} increases exponentially with separation. Thus, electron-hole pairs with lifetimes exceeding any reasonable experimental values must exist. Therefore, we account for an additional transport channel due to localized states in the disordered dielectric interconnecting Si crystallites. We consider that the electrons and holes can diffuse the corresponding distances x by hopping between localized sites as it is shown by dashed lines in Fig.2. The corresponding time-constants can be written as

$$\tau_2 = \frac{x^2}{D_h},\tag{5}$$

where the diffusion coefficient of the hopping process can be written as $D_h = \frac{1}{6\tau_p} b^2 \exp(-\frac{W}{kT})$ where b is the distance between the localized states, activation energy of the hopping process W is of the order of a few tenths of an electron-volt at room temperature and tends to zero for $T \rightarrow 0$. The above mentioned diffusion process is also a downward hopping. Below,we assume that energy loss in the transfer process between neighbouring crystallites is always smaller than the difference of corresponding band edges so that electrons and holes are thermalized in potential wells of silicon domains.

Our Monte-Carlo simulation of the radiative recombination takes into account both electron and hole transfer. We create electrons and holes randomly in crystallites with different values of r_i and let them relax by downward hopping (direct tunneling or diffusion) between the crystallites. Generation of electrons and holes in the Monte-Carlo procedure can take place in any Si crystallite. In the case where an electron is generated in a crystallite of a cluster consisting of several particles, it moves to the Si particle with minimum E_g. This is a rapid process compared with the time scale of the radiative recombination. We also assume that electrons and holes that get into one crystallite or cluster of crystallites in the process of initial distribution generation recombine nonradiatively before the onset of the radiative recombination. They do not influence the number of pairs existing at the beginning of light emission. As for the distant electron-hole pairs, we consider that a quantum of light with the wavelength given by (1) is emitted as soon as they meet in the crystallite r_i in the process of downward hopping. Backward activation hops of electrons with the probability proportional to $\exp(\frac{\Delta E_{gi} - \Delta E_{gj}}{2kT})$ are also taken into account in the process of simulation. A possibility exists that some of electrons or holes will experience hops to and from between neighboring crystallites that are equal in size. In this case, after the first hop between equal domains, electrons and holes are directed to the nearest larger size Si particle. As time evolves, electrons and holes, most of which were trapped in small crystallites, travel to deeper energy states (large domains). Therefore, a peak in the PL spectra corresponds to the intermediate values of wavelengths. There is an optimal value of r_i when electrons [3] and holes have already found each other and their concentration is still high. The electron-hole pairs were allowed to recombine until the pair density decreased to 30 % of its initial value. The simulation of recombination was repeated 10 times for a certain space configuration, then the configuration was changed. The results were averaged over all 150 configurations. Temporal dependencies of the number of emitted quanta were obtained for different wavelengths (values of

r_i). Fig.3 shows the results that illustrate the fate of 10^6 electron-hole pairs that re-combine radiatively at 300 K and 15 K correspondingly.

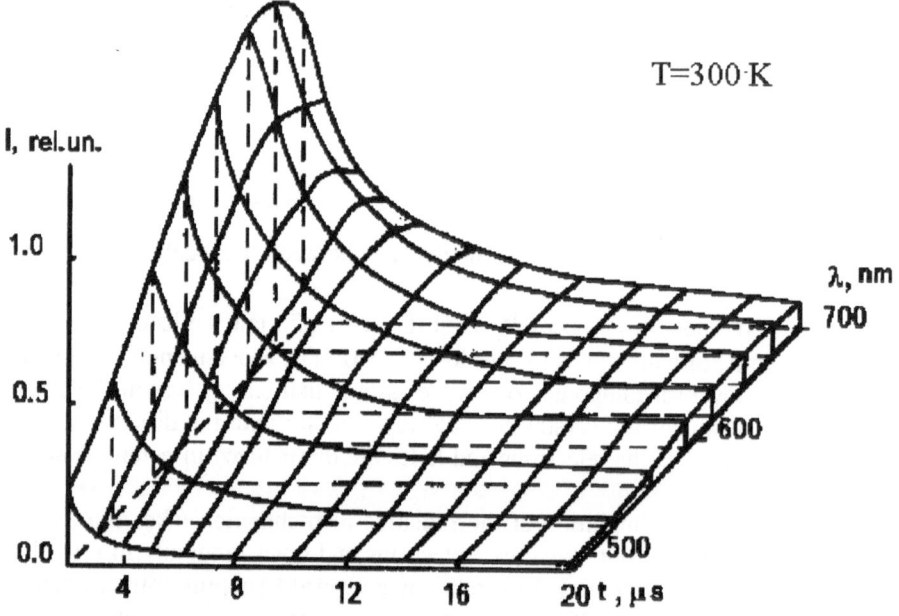

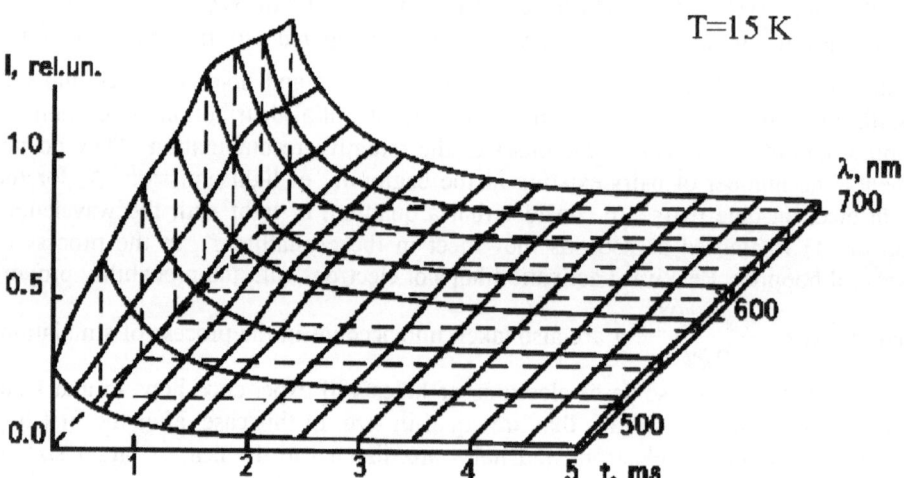

Fig.3. Calculated PL intensities as functions of time for different emission wavelengths and temperatures

The parameters of Gaussian size distribution and the values of diffusion coefficients D_h at room and helium temperatures were the only parameters that strongly influenced the final results. They were selected to obtain the best fits of the porous silicon experimental PL spectra (Fig.4) [16] and are given in the Table below.

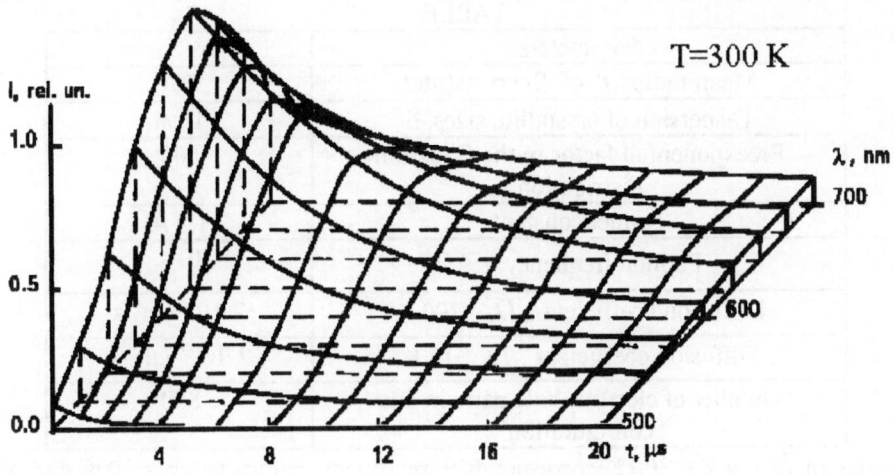

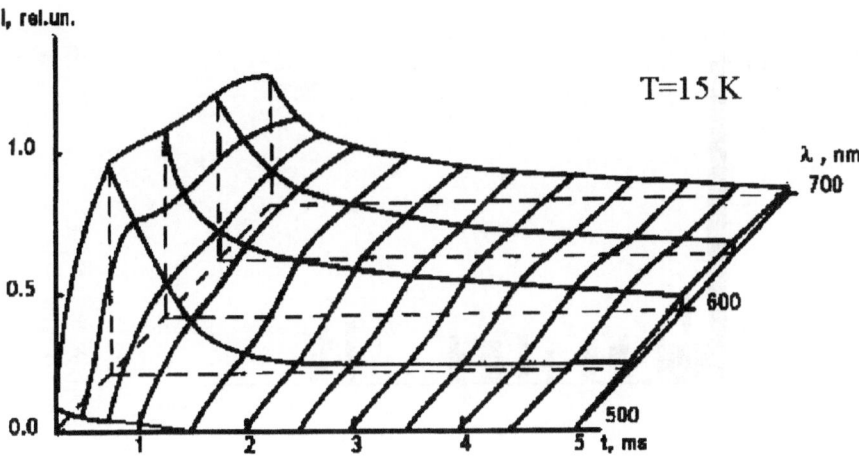

Fig.4. Experimental PL spectra [16]

We want to notice that the results shown in Fig.3 are rather insensitive to the value of initial electron-hole density N_i . Variations of N_i in the range from 100 to 1500 pairs in the cube selected for simulation did not result in significant changes of averaged $I(\lambda)$ dependencies. We attribute this fact to the existence of "favorite" recombination paths for electron-hole pairs. These mesoscopic ways of charge transfer between the Si crystallites lead to the peculiarities of $I(t,\lambda)$ dependencies for a certain configuration. These dependencies for one space configuration and two wavelengths are shown in Fig.5.

TABLE

Parameters	
Mean radius r of Si crystallites	7 Å
Dispersion of crystallite sizes, δ	10 Å
Preexponential factor in the Gaussian distribution	500
Tunnel constant	1.5 Å
Phonon frequency τ_p^{-1}	10^8 s^{-1}
Diffusion coefficient , D_h,, 300 K	$2 \cdot 10^{-9}$ cm^2/s
Diffusion coefficient , D_h,, 15 K	$2 \cdot 10^{-11}$ cm^2/s
Number of electron-hole pairs in one configuration	850

The value of D_h, at T = 300 K corresponds to reasonable parameters b = 10 Å and ΔE = 0.1 eV.

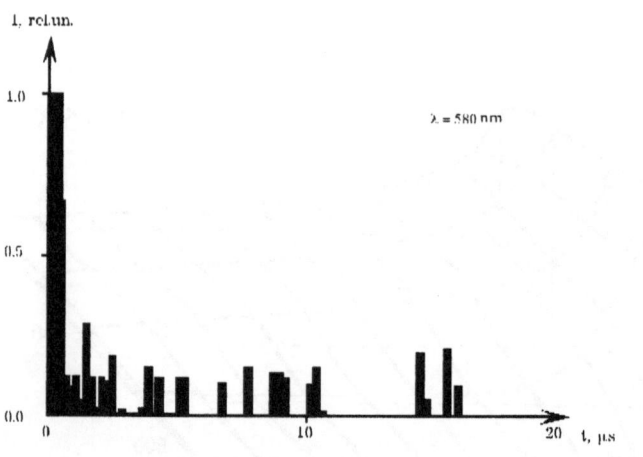

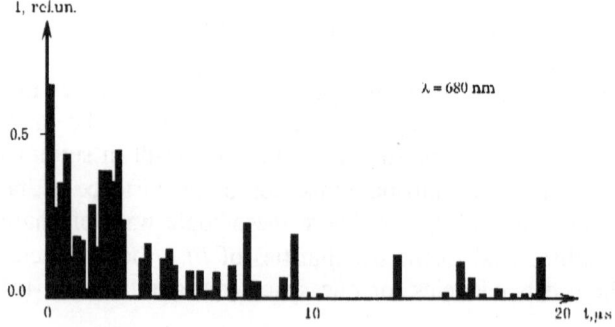

Fig.5. Time dependences of PL intensity for a single spatial configuration

For each type of crystallite (with λ_1 and λ_2) there exists a set of favorite transport paths with certain time constants. A large number of nonequilibrium charge carriers generated in PS recombine through "favorite" paths. It is clear that this effect will be more pronounced for smaller arrays of Si particles.

3. Discussion

As illustrated in the previous section, very good agreement can be obtained between the time-resolved PL spectra measured at 15 K and at room temperature, and the results of the Monte-Carlo simulation. We emphasize once again that we used minimum fitting parameters. The parameters that give the best fits have reasonable values typical for disordered solid state systems. The problem solved corresponds to the case of distant-pair recombination in PS. The necessary intermixing of electrons and holes prior to radiative recombination was ensured by high penetrability of intercrystallite barriers. The model developed also predicts the blue shift of the PL peak with increasing temperature registered in our experiments and in research by other authors. This shift can be attributed to thermal transitions from larger to smaller crystallites. These reverse transitions are more pronounced at elevated temperatures. Our model also predicts the experimental dependence observed of τ_e on temperature. This relation is associated with the temperature dependence of the effective diffusion coefficient that determines the rate of hopping between Si crystallites and clusters of crystallites.

In the case of oxidized or partly-oxidized PS, it is evident that interconnection of crystallites is infringed. Silicon dioxide films can grow at the surfaces of Si crystallites, and nonstoichiometric Si in the intercrystallite space which can also be transformed into SiO_2. This leads to two main consequences:
1. Electron-hole pairs are now thermalized mainly in the same cluster where they were born by the exciting quanta. Some of them travel very short distances and get into neighboring crystallites.
2. The hopping path between crystallites disappears due to the decrease of hopping site concentration.

As a result, mainly geminate pairs remain in PS after the end of the UV excitation pulse. It is clear that we can also find an intermediate situation for partly oxidized samples. Geminate pairs that have thermalized in the same cluster give rise to PL characterized by very short lifetimes.

4. Conclusions

We have proposed a novel model of radiative recombination processes in porous silicon that provides understanding of most aspects of experimental observations including spectral and temporal dependencies of PL for as-prepared and oxidized specimens. Our model which starts with a few very general assumptions, differs from analogous models for disordered semiconductors by existence of a certain disordered PS structure that was also modelled by a separate Monte-Carlo procedure. We have also taken into account the existence of an additional hopping path for electrons and holes travelling between silicon domains. With reasonable values of a small number of fitting parameters, we obtained good agreement between calculations and experimental $I(t,\lambda)$ dependencies at room and helium temperatures.

It cannot be claimed that we provide support for a certain model of the final microscopic process of radiative recombination in PS. In particular, both recombination of excitons confined in quantum clusters or recombination at the surfaces of quantum crystallites can be considered as plausible options that do not contradict our model. However, the results obtained give evidence that quantum confinement in silicon domains and charge transport between these domains must be taken into account in any interpretation.

5. References

1. Iyer, S.S., Collins, R.T., Canham, L.T. (1992) Light emission from porous silicon, MRS, Pittsburgh.
2. Takagahara, T., Takeda, K. (1992) Theory of the quantum confinement effect on excitons in quantum dots of indirect gap materials, Phys. Rev. B **46**. 15578-15581.
3. Teschke, O., Alvarez, F., Tessler, L., Kleinke, M.U. (1993) Nanosize structures connectivity in porous silicon and its relation to PL efficiency, Appl. Phys. Lett. **63**, 1927-1929.
4. Lin, C.H., Lee, Si.C., Chen, Y.F. (1993) Strong room-temperature photoluminescence of hydrogenated amorphous silicon oxide and its correlation to porous silicon, Appl. Phys. Lett. **63**, 902-905.
5. Suda, Y., Ban, T., Koizumi, T., Koyama, H., Tezuka, Y., Shin, S., Koshida, N. (1994) Surface structures and photoluminescence mechanisms of porous Si, Jpn. J. Appl. Phys. **33**, 581-585.
6. Proces, M. (1993) Light emission in thermally oxidized PS: Evidence for oxide-related luminescence, Appl. Phys. Lett. **62**, 3244-3246.
7. Takeda, Y., Hyodo, S., Suzuki, N., Motohiro, T., Hioki, T., Noda, S. An oligosilane bridge model for the origin of the intense visible photoluminescence of porous silicon, J. Appl. Phys. **73**, 1924-1928.
8. Stevens, P.D., Glosser, R. (1993) Anomalous photoluminescence behavior of porous Si, Appl. Phys. Lett. **63**, 803-805.
9. Lockwood, D.J. (1994) Optical properties of porous glasses, Solid State Commun. **92**, 101-112.
10. Kanemitsu, Y., Uto, H., Masumoto, Y.,Matsumoto, T., Fukagi, T., Mimura, H. (1993) Microstructure and optical properties of free standing porous silicon films, Phys. Rev. B **48**, 2827-2831.
11. Sinha, S., Baharjee, S., Arora, B.H. (1994) Photoluminescence excitation of porous silicon, Phys. Rev. B **49**, 5706-5719.
12. Tamura, H., Ruckschloss, M., Wirschem, T., Veprek, S. (1996) On the possible origin of the PL from oxidized nanocrystalline silicon, Thin Solid Films (in press).
13. Roy, A., Jayaram, K., Sood, A.K. (1994) Raman and photoluminescence studies of thermally annealed porous silicon, Solid State Commun. **89**, 229-233.
14. Tsybeskov, L., Vandyshev, Ju.V., Fauchet, P.M. (1994) Blue emission in porous silicon: Oxygen-related photoluminescence, Phys. Rev. B **49**, 7821-7824.
15. Matsumoto, T., Daimon, M., Futagi, T., Mimura, H. (1992) Picosecond luminescence decay in porous silicon, Jpn. J. Appl. Phys. **31**, L619-L621.
16. Roizin, Ya.O., Savin, D.P., Gevelyuk, S.A., Vorobyeva, V.A., Karpov, A.V., Korlyakov, A.B., Mugenski, E., Sokolska, I. (1996) Geminate and distant pair radiative recombination in porous silicon, Phys. Rev. B (in press).
17. Xie, Y.H., Hybertsen, M.S., Wilson, W.L., Ipri, S.A., Carver, G.E., Brown, W.L., Dons, E., Weir, B.E., Kortan, A.R., Watson, G.P., Liddle, A.I. (1994) Luminescence properties of nanometer-sized Si crystallites, Phys. Rev. B **49**, 5386-5391.
18. Dunstan, D.I., Boulitrop, F. (1984) Photoluminescence in hydrogenated amorphous silicon, Phys. Rev B **30**, 5945-5957.
19. Gutschker, O., Binderman, R. (1991) Monte-Carlo simulation of carrier relaxation and recombination in a-Si:H, Non-Cryst. Solids **137**, 579-583.
20. Tsybeskov, L., Peng, C., Duttagupta, S.P., Ettedgui, E., Gao, V., Fauchet, A.M., Curver, G.E. (1993) Comparative study of light-emission from porous silicon, Proc. MRS **298**, 307-311.
21. Landau, L.D., Lifshits, E.M. (1972) Quantum mechanics, Nauka, Moscow.
22. Sanders, G.D., Chang, Y.C. (1992) Theory of optical properties of quantum wires in porous silicon, Phys. Rev B **45**, 9202-9213.

QUANTUM MECHANICAL MODELLING OF EXCITON AND HOLE SELF-TRAPPING IN IONIC CRYSTALS

A. L. SHLUGER
Department of Physics, University College London,
Gower St., London WC1E 6BT, UK

V. E. PUCHIN
Institute of Chemical Physics, University of Latvia
Rainis blvd. 19, Riga, LV-1586, Latvia

1. Introduction

The self-trapping of holes and Frenkel excitons is the fundamental step in defect processes induced by electronic excitation in a wide class of insulating solids, including the formation of intrinsic defects and surface decomposition. It has been observed in many insulating crystals such as alkali-metal halides, quartz, rare-gas solids, alkaline-earth halides, perovskite-structured halides of the $KMgF_3$ family and in other systems. It drastically alters the optical, luminescent and energy transport properties of the crystal. Extensive experimental studies on self-trapped excitons (STEs) and holes in insulators have been carried out during the last few decades, using optical absorption, luminescence, ENDOR, resonant Raman scattering, and other techniques, as reviewed by Song and Williams [1].

However, the initial stages of the self-trapping process and even the structure of STEs are still unclear. Recently, femtosecond scale pulse optical spectroscopic techniques have been applied to the study of the hole polaron and exciton self-trapping in KI, RbI [2], KBr and RbBr [3]. The results of these pump-probe spectroscopic experiments were tentatively interpreted in terms of formation of some transient (lifetime < 5 ps) localized species preceding the formation of self-trapped holes (V_K centres), STEs and primary defect pairs.

Among plausible candidates for such a metastable hole state in alkali halides, the one-centre state has the largest relaxation energy, S, (about 1 eV or more) due to the polarization of the lattice by the strongly localized hole. The kinetic energy loss due to the hole localization in this state (localization energy), B, is determined by the structure of the upper valence band and may be roughly estimated as one-half of its width [4], which is more than 1 eV in many alkali halides. If the hole is equally delocalized between two or more lattice sites, the lattice polarization is smaller, and these small polaron states are less favourable than the one-centre state. However, in alkali halides and in alkaline-earth fluorides, possibly in alumina and other crystals, the

231

R. C. Tennyson and A. E. Kiv (eds.),
Computer Modelling of Electronic and Atomic Processes in Solids, 231–239.
© 1997 *Kluwer Academic Publishers.*

two-centre state is stabilized by chemical bonding between the two anions sharing the hole [1,5]. This leads to the formation of a quasi-molecular X_2^- state (X is the anion) where the distance between anions is much smaller than that in the perfect lattice. These are the only stable hole states observed so far in a pure lattice.

An exciton is an electron-hole pair which propagates in non-metallic solids. If the hole is already self-trapped as a V_K centre, the electron is also localized in the vicinity of X_2^-. Since the isolated X_2^- has no affinity to the second electron, the electron is trapped not by X_2^- itself, but by the field of the distorted lattice. The repulsion between the electron and X_2^- molecular ion can lead to the separation of the hole and electron components of the exciton. In particular, the most stable atomic configuration of the self-trapped exciton in some alkali halides is close to the nearest $F - H$ pair: X_2^- is shifted from its symmetrical position towards one of the anion sites (H centre), and the electron is localized in the adjacent anion site, which becomes vacant (F centre). However, STE in alkali halides may have up to three equilibrium atomic configurations having different optical absorption and luminescence spectra. One, which is called an on-centre configuration and corresponds to the electron bound to the V_K centre, has recently been proposed as a transient state in the exciton self-trapping in KBr and RbBr [3], and also as an equilibrium exciton configuration in NaBr and NaI [6]. However, most of the existing theoretical calculations predict the existence of only off-centre $F - H$ -like configurations [1].

Atomistic theories of the self-trapped excitons and holes in alkali halides have been reviewed in ref. [1, 7, 8]. In this paper, we discuss the results of recent studies of the transient hole and exciton states which highlight the importance of the lattice polarization and electron-hole correlation effects. In particular, we aim to assess how reliably cluster-type, many-electron calculation techniques are able to predict the structure and spectroscopic properties of defects in insulating crystals.

2. Embedded molecular cluster method

The embedded molecular cluster method used in these studies provides an adequate description of the point defects in the insulating crystals. The necessary condition for its applicability is the localization of the defect electronic wave function within the quantum cluster region. In the case of the hole or exciton, this means that it can be used only for those stages of the process when the quasi-particle is already trapped by the lattice deformation.

The detailed description of the method and corresponding software, which have been developed and applied to the excitons and point defects in alkali halides, can be found elsewhere [7, 9, 10]. Here we will describe qualitatively the main components of the theoretical model and the range of its applicability.

2.1. LOCALIZED ELECTRONIC GROUPS

The embedded cluster model employed in this study is based on the approximation that the perfect crystal can be divided into individual ions. It allows one to combine a quantum-mechanical treatment of a part of the crystal, including a defect (quantum cluster) with the classical description of the rest of the crystal. This is made by substituting a number of classical ions by a quantum cluster and by using a "self-consistency" procedure based on consecutive iteration of two computational methods as discussed in ref. [7, 9, 10]. To be used together, both methods must give the same lattice constants of the perfect lattice and, in the more general case, must yield the same optimized structure.

The embedded cluster model is based on the method of localized electron groups as discussed in detail in ref. [11, 13]. If all the ions except those in the cluster region are only slightly perturbed by the defect, the problem is reduced to the solution of the equation for the cluster wave function in the self-consistent field of the rest of the lattice. This external field, or embedding potential, consists of the Madelung potential, pseudopotentials of the ions surrounding the cluster, and terms due to the polarization of the crystal induced by the defect.

2.2. MADELUNG FIELD

The Madelung potential of the crystalline infinite (or semi-infinite in the case of the surface) lattice of point charges can be obtained by summation of conditionally converged series. The Ewald technique can be used for this purpose. Alternatively, the summation can be performed over the sufficiently large, but finite array of the point charges. In the latter case, the calculated potential in the cluster region differs from the exact value by a constant, which should be subtracted afterwards. This is especially important for the modelling of the surface defects, where the asymptotics of the electrostatic potential define the energetics and the decay of the cluster wave function into the vacuum.

In the case of completely ionic crystals (NaCl, MgO), the charges of ions can be taken as integers. When the ionic charges differ from the formal valence charges, partition of the total wave function into the ionic wave functions is not strictly justified. However, the cluster model with fractional ionic charges has been applied successfully to the defects in the bulk and molecular adsorption on the surfaces of the crystals, which are known to be partially ionic (SiO_2, Al_2O_3, TiO_2). In this case, the cluster should be chosen to contain ions in stoichiometric proportion, in order to maintain at least the integer number of electrons in the cluster.

2.3. PSEUDOPOTENTIALS

The point ion approximation is not valid for those ions in the rest of the lattice, which are close to the cluster boundary and whose electron densities overlap with the cluster one. Finite ion size effects should be taken into account. The potential experienced by the cluster electrons, due to these ions besides the bare Coulomb potential, includes non-point electrostatic correction, exchange and orthogonalization (or Pauli repulsion) terms. The former is local and depends only on the ion electron density. Two others are nonlocal, i.e. depending also on the cluster electron state, and can be expressed in terms of projection operators to the ionic wave function. The effective Pauli repulsion term keeps the cluster wave function orthogonalized to that of the remainder of the lattice, and thus localized in the quantum cluster region.

For technical reasons, the norm-conserving semi-local pseudopotential method is the most efficient for the treatment of the ions in the cluster neighbourhood. It allows the significant reduction of the basis set expansion of the cluster wave function calculation, because the pseudo wave function, which is replacing the wave function in this method, is smoothed and has no nodes in the frozen core region. Standard pseudopotential parameters, which were generated for the atomic cores from the atomic wavefunctions, can be used for the positive ions. For the negative ions, the pseudopotentials could be derived from the corresponding ionic wave functions. However, the non-point electrostatic and exchange contributions, which are attractive in the case of negative ions have the same order of magnitude and exponential asymptote as the repulsive Pauli term. Thus, these contributions cancel each other to some extent and the point charge approximation remains valid for the negative ions.

2.4. PAIR POTENTIALS

If *ab initio* methods are used to calculate the cluster electronic wavefunction, the total energy, and the equilibrium atomic structure, the size of the cluster is strictly limited by computer performance. Thus, it is feasible to restrict the number of electrons treated by the quantum chemistry method only to those which are involved explicitly in the chemical bonding or excitation process. In this case, the ions, which appear to be on the crystal boundary and are replaced by pseudopotentials, may still have significant displacements with respect to their perfect lattice positions. For those boundary ions, the pair potential interaction with ions in the remainder of the lattice and cores in the quantum cluster region should be introduced. Besides bare Coulomb interaction the pair potentials contain the short-range repulsion and Van der Waals or dispersion terms.

The parameters of the pair potential are the important part of the embedded molecular cluster model and can be derived self-consistently from the perfect crystal calculations. They satisfy the following criteria:

(a) the equilibrium geometry of the cluster simulating the perfect lattice has to coincide with the corresponding fragment of the infinite lattice;

(b) the total energy of this cluster has to behave symmetrically with respect to the displacement of the ions both inward and outward from the border of the cluster.

Such an approach to the calculation of the interactions between the "quantum-mechanical" ions and the rest of the ions has a substantial advantage provided that the ionic approximation holds with a high accuracy. Indeed, it makes cluster size very flexible and enables changing it without inducing jumps in the total energy of the system. In particular, in the process of calculating the adiabatic potential energy surface, some of the frozen ions may easily be included in the geometry optimization, whereas some may be fixed in their lattice sites if their displacements appear to be negligibly small. Furthermore, more ions may be made quantum mechanical if their electronic structure is expected to change appreciably during the next step of the defect diffusion process. Simultaneously, some of the other ions may be frozen as whole-ion pseudopotentials if they are not perturbed by the defect.

2.5. LATTICE POLARIZATION

The polarization within the quantum cluster is treated within the accuracy of the particular method of the wave function calculation and chosen basis set of atomic orbitals. The displacements of the ions in the vicinity of the quantum cluster, calculated via the pair potentials described above, partially describe also the polarization of the rest of the lattice. Thus, for the neutral defects, such as the self-trapped exciton, F and H centre, the further improvement in the treatment of polarization effects does not change the results.

The polarization of the rest of the lattice is most important for the charged defects, such as V_K centre, cation or anion vacancy, unless we restrict ourselves to the calculation of the energy differences for the defect in the same charge state.

The polarization energy of the lattice which depends on the particular charge distribution within the cluster can be taken into account using the Mott-Littleton method and the shell model [14]. In this model, each ion within the finite region outside the cluster is represented by the negatively-charged shell connected to its core by a spring. The outermost part of the crystal is treated as a dielectric continuum. The parameters of the model including shell and core charges, elastic constant of the spring and interionic interaction potentials are fitted to the experimental data or to the *ab initio* calculations.

The response of the polarized lattice is then given in a form of the lattice polarization energy and an electrostatic potential ϕ_i produced by the cores and shells outside the cluster at the position of each nuclei, i, inside the quantum cluster. This potential, however, is calculated using not the formal ionic charges employed in the parametrization of the shell model, but those obtained for the perfect lattice using the periodic model and the same quantum-chemical method. Since the quantum-mechanical charges are different from the formal ionic charges used in the Mott-Littleton model, this ensures a homogeneous charge distribution across the cluster border. Note that ϕ_i in this approach includes both the Madelung term and the dipole polarization term. The diagonal matrix elements of this potential calculated on atomic orbitals μ, $\langle \mu | \phi_i | \mu \rangle$ are then added to the Fock matrix. The total energy of the whole system including the quantum cluster embedded in the infinite polarizable lattice is minimized with respect to the LCAO coefficients, positions of the nuclei inside the cluster and of the cores and shells of the rest of the crystal. This method is implemented in the ICECAP and CLUSTER95 computer codes [7,10].

2.6. ELECTRON CORRELATION

In the localized electron groups approach, the electron correlation effects are naturally separated in two parts: intragroup, and intergroup, correlation. The former can be treated within the quantum cluster using standard quantum chemistry methods, such as Møller-Plesset perturbation theory or the configuration interaction method including the electron excitations within a single electron group. Intergroup correlation can be described in terms of the charge transfer excitations between the different electron groups. In this case, the total wavefunction is taken as the linear combination of the antisymmetrized products of the localized electron group wave functions.

3. Application to the hole and exciton self-trapping problem

The results of extensive applications of this technique are described in ref. [7, 8, 15]. In this paper, we will summarize the results of recent studies which concern two particular problems related to the initial stages of self-trapping of holes and excitons.

3.1. ONE-CENTRE TRAPPING OF THE HOLES IN ALKALI HALIDE CRYSTALS

Spectroscopic measurements were recently performed in KI on the time domain from 0.3 ps to 100 ps after the excitation pulse at both room and liquid nitrogen temperatures [3]. The two-photon excitation with an energy of about 8 eV employed in these experiments first produces an electron-hole pair in the bulk of KI. The electron is quickly trapped by the NO_2^- impurity or is delocalized in the lattice. Optical absorption spectra in the energy range from 1.5 to 3.2 eV have been observed both in pure samples of KI and those doped with the electron-trapping impurity (NO_2^-). Transformation of the optical absorption spectrum in KI and RbI is attributed to the hole and takes place via three distinct stages:

  (1) At least two intense optical absorption bands with maxima near 2.3 and >3.2 eV were already observed at 0.3 ps after the excitation pulse in doped crystals. Their intensity rises within approximately 1 ps. With some delay (about 0.5 ps) a third absorption band at about 2.6 eV begins to rise and broadens.
  (2) About 3 ps after the pulse, the whole spectrum transforms into a featureless broad band.
  (3) Subsequently, the well-known optical absorption of the V_K centre appears and rises during 10 ps with a time constant of about 3 ps. Similar transient optical absorption has been observed in RbI.

In a recent paper [10], we have demonstrated that these results can be reasonably understood if we consider that the hole first localizes in the one-centre state which then transforms into the two-centre state, and finally into the V_K centre. This conclusion is based on the following calculations performed using the CLUSTER95 code.

One simple argument is based on the static energetic criterion that the self-trapped state should have lower energy than the bottom of the free hole band, i.e. E_{st} should be negative. The band structure calculation of the KI using a relativistic technique [16] has demonstrated that spin-orbit effects split the upper p valence band into two non-overlapping sub-bands. The two-photon excitation with an energy of 8 eV produces holes in the upper sub-band with the angular momentum $J = 3/2$. The width of this sub-band, calculated using a relativistic mixed-basis method and a 'muffin tin' potential [16], was found to be equal to about 1 eV, whereas the width of the spin-orbit split valence band is 1.82 eV. This is about 1 eV smaller than the experimental value [17] of 2.8 eV. The valence band width obtained in our calculation without the spin-orbit interaction is 2.05 eV which, with the addition of the spin-orbit splitting, gives 2.95 eV.

The localization energy B for the localized one-centre state in the Wannier representation is determined by the position of the "centre of mass" of the density of states (DOS) in the valence band. For a symmetrical DOS, it can be calculated as a half of the valence band width [4]. A more accurate value is usually smaller than this estimate by about 20%. This is due to an angular dependence of transfer integrals for p orbitals which is neglected in this simple approximation. Thus, a conservative estimate for the upper limit of the localization energy B of the one-centre hole is half of one of the two split sub-bands < 1.0 eV.

To find the geometric and electronic structures of the hole in the one-centre state, we first fixed the position of the I atom carrying the hole in the lattice site, whereas all other crystal ions were allowed to relax. The nearest neighbour cations are displaced outwards by about $0.1a$ and the next nearest-neighbour anions are displaced inwards by about $0.015a$ (a is the inter-atomic distance). The relaxation energy, S, as a difference between the energy of the completely-relaxed state and that without core displacements from their perfect lattice sites, when only the electronic polarization responds to the presence of the hole, is equal to 1.3 eV. Then, the self-trapping energy for the one-centre hole state can be estimated as $B - S \leq -0.3$ eV.

Calculations of the optical absorption energies and the matrix elements of the corresponding dipole electronic transitions for the relaxed state of the one-centre hole were made using the configuration interaction for single electron excitations technique (CIS) [18]. In the completely symmetrical configuration, the allowed transitions are from E_{1g} and A_{1g} states to the single occupied E_{1u} hole state and are accompanied by the hole delocalization. The latter should lead to the change in the lattice polarization which was not taken into account in our CIS calculations. To check how this can affect the calculated optical absorption energies, we calculated some of the transitions as the difference between the total energies of the ground and excited hole states with the self-consistent account of the lattice polarization in both states. With this correction, transitions calculated using CIS in the same cluster become 1.9 eV, 2.1 eV and 3.3 eV. They are in the range of the splitting and positions of the maxima of the absorption bands observed at 0.3 ps after the excitation pulse at both room and liquid nitrogen temperature in KI.

Further, we have considered the adiabatic potential for transformation of the one-centre into the two-centre state. The former state is not stable with respect to the displacement of the two anions closer to each other. However, once formed, the one-centre hole polaron state does not collapse into the two-centre state immediately, but this requires the iodine ions to come closer than a critical distance of about 4.1 Å. Before that happens, the hole remains localized on one of the ions. This behaviour of the adiabatic potentials results from the competition between two main factors: the chemical bonding between two iodines, and the lattice polarization.

These results suggest that the one-centre polaron state in KI can exist as a transient state. However, this does not answer the question of why holes produced by crystal excitation prefer to be trapped first in this state but not, for instance, in the two-centre state. We are presently working on this problem.

3.2 SELF-TRAPPED EXCITON

Although the STE in some alkali-halides may have up to three equilibrium configurations, most of the theoretical studies predict only the configuration corresponding to the nearest pair of the F and H centres [1, 7]. As has been demonstrated in our recent calculations for the NaCl crystal [8], the electron-hole correlation is an important factor which strongly affects the structure of the STE.

In the lowest energy configuration of the STE in NaCl, the hole component of the STE, X_2^-, is oriented along the <110> axis and its position is intermediate between those in H and V_K centres, such that one of the halogens is almost in the anion lattice site. The electron is localized around the adjacent vacant anion site. The X_2^- bond length is only 2% larger than that in the H centre, which would give almost the same vibration frequency of the stretching mode for both defects in agreement with the Resonant Raman Spectroscopy study. The calculated exciton luminescence energy is 2.9 eV, which is in satisfactory agreement with the experimental value of 3.36 eV. The optical absorption spectrum does not correspond to that of the superposition of isolated F and H centres. The F centre band is split into three, due to the lower symmetry. The hole excitations within the X_2^- are mixed with the electron ones of the same symmetry. However, the lowest excited state of Π symmetry and "mostly hole" character has a transition energy of 1.9 eV, close to the H centre excitation, and the lowest state of Σ symmetry and the predominantly electron character is 2.8 eV higher than the ground state, as for the isolated F centre. According to the optical adsorption and luminescence spectra, this configuration can be classified as type III.

Another equilibrium configuration of the STE found in this study has the D_{2h} symmetry and the V_K centre as its hole component. However, the X_2^- bond is longer than that in the V_K centre because an excited electron occupies the bonding σ_g atomic orbital. In the Hartree-Fock approximation this configuration corresponds to the saddle point, but is stabilized as a result of the strong interaction of several electronic configurations corresponding to the electron and hole components of the STE. This local minimum of the adiabatic potential energy surface is higher by 0.56 eV than the off-centre one. High luminescence energy (6.2 eV) and low absorption energy of the electron component (0.6 eV) suggests that this configuration corresponds to the type I STE. Note that similar configuration has recently been proposed as a transient state for the exciton self-trapping in KBr.

Other calculated equilibrium structures of the STE correspond to the nearest $F - H$ pair, but with the H centre rotated by 90°, or to the next nearest and further separated $F - H$ pair. For all these configurations, the energy difference between the lowest STE state and the ground singlet state of the distorted lattice is either small or negative, which indicates that they cannot be considered as an initial state of some luminescence band. The activation energy for the separation of the F and H centres from the lowest STE configuration was estimated to be 0.65 eV in the Hartree-Fock approximation.

4. Conclusion

The results of the theoretical simulation of the self-trapped exciton and elementary intrinsic defects are in broad overall agreement with the experimental data provided by optical, Raman and magnetic resonance spectroscopies. In most cases, the Hartree-Fock approximation gives a reasonable estimate for the defect atomic structures. However, the electron correlation correction is sometimes critical. It is especially important for the calculation of excited states and activation energies for defect diffusion and reactions. In the case of charged defects, the lattice polarization should be taken into account.

5. Acknowledgements

The authors are grateful to N. Itoh, K. Tanimura, A. M. Stoneham, K. S. Song, R. T. Williams, L. N. Kantorovich, E. A. Kotomin for numerous valuable discussions.

6. References

1. K. S. Song and R. T. Williams, *Self-Trapped Excitons* (Springer-Verlag, Berlin, 1993).
2. S. Iwai, T. Tokizaki, A. Nakamura, *et al.*, Phys. Rev. Lett. **76**, 1691 (1996).
3. T. Sugiyama, H. Fujiwara, T. Suzuki and K. Tanimura, Phys. Rev. B (at press).
4. L. Kantorovich, A. Stashans, E. Kotomin, *et al.*, Int. J. Quant. Chem. **52**, 1177 (1994).
5. A. M. Stoneham, *Theory of Defects in Solids* (Oxford University Press, Oxford, 1985).
6. K. Edamatsu, M. Sumita, S. Hirota, *et al.*, Phys. Rev. B **47**, 6747 (1993).
7. A. L. Shluger, A. H. Harker, V. E. Puchin, *et al.*, Modelling Simul. Mater. Sci. Eng. **1**, 673 (1993).
8. V. E. Puchin, A. L. Shluger, and N. Itoh, Phys. Rev. B **52**, 6254 (1995).
9. J. M. Vail, R. Pandey, and A. B. Kunz, Rev. Solid State Sci. **5**, 241 (1991).
10. A. L. Shluger and J. D. Gale, Phys. Rev. B (at press) (1996).
11. L. N. Kantorovich, J. Phys. C: Solid State Phys. **21**, 5041 (1988).
12. L. N. Kantorovich and B. P. Zapol, J. Chem. Phys. **96**, 8420 (1992).
13. L. N. Kantorovich and B. P. Zapol, J. Chem. Phys. **96**, 8427 (1992).
14. These techniques have been reviewed in a special issue: C. R. A. Catlow and A. M. Stoneham (ed.) J. Chem. Soc. Faraday. Trans. II 85 (1989).
15. V. E. Puchin, A. L. Shluger, K. Tanimura, *et al.*, Phys. Rev. B **47**, 6226 (1993).
16. A. B. Kunz, J. Phys. Chem. Solids **31**, 265 (1970).
17. R. T. Poole, J. G. Jenkin, J. Liesegang, *et al.*, Phys. Rev. B **11**, 5179 (1975).
18. J. B. Foresman, M. Head-Gordon, and J. A. Pople, J. Phys. Chem. **96**, 135 (1992).

CORRELATION BETWEEN ELECTRONIC STRUCTURE AND ATOMIC CONFIGURATIONS IN DISORDERED SOLIDS

YU.N.SHUNIN* and K.K.SCHWARTZ
Gesellschaft für Schwerionenforschung mbH, Darmstadt, Deutschland

1.Introduction

Correlations between atomic and electronic properties of solids must be treated through electronic, phononic and other excitation spectra. Developed in recent years, the cluster approach for disordered solids is based on the cluster concept and effective medium approximation (EMA). This approach utilizes in a practical manner some representations of the atomic structures of disordered solids, multiple scattering theory and coherent potential approximation (CPA). The main goals of numerical investigations by means of the cluster approach are atomic structure modelling, and electronic and phononic spectra calculations. Structurally-correlated electronic and phononic spectra changes allow the simulation of photo- and radiationally-induced processes in solids. Special attention is paid to photo-induced processes in amorphous semiconductors (chalcogenides, a-Si etc), which are important for optical recording systems, and elementary track excitations induced by swift heavy ion passages in amorphous metals and alkali halides (e.g., LiF).

2. Mathematical and computer modelling of fundamental properties of disordered condensed media

The general scheme of computer modelling of the fundamental properties of disordered condensed media is represented in Figure 1. An origin modelling point is established in the introduction of initial atomic structures and atomic potentials. Then, the electronic structure is calculated in a self- consistent procedure (in CPA or in EMA). The next step is the estimation of total energy for the atomic configurations defined above and the minimization of the total energy functional over atomic positions. Lastly, the atomic distribution functions, pair potentials and force constants are estimated, and the first cycle of iteration is over. We may finish our calculation or continue it with a new atomic configuration (see Figure 1).

R. C. Tennyson and A. E. Kiv (eds.),
Computer Modelling of Electronic and Atomic Processes in Solids, 241-257.
© 1997 *Kluwer Academic Publishers.*

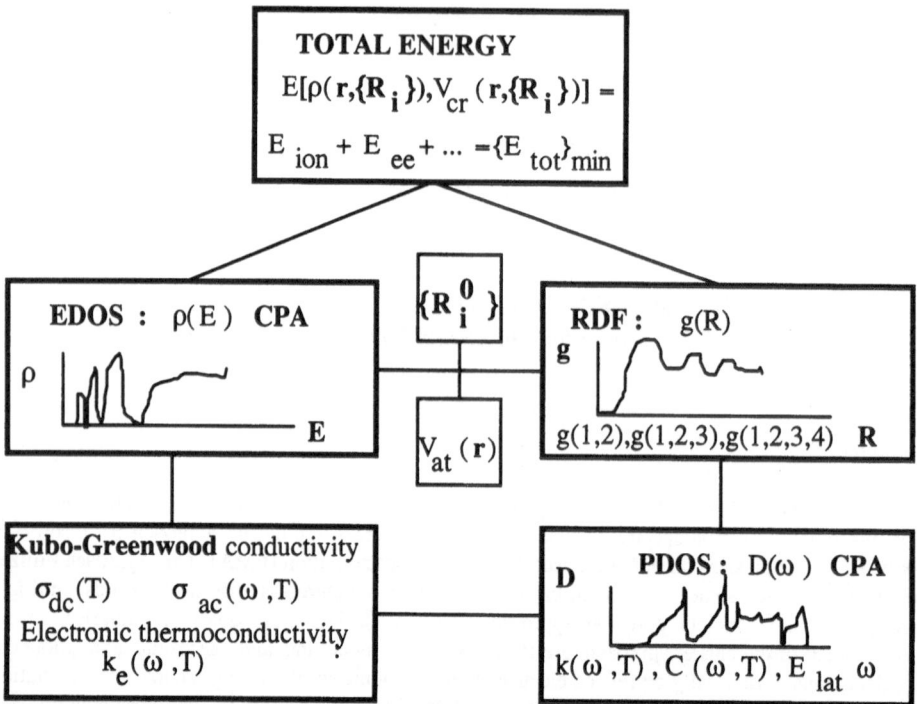

Figure 1.
Strategy of computer simulation of correlation between
electronic structure and atomic configurations.

The procedure described above is confined to an s.c. small iteration cycle and corresponds to the well-known adiabatic approximation (atomic positions are frozen during electronic structure calculations). From this cycle, two other branches emerge, namely conductivity and phonon spectra calculations, with possible considerations of electron-phonon correlations. We can close the large cycle of iterations, taking into account all-around correlations between the atomic structure, electronic and phononic subsystems. The methodological foundation of this general scheme is the cluster approach.

3. The general approach to the structural disorder description

The structural disorder description is based on the formalism of atomic distribution functions. The correlation lengths of the atomic structure define the scales of order in a disordered condensed medium. First, we must consider the scales of ordering, introducing a spatial correlation function of mass density fluctuation:

$$\langle \Delta\rho_1 \cdot \Delta\rho_2 \rangle = \overline{\rho_1\rho_2} - \overline{\rho}^2. \tag{3.1}$$

An essential decreasing of $\langle f(\mathbf{r}_1)f(\mathbf{r}_2)\rangle$ -like correlations (i.e.,→0) defines the scale of a structural order. Thus, one considers these scales as follows:

$|\mathbf{r}_1 - \mathbf{r}_2| \sim L_1 \approx 0.1 - 0.5nm$ (the short range order - SRO);

$|\mathbf{r}_1 - \mathbf{r}_2| \sim L_2 \approx 0.5 - 5.0nm$ (the medium range order - MRO)

$|\mathbf{r}_1 - \mathbf{r}_2| \sim L_3 > 5.0nm \rightarrow \infty$ (the long range order - LRO).

For the next step, we need distribution functions. The s-particle distribution function is as follows [17,18] :

$$g(1,2,3,...s) = n(1,2,3,...,s)/n^s. \tag{3.2}$$

A pair-distribution function can be written in the form:

$$g(1,2) = g(\mathbf{R}_{12}) = n^{-1}\delta(\mathbf{R}_{12} - \mathbf{l}), \tag{3.3}$$

where l-a lattice vector. In the case of an isotropic medium, one can obtain:

$$g(\mathbf{R}_{12}) = g(R) = n^{-1}N(l)\delta(|\mathbf{R} - \mathbf{l}|). \tag{3.4}$$

Simple liquids can be described by 3-particle functions:

$$g(1,2,3) = n^{-2}\delta(\mathbf{R}_{12} - \mathbf{l})\delta(\mathbf{R}_{13} - \mathbf{l}'). \tag{3.5}$$

Using, for example, well-known superpositional and over-superpositional approximations one can represent the high-order distribution functions by means of low-order ones. Thus,

$$g(1,2,3) \approx g(1,2)g(2,3)g(3,1) \tag{3.6}$$

and

$$g(1,2,3,4) = \frac{g(1,2,3)g(1,2,4)g(1,3,4)g(2,3,4)}{g(1,2)g(1,3)g(1,4)g(2,3)g(2,4)g(3,4)}. \tag{3.7}$$

A system of coupled equations consists of equations of the type:

$$\frac{1}{kT}\nabla_1 \ln g(1,2,...,s) = \nabla_1 g(1,2,...,s) + \frac{n}{g(1,2,...,s)}\int_{V_s} G(1,2,...,s+1)\Phi(1,s+1)dV_s. \tag{3.8}$$

All many-particle interaction potentials can be approximated by 2-particle potentials:

$$\Phi(1,2,...,N) \cong \frac{1}{2}\sum_{i,j}^{N}\Phi(i,j). \tag{3.9}$$

Many-particle distribution functions are expressed as:

$$g(1,2,...,N) = \exp(-\frac{1}{kT}\Phi(1,2,...N)). \tag{3.10}$$

To return to the pair approximation, we must use the integration procedure as follows:

$$g(1,2) = C\int...\int \exp(-\frac{1}{kT}\Phi(1,2,...,N)dV_3...dV_N, \tag{3.11}$$

which for gases can be presented simply:

$$g(1,2) = C\exp(-\frac{1}{kT}\Phi(1,2)). \tag{3.12}$$

For condensed media, it is necessary to use an averaged pair-potential $\overline{\Phi}(1,2)$, which is similar to pseudopotentials for electronic spectra calculations. Thus, the averaged pair distribution function can be expressed as:

$$g(1,2) = \exp(-\frac{1}{kT}\overline{\Phi}(1,2)).$$

(3.13)

The total correlation function

$$h(R) = g(R)-1$$

(3.14)

may be used as a measure of structural order, where $g(R) = g(1,2)$. It means that if $h(R) \to 0$ for R>L, then L is the scale of order. A correlation length L is a measure of order (or disorder).

4. Electronic structure calculation of disordered material

4.1. POTENTIAL CONSTRUCTIONS:ATOMIC AND CRYSTALLINE

An electronic structure calculation is considered here as a scattering problem, where centers of scattering are atoms of clusters. So, the first step of modelling is the construction of potentials, both atomic and crystalline .

The potential of a screened atomic nucleus looks as follows:

$$V^G(r) = -\frac{2Z}{r}\frac{\exp(\frac{\lambda r}{\mu})}{1+\frac{Ar}{\mu}},$$

(4.1)

where $\lambda = 0.1837, A = 1.05, \mu = 0.8853Z^{-1/3}$. Using a statistical approximation of atoms, one usually applies $X\alpha$ - or $X\alpha\beta$ -approximations for the electronic exchange and correlation:

$$V_{X\alpha}(r) = -6\alpha[\frac{3}{8\pi}\rho(r)]^{1/3},$$

(4.2)

where α depends on the charge number Z, and

$$V_{X\alpha\beta}(r) = -6[\alpha + \beta G(\rho)](\frac{3}{8\pi}\rho)^{1/3},$$

(4.3)

$G(\rho) = \frac{4}{3}(\frac{\nabla\rho}{\rho})^2 - 2\frac{\nabla^2\rho}{\rho}$, $\alpha \approx 0.67, \beta \approx 0.003$, $\rho(r) = \frac{\Delta_r V_e(r)}{8\pi}$- the electronic charge density,

$V_e(r) = \frac{2Z}{r} - V^G(r)$- the electronic part of the potential. Thus, the atomic potential of a neutral atom can be expressed as:

$$V_{at}(r) = V^G(r) + V_{ex-corr}(r),$$

(4.4)

where we consider $V_{ex-corr}$ as potentials $V_{X\alpha}$ or $V_{X\alpha\beta}$.

The "crystalline" potentials can be estimated by expressions:

$$V_{coul}(\mathbf{r}) = V^G(\mathbf{r}) + \sum_{\gamma,\mathbf{n}_\gamma}V_\gamma^G(\mathbf{r} - \mathbf{R}_{\mathbf{n}_\gamma}^\gamma),$$

(4.5)

$$\rho_{cryst}(\mathbf{r}) = \rho(\mathbf{r}) + \sum_{\gamma, n_\gamma} \rho_\gamma (\mathbf{r} - \mathbf{R}_{n_\gamma}^\gamma). \tag{4.6}$$

Then we apply the so-called MT-approximation (MTA, MT-muffin tin):

$$V_{MT}(r) = \langle V_{cryst}(\mathbf{r}) \rangle - V_{MTZ}, \tag{4.7}$$

$$V_{cryst}(\mathbf{r}) = V_{coul}(\mathbf{r}) + V_{ex-corr}(\mathbf{r}), \tag{4.8}$$

$$V_{cryst}(\mathbf{r}) = -6\alpha [\frac{3}{8\pi} \rho(\mathbf{r})]^{1/3}, \tag{4.9}$$

where V_{MTZ}-MT-zero estimation of potential calculation, and <...> means space averaging. To obtain the electronic structure, the calculation of scattering properties is necessary. Scattering properties of these potentials are calculated in the form of phase shifts, or generally, in the form of T- and S-matrices of scattering. A very important case of MT-appoximation of potentials has been widely investigated, but much attention was paid to the spherically nonsymmetric potentials. The results of potential modelling and phase shift calculations in the framework of the MT-approximation are presented in [1, 3] .

4.2. TOTAL ENERGIES, INTER-ATOMIC POTENTIALS

The total energy estimations, inter-atomic potentials and force constant calculations are based on the same analytical potential functions as atomic and crystalline potentials:

$$\mathbf{F}_i = -\nabla_{\mathbf{R}_i} E^{CL} = 2 \sum_{i \ne j} \frac{Z_i' Z_j' (\mathbf{R}_i - \mathbf{R}_j)}{|\mathbf{R}_i - \mathbf{R}_j|^3} - \nabla_{\mathbf{R}_i} E_{ee} = \mathbf{F}_i^{ion} + \mathbf{F}_i^{el}, \tag{4.10}$$

$$\mathbf{F}_i^{el} = -\int d\mathbf{r} \frac{\partial V_{ext}(\mathbf{r}, \mathbf{R})}{\partial \mathbf{R}_i} \rho(\mathbf{r}, \mathbf{R}) - \int d\mathbf{r} \frac{\delta E_{ee}}{\delta \rho} \frac{\partial \rho(\mathbf{r}, \mathbf{R})}{\partial \mathbf{R}_i} = \mathbf{F}_i^{el(1)} + \mathbf{F}_i^{el(2)}, \tag{4.11}$$

$V_{ext}(\mathbf{r}, \mathbf{R})$ - the external crystalline potential,

$$\mathbf{F}_i^{HF} = \mathbf{F}_i^{ion} + \mathbf{F}_i^{el(1)} - \tag{4.12}$$

is the Hellmann-Feynman force.
In simple cases, the pair-interaction potential can be estimated as:

$$\Phi(R) = \int \rho_1(r) V_2(R - r) 4\pi r^2 dr, \tag{4.13}$$

and then the inter-atomic force:

$$F(R) \approx -\frac{d\Phi}{dR} = -\int \rho_1(r) [\frac{\partial V_2(R-r)}{\partial R}] 4\pi r^2 dr, \tag{4.14}$$

where $\rho_1(r)$-the averaged density of electronic charge of a "central" atom, and $V_2(R-r)$ - the potential of the "second" atom (the quasi-atom or averaged crystalline potential).

The total energy calculations for a neutral atom are based on the relationship:

$$E_{tot}[\rho(r)] = \int [\frac{3}{5}(3\pi^2)^{2/3} \rho^{5/3}(r) + \frac{1}{2}\rho(r)V_e(r) - \frac{2Z}{r}\rho(\pi) - \frac{9}{2}\alpha(\frac{3}{8\pi})^{1/3}\rho^{4/3}(r)]4\pi r^2 dr \quad .(4.15)$$

The next procedure looks as follows:

$$E_{tot}^{CL} = E_{ion-ion} + E_{ee}[V_{ext}, \rho], \tag{4.16}$$

$$\rho^{CL}(\mathbf{r}, \mathbf{R}_1, \mathbf{R}_2, ..., \mathbf{R}_N) = \sum_i \rho^i(\mathbf{r} - \mathbf{R}_i),$$ (4.17)

$$V_{ext-ion}^{CL}(\mathbf{r}, \mathbf{R}_1, ..., \mathbf{R}_N) = 2\sum_i \frac{Z_i}{|\mathbf{r} - \mathbf{R}_i|},$$ (4.18)

$$V_{ee}^{CL}(\mathbf{r}, \mathbf{R}_1, ..., \mathbf{R}_N) = \sum_i V_e^i(\mathbf{r} - \mathbf{R}_i).$$ (4.19)

Using the MT-approximation, these relationships may be rewritten as:

$$E_{tot}^{CL} \approx \sum_i E_i^{MT} + \sum_{i \neq j} \frac{Z_i'' Z_j'}{|\mathbf{R}_1 - \mathbf{R}_J|} + E_{ee}'[V_{ext}', \rho'].$$ (4.20)

Then, the total energy of a solid is estimated by :

$$E_{tot}^{sol} \approx E_{tot}^{CL} + E_{rest \cdot medium}$$ (4.21)

and the total energy change is

$$\Delta E_{tot}^{sol} \approx \Delta E_{tot}^{CL}.$$ (4.22)

4.3. ELECTRONIC STRUCTURE

Electronic structure calculations start from the definition of the initial atomic structure to produce a medium for solution of the scattering problem for a trial electronic wave (see Figure 2) [1-4].

The formalism we use for electronic structure calculations is based on the CPA, the multiple scattering theory and cluster approach. As the first step in the modelling procedure, one postulates the atomic structure on the level of short- and medium-range orders. The next step is to construct a "crystalline" potential and introduce the MTA. This is accomplished by using realistic analytical potential functions. Then, the electronic wave scattering problem is solved, and the energy dependence of the scattering properties for isolated muffin-tin scatterers is obtained in the form of phase shifts $\delta_{lm}(E)$

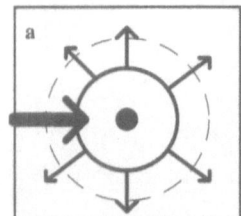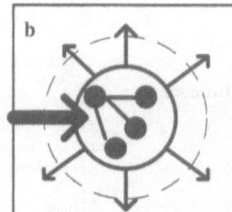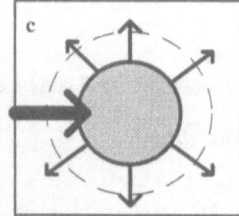

Figure 2.
Models of scattering clusters in: a) a single atom cluster; b) a many-atomic
cluster; c) a cluster with general type potentials.

and the T-matrix of the cluster as a whole is found. The indices l and m arise as a result of expansions of the functions as Bessel functions j_l Hankel functions h_l and spherical harmonics Y_{lm}. We note that the introduction for the potentials simplifies the solution of the scattering problem considerably, since this procedure symmetrizes the potentials. However, the analysis of the scattering problem for potential functions of a general type not possessing spherical symmetry within the confines of the cluster volume has demonstrated that the MTA is applicable for a wide class of disordered materials, both metals and semiconductors. In the procedure of constructing the potentials, the $X\alpha\beta$-approximation for the exchange-correlation interaction has also been used. This approximation has important advantages for describing binary materials, since its parameters α and β are practically constant for different elements.

As we have said, the modelling of disordered materials involves representing them as a set of atoms or clusters immersed in an effective medium with the dispersion relation $E(\mathbf{K})$ and a complex energy-dependent coherent potential $\Sigma(E)$, which is found self-consistently in the framework of the CPA. The basic relations are:

$$\Sigma(E) = V_e + \langle T \rangle (1 + G_e \langle T \rangle)^{-1}, \tag{4.23}$$

$$N(E) = -(2/\pi) \ln \det \| G(E) \|, \tag{4.24}$$

$$G(E) = G_e + G_e \langle T \rangle G_e = \langle G \rangle, \tag{4.25}$$

$$\langle T(E, \mathbf{K}) \rangle = 0, \tag{4.26}$$

$$\Sigma(E) = V_e, \tag{4.27}$$

$$\langle G \rangle = G(E) = G_e. \tag{4.28}$$

Here <...> denotes configurational averaging , V_e and G_e are the potential and the Green's function of the effective medium ,$N(E)$ is the integral density of electronic states, and $T(E, \mathbf{K})$ is the T matrix of the cluster. Condition (4.26)) can be rewritten in the form

$$\langle T(E, \mathbf{K}) \rangle = \mathrm{Sp} T(E, \mathbf{K}) = \int \langle \mathbf{K} | T(E, \mathbf{K}) | \mathbf{K} \rangle d\Omega_{\mathbf{K}} = 0, \tag{4.29}$$

where

$$|\mathbf{K}\rangle = 4\pi \sum_{l,m} i^l j_l(Kr) Y_{lm}^*(\mathbf{K}) Y_{lm}(\mathbf{r}), \tag{4.30}$$

and the integration is over all angles of $\mathbf{K}$. Relation (4.29) is the self-consistent condition and enables one to obtain the dispersion relation $E(\mathbf{K})$ of the effective medium. The calculation of the density of electronic states on the basis of relation (4.24) is done by the variational procedure

$$\rho(E) = \frac{\delta N(E)}{\delta E}. \tag{4.31}$$

The cluster idea has been successfully implemented both for elemental and binary semiconductors (Si, Ge, As, Se, Te) [1-3]. Special attention was paid to binary semiconductors such as $As_x Se_{1-x}$ and $Sb_x Se_{1-x}$ [2,12] (prospective materials for systems for optical recording). The energy gap calculations were based on the cluster approach and on using the concept of statistical weighting of basic clusters .

In order to reproduce the specified "x" for a system $A_x B_{1-x}$, it is necessary to find the most probable statistical weights α_i of the clusters in this system, specifically

$$A_x B_{1-x} \sim \sum_i^N [A_{k_i} B_{l_i}], \tag{4.32}$$

where $[A_{k_i} B_{l_i}]$ is the symbolic notation for a cluster, k_i and l_i are the numbers of atoms of types A and B in the ith cluster, and N is the number of basic clusters. Further, to obtain the averaged physical characteristics of the system as a whole (e.g., for the density of electronic states) it is necessary to use relations of the form

$$\rho(E) = (\sum_{i}^{N} \alpha_i \rho_i(E)) / (\sum_{i}^{N} \alpha_i). \tag{4.33}$$

4.4. CONDUCTIVITY

The calculations of conductivity are based on the Kubo-Greenwood formula

$$\sigma_E(\omega) = (\pi\Omega/4\omega) \int [f(E) - f(E + \hbar\omega)] |D_E|^2 \rho(E)\rho(E + \hbar\omega)dE, \tag{4.34}$$

where $f(E)$ is the Fermi-Dirac distribution function, $D_{E',E} = \int_{\Omega} \psi_{E'}^* \nabla \psi_E d\mathbf{r}$,

$\psi_{E(\mathbf{K})} = A\exp(i\mathbf{Kr})$, $\mathbf{K}$ is the complex wave vector of the effective medium. The dispersion relation $E(\mathbf{K})$ determines the properties of the wave function $\psi_{E(\mathbf{K})}$ on the isoenergy surface in $\mathbf{K}$ space. The imaginary part of $\mathbf{K}$ causes damping of the electron wave and is due to the absence of long-range structural order.

The second possibility for estimating the conductivity comes from the Thouless model [20], where the loss of phase of the electron wave is linked with the structural disorder and is taken into account in a purely phenomenological way through the coherence length λ. The complex wave number $K_R + i(1/2\lambda)$ in this model can now be related to the calculated dispersion relation $E(\mathbf{K})$. Thus, if we calculate the coherence length

$$\lambda = \frac{1}{2}(\text{Im}\, K(E))^{-1}, \tag{4.35}$$

then the conductivity can be estimated as

$$\sigma_E(\omega) = \frac{16}{3}\pi^2\lambda\rho^2(E_F)[1 + \frac{\omega^2\lambda^2}{4E_F}]^{-1}. \tag{4.36}$$

Finally, one must also devote some attention to the effective mass of the electron, which is an extremely nontrivial dynamical characteristic for a disordered structure. Formally, one can obtain an estimate for the effective mass on the basis of the known dispersion relation $E(\mathbf{K})$, in the form

$$m^* = (\partial^2 E / \partial K_R^2)^{-1}. \tag{4.37}$$

The same idea of statistical weighting is used for conductivity calculations of As_xSe_{1-x} and Sb_xSe_{1-x}. The calculations of effective masses should be particularly noted [2]. Many singularities produce the conditions of localization and delocalization energy points and bands in these complicated highly disordered systems.

5. Vibrational spectra calculations: quasi-one-dimensional chains

The cluster approach is developed for vibrational spectra calculations of quasi-one-dimensional chains. Amorphous Se-like systems can be considered as an assembly of finite size chains and rings. Spectral pecularities are formed on the basis of chain and interchain vibration. The main problem here is the quasi-elastic medium properties calculation. Possible variants for such a calculation are CPA or quasi-elastic force constant estimations. In particular, some successful estimations of the force constant for S , Se and Te have been carried out [19].

6. Elementary excitations in solids after heavy ion passages

Various elementary excitations take place in solids after heavy ion passages. Swift heavy ions induce different types of excitation waves (electronic, phononic, excitonic etc) and phase transition in solid targets. It is worth discussing here a large scale of *in situ* and post-experimental measurements which demonstrate relaxations processes and new phase states in track volumes and in whole matter.

6.1. A MODEL OF CORRELATED TRACK EXCITATIONS

We formulate now the basis of a model of **correlated track excitations (CTE):**
a) an undisturbed surrounding defines the times of relaxation processes within the track volume;
b) any excitation energy within the excited track is taken out of a track "cylinder" by a corresponding
       channel of excitations in undisturbed matter;
c) dispersion laws of solids give the velocities of excitation propagation in the undisturbed matter;
d) any mode of considered excitation is the channel for the energy transfer.

A heavy ion passage within matter produces various scales of excitations which originate due to a non-steady, non-equilibrium short-life plasma in the track after the ion passes. The instant energy density within the elementary track volume is defined by the energy losses (dE/dx) and equals:

$$w = w(\mathbf{r}, t) = (dE / dx) / A, \qquad (6.1)$$

where A is a cross-section of a track, but w(r,0) is the starting point for calculations of the energy density relaxation function w(r,t).

The energy exchanges between the track volume and the "rest" medium are dictated by the medium through its dispersion laws. The first step of our investigation is the problem of cooling of the track fragment after the swift heavy ion passage. Moreover, we can get important information on cooling rates which are responsible for the post-track matter state, because the cooling rates determine the conditions of a matter state within the track volume.(As it is known, a cooling rate define final states of a matter and is a determinant in prepairing technologies of glasses). This analysis is based on the local energy conservation law:

$$\frac{\partial w}{\partial t} + \operatorname{div} \mathbf{S} = \sum_{i,j} \sigma_{ij}^{\pm}(\mathbf{r}, t), \qquad (6.2)$$

where S is the vector energy flux intensity, $\sigma_{ij}^{\pm}(\mathbf{r}, t)$ are the point energy sources within the track volume ("+" - the energy generators,"-" - the energy annihilators). Thus, the main goal of modelling is the energy loss estimations. We can factorize the problem of tested track volume cooling, considering partial contributions of energy flux intensity into the total intensity as:

$$\mathbf{S} = \mathbf{S}_{electron} + \mathbf{S}_{phonon} + \mathbf{S}_{solitons} + \mathbf{S}_{plasmon} + \mathbf{S}_{magnon} + \dots \quad , \qquad (6.3)$$

where

$$\mathbf{S}_{electron} = w_{electron} \mathbf{v}_{electron} \, , \qquad (6.4)$$

$$\mathbf{S}_{phonon} = w_{phonon} \mathbf{v}_{phonon} \, , \qquad (6.5)$$

and so on. The velocities can be found from the dispersion relations:

$$\mathbf{v}_{electron}(E) = \frac{1}{\hbar} \frac{\partial E}{\partial \mathbf{k}} \, , \qquad (6.6)$$

$$\mathbf{v}_{phonon}(\omega) = \frac{\partial \omega}{\partial \mathbf{k}}.$$ (6.7)

These relationships (6.1)-(6.7) show connections between track excitations and solid states. The velocities of excitation propagation define the times of relaxation processes within a matter.

6.1.1. Stopping energy calculations.

The right side contribution $(-\frac{\partial w}{\partial t})_{point}$ in equation (6.2) is defined by the relaxation mechanisms of different excitations, w-the local energy density, S=wv - the intensity of energy flux. Then,

$$div(w\mathbf{v}) = wdiv\mathbf{v} + \mathbf{v}\nabla w,$$ (6.4)

where **v** - the ion (or energy carrier) velocity. We assume that $div\mathbf{v} \approx \frac{\partial v}{\partial x}$, but the estimation of grad w not so evident.

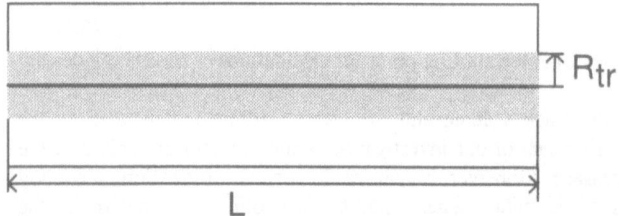

Figure 3. A model of a primary excited track volume

The first step of our investigation is connected with the consideration of the primary cylindrical track volume, where the lost ion energy is concentrated. So, our goal is an estimation of the energy density w(r, x, t) distribution along the ion track as the initial condition for the excitation relaxations after an ion passage.

$$E \approx E(\tau) = \int_0^L \int_0^{R_r} w(r,x,\tau)2\pi r dx,$$ (6.5)

where τ - the relaxation time. This energy initiates the secondary perturbation in a lattice of an irradiated sample over the primary track.

6.2. PERCOLATION OF CONDUCTIVITY OF IRRADIATED METAL TARGETS

This section is devoted to some problems of the glassy state of metals and its ion beam induced modification. We pay special attention to Fe-B-Si-C systems. Glassy metals as objects of radiation damage investigations are very convenient because practically all solid state experimental methods (EXAFS, PES, conductivity measurements, magnetic investigations etc) can be applied. In particular, there are well-known measurements of resistance changes of glassy metallic films irradiated by heavy ion beams.

From the other point of view, glassy metals are ideal objects for the above-mentioned cluster calculations of electronic structure and conductivity. Thus, it is possible to simulate the resistance changes by means of the developed cluster approach.

Another simulation idea is based on the percolation theory concept [15,16]. The irradiation process produces regions with new specific resistivities as a result of phase transitions within ion tracks. In particular, numerical estimations of resistance changes of Fe-B-Si-C are in good agreement with experiments [5], where the growth of resistance difference is demonstrated.

6.3. DEFECT CREATION IN ALKALI HALIDES

High-energy ions produce a wide scale of different defects in ionic crystals. Among the above-mentioned defects, we can define some types of point defects: vacancies (defects of Shottky and Frenkel), and colour centers. F-centers and their aggregates are the most significant ones which define luminescence spectra [10,11].

6.3.1. On the generation of F-centers and their aggregates in LiF within the track volume after a heavy ion passages

We now pay special attention to LiF. A passage of a high-energy heavy ion through a crystal allows us to consider it initially as a dense molecular gas in which some chemical reactions take place. We note only the main reactions:

$$LiF \leftrightarrow Li^+ + F^- - \varepsilon_q \; , \tag{6.6}$$

$$LiF \leftrightarrow Li + F - \varepsilon_d \; , \tag{6.7}$$

$$F^- \leftrightarrow F + e - \varepsilon_a \; , \tag{6.8}$$

$$Li \leftrightarrow Li^+ + e - \varepsilon_i \; . \tag{6.9}$$

Thus, a primary excited state of a track volume is a high energy plasma which relaxes to a defect saturated state which is "frozen" in the destroyed crystal after fast cooling of an ion trace. The reactions (6.6)-(6.9) form the basis of plasma-like states of excited track matter in LiF. Of course, we wish to take into account more complicated kinetic processes, e.g., reactions of deeper ionizations of atoms.

Now, let us define activation energies in reactions (6.6)-(6.9): ε_c -the cohesion energy per unit molecule LiF in crystal, ε_d -the dissociation energy per unit molecule, ε_i -the energy of ionization of Li, ε_a -the energy of activation of F. These values are presented in Table 1.

TABLE 1. Activation energies of LiF [7]

$\varepsilon_{c,}eV$	$\varepsilon_{d,}eV$	$\varepsilon_{j,}eV$	$\varepsilon_{a,}eV$
10.52	8.52	5.44	3.44

6.3.2. Theoretical estimations of partial particle concentrations of track plasma components

We start our calculations assuming that the partial distributions of track plasma components (i.e., LiF molecules, neutral atoms Li and F, ions Li$^+$ and F$^-$, electrons (holes)) produce after relaxation and cooling, all possible defects in a track volume. We also consider the existence of local equilibrium of plasma components at the moment of the ion passage. This equilibrium is practically conserved (or frozen) due to very short times of cooling. In general, our concept is based on the balance of statistical sums of all the components in any reaction [7]. Thus, for a general reaction,

$$v_A A + v_B B \leftrightarrow v_C C + v_D D \tag{6.10}$$

we can write the relationship:

$$\prod_A [\frac{1}{n_A}(\frac{2\pi m_A kT}{h^2})^{3/2}(Q_i)_A]^{v_A} = 1, \tag{6.11}$$

where A, B, C, D-reaction components, v_A, v_B, v_C, v_D- integer numbers, n_A, m_A, Q_A -the concentration, the mass and the statistical sum of A-component, respectively. We begin our consideration with the reaction (6.9). Using the expression (6.11), we obtain immediately:

$$\frac{n_{Li^+} n_e}{n_{Li}} = (\frac{2\pi kT}{h^2})^{3/2}(\frac{m_{Li^+} m_e}{m_{Li}})^{3/2}\frac{Q_{Li^+} Q_e}{Q_{Li}}, \tag{6.12}$$

where

$$Q_{Li} \approx (2I+1)2, \tag{6.13}$$

$$Q_{Li^+} = (2I+1)\exp(-\frac{\varepsilon_i}{kT}), \tag{6.14}$$

$$Q_e = 2, \tag{6.15}$$

$(2I+1)$- the generation factor on the nuclear spin. Then, instead of (6.12) we can write

$$\frac{n_{Li^+} n_e}{n_{Li}} = (\frac{2\pi m_e kT}{h^2})^{3/2}(\frac{g^+ g_e}{g})^{3/2}\exp(-\frac{\varepsilon_i}{kT}), \tag{6.16}$$

where $g_e g^+ / g = 1$ and g_e -the g-factor of electron. In a similar method, we consider other reactions. Thus,

$$\frac{n_F n_e}{n_{F^-}} = (\frac{2\pi kT}{h^2})^{3/2}(\frac{m_F m_e}{m_{F^-}})^{3/2}\frac{Q_F Q_e}{Q_{F^-}},$$ (6.17)

$$Q_F \approx (2I+1)2 ,$$ (6.18)

$$Q_{F^-} = (2I+1)\exp(-\frac{\varepsilon_a}{kT}),$$ (6.19)

$$Q_e = 2 ,$$ (6.20)

$$\frac{n_F n_e}{n_{F^-}} = (\frac{2\pi m_e kT}{h^2})^{3/2}(\frac{gg_e}{g^-})^{3/2}\exp(-\frac{\varepsilon_a}{kT}).$$ (6.21)

For a LiF -molecule's decomposition according to reactions (6.6) and (6.7), we have two main possibilities of concentration redistributions:

$$\frac{n_{Li^+} n_{F^-}}{n_{LiF}} = (\frac{2\pi kT}{h^2})^{3/2}(\frac{m_{Li^+} m_{F^-}}{m_{LiF}})^{3/2}\frac{Q_{Li^+} Q_{F^-}}{Q_{LiF}},$$ (6.22)

$$\frac{n_{Li} n_F}{n_{LiF}} = (\frac{2\pi kT}{h^2})^{3/2}(\frac{m_{Li} m_F}{m_{LiF}})^{3/2}\frac{Q_{Li} Q_F}{Q_{LiF}} .$$ (6.23)

However, we must take into account the more complicated character of molecular statistic sums. The general expression of a molecular statistical sum looks as follows:

$$Q = Q_t Q_r Q_v \ldots$$ (6.24)

where Q_t, Q_r, Q_v - are the translational, rotational and vibrational parts, respectively. Thus, the LiF - molecule statistical sum can be presented as:

$$Q_{LiF} = Q_t Q_r Q_v = (\frac{2\pi kT}{h^2})^{3/2}(\frac{m_{Li} m_F}{m_{LiF}})^{3/2}(\frac{8\pi^2 I_m kT}{h^2})(1-\exp(-\frac{h\nu_0}{kT}))^{-1} ,$$ (6.25)

where $I_m = \frac{m_{Li} m_F}{m_{Li} + m_F} r_e^2$ -the moment of inertia of molecule LiF, ν_0 - the eigen vibration frequency ($h\nu_0 \approx 0.0345 eV$, [9]). Now we can rewrite (6.22) and (6.23) in the form:

$$\frac{n_{Li^+} n_{F^-}}{n_{LiF}} = (\frac{2\pi kT}{h^2})^{3/2}(\frac{m_{Li^+} m_{F^-}}{m_{LiF}})^{3/2}\exp(-\frac{\varepsilon_d}{kT})(1-\exp(-\frac{h\nu_0}{kT}))^{-1},$$ (6.26)

$$\frac{n_{Li} n_F}{n_{LiF}} = (\frac{2\pi kT}{h^2})^{3/2}(\frac{m_{Li} m_F}{m_{LiF}})^{3/2}\exp(-\frac{\varepsilon_q}{kT})(1-\exp(-\frac{h\nu_0}{kT}))^{-1} .$$ (6.27)

Using equations (6.26) and (6.27), we can estimate a new balance of concentrations

$$\frac{n_{Li^+} n_{F^-}}{n_{LiF}} = C_{charge},$$ (6.28)

$$\frac{n_{Li} n_F}{n_{LiF}} = C_{neutral}.$$ (6.29)

However, we must pay more attention to reactions which can produce free electrons and neutral F atoms, following our interest to F-centers and their aggregates. In this respect, our primary interest is connected with the expression (6.26). In this way, decomposition of LiF leads us to the estimation of concentration n_F. Using equations (6.17) and (6.23), we obtain

$$\frac{n_{Li^+} n_e}{n_{Li}} \frac{n_{F^-}}{n_F n_e} = \frac{n_{Li^+} n_{F^-}}{n_{Li} n_F} = C_{in},$$ (6.30)

$$\frac{n_{Li} n_F}{n_{LiF}} = \frac{C_{charge}}{C_{in}},$$ (6.31)

$$\frac{n_F}{n_{Li}} = \exp(\frac{\varepsilon_a - \varepsilon_i}{kT}),$$ (6.32)

$$n_{F_0} = \sqrt{(n_{LiF})_{new} \exp(-\frac{\varepsilon_a - \varepsilon_i}{kT})}.$$ (6.33)

Here we must point out the difference between old and new concentrations of LiF, namely $(n_{LiF})_{old}$ and $(n_{LiF})_{new}$, respectively. The $(n_{LiF})_{old}$ is the initial concentration of LiF molecules before an ion passage.

6.3.3. Defect creation in heavy ion tracks.

The defect creation in solids by irradiation with heavy ions with specific energy of 10 MeV/u occurs under a high excitation energy (dE/dx) of several kiloelectronvolt per Ångström [21, 22]. In LiF crystals the lateral energy loss distribution according to [23, 24] is concentrated in a cylindrical zone of the radius (r_{ex}) of 10-20 Å (90%). In this central cylindrical region around the ion path, the energy transfer to each ion exceeds the energy of self-trapped excitons and an exciton plasma state can be created [21,25]. Therefore, the defect creation in the central zone with $r_{ex} \approx 10$ Å can be different from that at a lower excitation density in the lateral track region with $r_{lat} > r_{ex}$.

 Radiation damage creation (F-,F$_2$- and colloidal centers) were studied in LiF crystals irradiated with Zn, Se, Xe, Au, and U (the specific energy of the ions was 11.4 MeV/u). The concentration of F- and F$_2$-centers (N_F, N_{F_2}) was estimated by optical spectroscopy and the presence of macroscopic aggregates was detected by chemical etching [25] and by small angle X-ray scattering (SAXS) [10]. The energy to create on F-center (or more correctly, one Frenkel pair) ΔE_F was estimated according to $\Delta E_F = E_0 \Phi / N_F R_p$, where E_0 is the total energy of the ion, Φ is the fluence (ions/cm^2), and R_p is the range of the ion. The efficiency of F- and F_2-center creation increases with the the increasing dE/dx from Zn to Xe (ΔE_F are 1200 eV (Zn), 1000 eV (Se), and 900 eV (Xe)). A further increase of dE/dx leads to a decrease of the efficiency of F- and F_2- center creation (ΔE_F is 1200 eV (Au) and 1400 eV (U)). If the value of dE/dx exceeds a critical (dE/dx)$_{crit}$ $\approx$ 1.2 keV/ Å macroscopic aggregates (Li colloids) are created in the central cylindrical region around the ion path with radius $r_{1,lat}$ from 10-15 Å [10]. Such macroscopic aggregates were observed in LiF crystals irradiated with Se, Xe, Au, and U.

The diffraction contrast and the magnitude of $r_{1,lat}$ increase with the magnitude of dE/dx. For Se only a weak difraction contrast by SAXS was detected due to the small region of the aggregate center formation along the ion track (only in the range of ΔR_p $\approx$ 7 Å around the Bragg maximum, the magnitude of R_p= 92 µm). In LiF irradiated with Au and U ions, the difraction contrast of SAXS was the strongest and the value of $r_{1,lat}$ $\approx$ 15 Å. The correlation of the increase of ΔE_F with the formation of macroscopic aggregate centers (Li colloids) can be explained as the decrease of the efficiency of single F-center creation due to their participation in the colloid center formation. The estimation of the distribution radius $r_{2,lat}$ of single F - and F_2-centers according to the model of Thevenard (see [22]) leads to a value of $r_{2,lat}$ $\geq$ 400 Å which is much larger than $r_{1,lat}$. The concentration of F-centers per single track is of the order from 10^5 (Zn, Se) to 10^6 (Xe, Au, U) independently from the fluence (in the range of 10^9- $5 \cdot 10^{10}$) which demonstrates that no overlapping of single tracks take place.

The macroscopic aggregate center fomation is possible if the local concentration of F-centers exceeds a critical value and if the hole centers complementary to F-centers are spatially separated [25-27]. The critical concentration depends on the magnitude of dE/dx and the spatial separation probably occurs by the diffusion of hole centers which are much more mobile than F-centers.

Thus, the track damage morphology of heavy ion tracks is complicated. For numerical simulation, additional experimental and theoretical investigations are necessary.

6.3.4. Relaxation processes within the track plasma

The second step of our numerical investigations is the study of relaxation processes which take place as a result of a heavy ion passage. There are some competing processes in these conditions. We must mention here the primary plasma recombination, Coulomb explosion and ambipolar diffusion. But the role of different relaxation processes, which have their own specific relaxation times, is not the same.

With alkali halides, the main channels of energy exit are connected with excitonic and phononic waves. This leads to the creation of definite concentrations of charge centers and dislocations around the single track volume. The situation changes qualitatively by increasing ion fluences when saturation can provide a threshold-like transformation of matter properties, similar to percolation mechanisms.

We consider here the plasma decay after a heavy ion passage through matter. Suppose a heavy ion track in matter is considered as a linear plasma source, then we must solve the problem of charge diffusion within the track plasma. The main difficulties are connected with the definition of the kinetic coefficients of the plasma states [13-14].

Diffusion Model of Plasma Decay. Main relationships of ambipolar plasma diffusion are as follows:

$$\frac{\partial n}{\partial t} + \nabla \Gamma_j = 0, \tag{6.34}$$

$$\Gamma = -D\nabla n , \tag{6.35}$$

$$\Gamma = \mu_i n\mathbf{E} - D_i\nabla n = -\mu_e n\mathbf{E} - D_e\nabla n , \tag{6.36}$$

$$\mathbf{E} = [(D_i - D_e)/(\mu_i - \mu_e)](\nabla n / n) , \tag{6.37}$$

$$\Gamma = -\nabla n(\mu_i D_e + \mu_e D_i)/(\mu_i + \mu_e) , \tag{6.38}$$

the ambipolar diffusion coefficient:

$$D_a \equiv (\mu_i D_e + \mu_e D_j)/(\mu_i + \mu_e) \tag{6.39}$$

Assume, the charge concentration as,

$$n(\mathbf{r},t) = T(t)S(\mathbf{r}) \tag{6.40}$$

then we obtain the equation of the cylindrical region for the spatial part -

$$d^2S / dr^2 + (1/r)dS / dr + (1/D\tau)S = 0. \tag{6.41}$$

However, in any case, we need some hypothesis on recombination effects , e.g. , in the form

$$\frac{\partial n}{\partial t} = -\alpha n^2 \tag{6.42}$$

as an additional part of a diffusion equation.

The third step is connected with the solutions of plasma equations (e.g., Fokker-Plank-like equations) self-consistently with undisturbed matter to obtain a spectrum of the nonequilibrium temporal plasma and spatial-time distribution of energy density w(**r**, t). The main goal of this step is the formulation of self-consistent conditions for the waves of excitation within and out of the track. Namely, we must sew the "internal" and "external" wave functions of any excitation on the boundaries of a track volume. The external wave functions are found from the solution of an undisturbed solid state problem (e.g., from the cluster approach). Thus, we must "precisely" solve the problem within the track, and the undisturbed matter plays the role of the effective medium.

7. Conclusions.

1. The cluster approach is the conceptual foundation for the description of fundamental properties of disordered solids.
2. The method of potential calculation (atomic, crystalline, inter-atomic) is developed on the basis of analytical functions.
3. The electronic structure calculations of disordered semiconductors in the framework of the cluster approach showed the principal role of a short range order (cluster with the sizes of 1-2-3 coordinating spheres).
4. The cluster model of binary covalent compounds is based on the concept of basic clusters, presented in the atomic structure with different statistical weights.
5. A modelling of photo-induced processes is carried out by the EDOS calculations of different atomic configurations.
6. An application of the cluster approach for a direct atomic structure modelling, vibrational spectra and conductivity calculations is possible and has provided known results.

7. The developed cluster approach and the model of correlated track excitations allow one to simulate a large scale electron-atomic correlation in solids induced by heavy ion passages.

Acknowledgements

This work is partly supported by scientific grant No. 93.197 of Latvian Counsel of Science. The authors also express their thanks to the Head of Material Research Division of GSI (Darmstadt) Dr. N.Angert, Prof. R. Neumann (GSI), Dr. C.Trautmann (GSI), and Prof. H. Fue, (Technical University of Darmstadt) for support of this work and fruitful discussions.

References

1. Shunin, Yu.N., Shwartz K.K.(1986) Calculation of the electronic structure in disordered semiconductors, *phys.stat.sol(b)* **135**, 15-36.
2. Shunin,Yu.N., Shwartz K.K. (1994) Electronic structure and conductivity of disordered binary semiconductors, *Tech.Phys.* **39**, 1025-1031.
3. Shunin Yu.N., and Shwartz (1991) Cluster approach for disordered solids, *Latv. J. Phys. Tech. Sci.* No.3, 3-29.
4. Shunin, Yu., Schwartz, K.K., Trautmann, C. (1996) Elementary excitations and radiation defects in glassy metals induced by swift heavy ions, *GSI scientific reports*, Darmstadt. 25 p. (in print)
5. Audouard A., Balanzat E., Jousset J.C., Lesueur D., Thome (1993) Atomic displacements and atomic motion induced by electronic excitation in heavy-ion- irradiated amorphous metallic alloys , *J.Phys.:Condens. Matter* **5**, 995-1018
6. Fischer, B.E., and Spohr, R. (1983) Production and use of nuclear tracks:imprinting structure on solids, *Rev.Mod.Phys.* **55**, 907-948.
7. Heer, C.V. (1972) *Statistical Mechanics, Kinetic Theory, and Stochastic Processes*, Ac.Press, N.-Y.-London.
8. Evarestov, R., Kotomin, E.A., Ermoshkin, A.N.(1983) *Molecular models of point defects in wide-gap solids*, Zinatne, Riga (in Russian)
9. Kittel, Ch. (1975) *Introduction to the Solid State Physics*, John Wiley and Sons Inc., N.Y.-London-Sydney- Toronto.
10. Schwartz, K., Trautmann, C., Steckenreiter, T. (1995) Ion track in LiF crystals, *GSI Jahresbericht*, Darmstadt, 221.
11. Schwartz, K., and Ekmanis, Yu. (1989) *Radiation damage and radiation sensitivity of dielectric materials*, Zinatne Publ. House, Riga.
12. Shunin, Yu.N. (1989) Composition influence on the electronic structure of amorphous As_xSe_{1-x}, *Fiz.Tekh.Polupr.* **23**, 1049-1053
13. Chen, F.F. (1984) *Introduction to Plasma Physics and Controlled Fusion.Vol.1. Plasma Physics*, Plenum Press, N.-Y..
14. Birdsall, Ch.K., Langdon, A.B. (1985) *Plasma Physics , via Computer Simulation*, McGraw Hill, N.-Y.
15. Kirkpatrick, S. (1973) Percolation and conduction , *Rev.Mod.Phys.* **45**, 574-588
16. Isichenko, M.B. (1992) Percolation, statistical topography, and transport in random media, *Rev.Mod.Phys.* **64**, 961-1043.
17. Ziman, J.M. (1979) *Models of disorder*, Cambridge University Press, Cambridge- London-N.-Y.-Melbourne
18. Kuni, F.M. (1981) *Statistical Physics and Thermodynamics*, Nauka, Moscow (in Russian)
19. Shunin, Yu.N. (1991) *Modelling of atomic and electronic structure of disordered semiconductors* , Dr.Sc.Hab.Thesis, Physics Inst., Riga-Salaspils.
20. Thouless, D.J. (1972) *The quantum mechanics of many-body systems*, Academic Press, N.-Y. -London
21. Itoh, N., Tahimura, K. (1986) Radiation effects in ionic solids, *Rad.Eff.* **98**, 269-287
22. Perez ,A., Balanzat E., Dural, J. (1994) Experimental study of point defect creation in high energy heavy-ion tracks, *Phys.Rev. B* **41**, 3943-3950
23. Krämer, M., Kraft, G. (1994) Monte-Carlo calculations of the lateral energy distribution, *Radiat.Environ.Biophys.* **33**, 91-101.
24. Katz, R., Loth, K.S., Daling, L., Huang, G.R. (1990) An analytic representation of the radial distribution of dose from energetic ions in water, Si, LiF, NaI, and SiO_2, *Rad.Eff.* **114**, 15-20
25. Schwartz, K. K. (1996) Electronic excitations and defect creation in LiF crystals, *Nucl. Inst.Meth. B* **107**, 128-132
26. Davidson, A.T., Comins, J.D., Derry, T.E., Khumalo, F.S. (1986) The production of defects and colloids in lithium fluoride crystals by implantation with rare gas ions, *Rad.Eff.* **98**, 305-312.
27. Hudges, A.E. (1979) Metal colloids in crystals, *Adv. Phys.* **28**, 717-828.

Structure and Properties

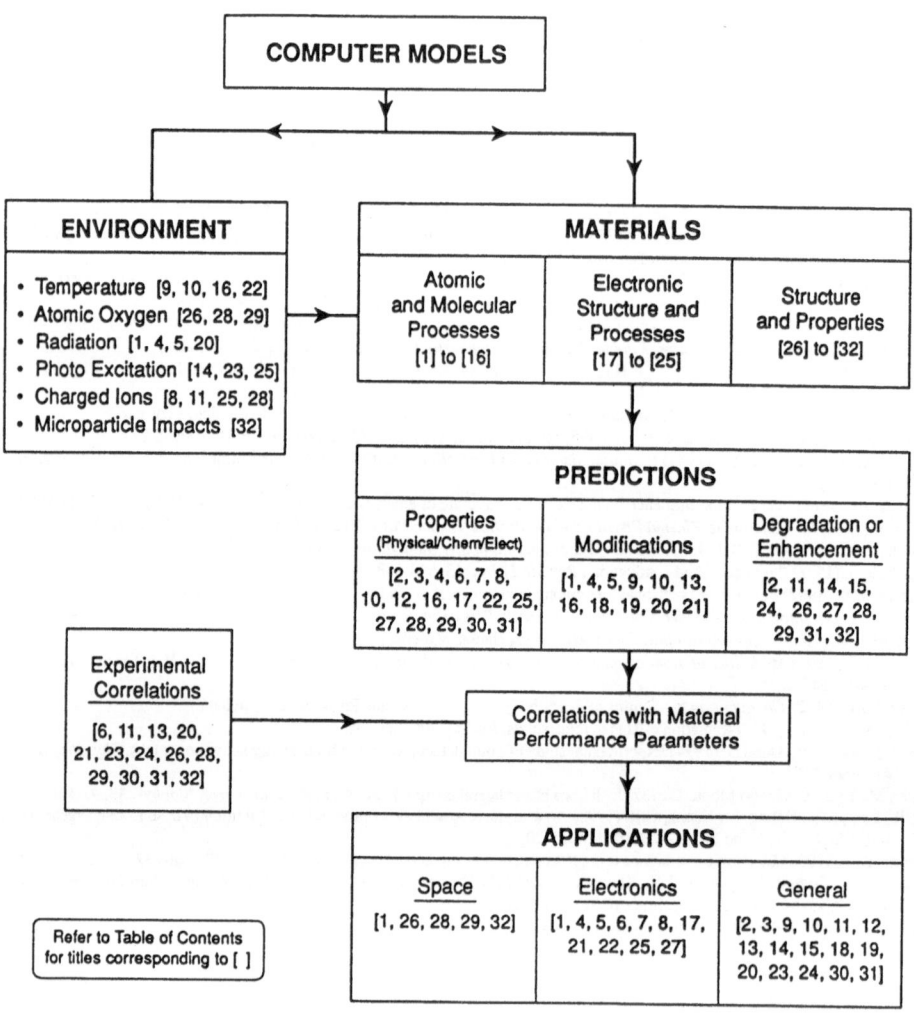

COMPUTER MODELS

ENVIRONMENT

- Temperature [9, 10, 16, 22]
- Atomic Oxygen [26, 28, 29]
- Radiation [1, 4, 5, 20]
- Photo Excitation [14, 23, 25]
- Charged Ions [8, 11, 25, 28]
- Microparticle Impacts [32]

MATERIALS

| Atomic and Molecular Processes [1] to [16] | Electronic Structure and Processes [17] to [25] | Structure and Properties [26] to [32] |

PREDICTIONS

| Properties (Physical/Chem/Elect) [2, 3, 4, 6, 7, 8, 10, 12, 16, 17, 22, 25, 27, 28, 29, 30, 31] | Modifications [1, 4, 5, 9, 10, 13, 16, 18, 19, 20, 21] | Degradation or Enhancement [2, 11, 14, 15, 24, 26, 27, 28, 29, 31, 32] |

Experimental Correlations

[6, 11, 13, 20, 21, 23, 24, 26, 28, 29, 30, 31, 32]

Correlations with Material Performance Parameters

APPLICATIONS

| Space [1, 26, 28, 29, 32] | Electronics [1, 4, 5, 6, 7, 8, 17, 21, 22, 25, 27] | General [2, 3, 9, 10, 11, 12, 13, 14, 15, 18, 19, 20, 23, 24, 30, 31] |

Refer to Table of Contents for titles corresponding to [ ]

MONTE CARLO COMPUTATIONAL TECHNIQUES FOR PREDICTION OF ATOMIC OXYGEN EROSION OF MATERIALS

BRUCE A. BANKS, THOMAS J. STUEBER,
NASA Lewis Research Center, Cleveland Ohio
NYMA Inc., Brook Park, Ohio

Abstract

Materials on the surface of spacecraft in low Earth orbit (LEO) are exposed to the remnants of the Earth's upper atmosphere. Energetic solar photons cause photodissociation of O_2 to produce highly reactive atomic oxygen. As spacecraft orbit through the atomic oxygen, impact energies of 4.5 $\pm$ 1 eV result with an arrival flux sufficient to cause polymeric materials to be oxidized at rates high enough to cause durability concerns. To increase materials durability adequate to meet spacecraft mission lifetime requirements, atomic oxygen protective coatings have been applied over polymers. Such coatings typically consist of metal oxide thin films. The durability of such protected polymers used for solar array blankets and thermal control is limited as a result of microscopic defects in the protective thin films.

Ground laboratory thermal energy ($\sim$ 0.04 eV) atomic oxygen plasmas are typically used to evaluate the protective quality of metal oxide coated polymers. However, the projection of in-space durability of these materials based on thermal energy ground laboratory testing is not obvious as a result of differences in atomic oxygen arrival directions and energies. A Monte Carlo computational model has been developed which is capable of modeling the atomic oxygen erosion effects at defect sites in protective coatings for both thermal energy plasma and LEO environments. This two dimensional model is based on detailed mechanistic interaction information such as: reaction probability dependence on oxygen atom impact energy and angle of attack, probability of recombination, mix of specular and diffuse ejection of unreacted oxygen atoms, degree of thermal

R. C. Tennyson and A. E. Kiv (eds.),
Computer Modelling of Electronic and Atomic Processes in Solids, 259–270.
© 1997 *Kluwer Academic Publishers.*

accommodation of unreacted oxygen atoms, and energy and angular-arrival distribution of impacting oxygen atoms. The simulated effects of atomic oxygen erosion are produced by allowing thousands of model oxygen atoms to enter through a model defect in a protective coating to oxidize the underlying polymeric material. The polymer is represented by an orthogonal array of oxidizable (removable) cells. Atoms entering the computational model defect can react with the polymer, recombine on the surface to produce non-reactive O_2 or not react and scatter based on assigned probabilities and mechanistic interaction behaviors previously outlined.

Although optimal quantification of the various possible interaction parameters has not been completed, reasonable assumptions for the values have produced computational predictions in close agreement with experimentally observed results from both ground laboratory plasma and in-space testing. Figure 1 shows the results of atomic oxygen attack at the site of a wide crack in a protected coating over a hydrocarbon polymer exposed to a thermal energy isotropic oxygen plasma (Figure 1a) and to normal incident attack in LEO (Figure 1b). As can be seen from Figure 1, the resulting surface texture from isotropic plasma exposure is much smoother than the carpet-like texture from directed exposure in LEO. These results are very consistent with those experimentally observed in RF plasma ashers and in space. Other computational results assuming limiting values of interaction parameters as well as reasonable values for these parameters have produced credible replication of observed atomic oxygen undercutting phenomena.

1. Introduction

Materials on spacecraft in low Earth orbit (LEO) are exposed to atomic oxygen with impact energies of 4.5 ± 1 eV and an arrival flux sufficient to cause polymeric materials to be oxidized at rates high enough to cause durability concerns. The durability of polymers protected with metal oxide films is limited as a result of microscopic defects in the protective thin films.

Ground laboratory thermal energy (~ 0.04 eV) atomic oxygen plasmas are typically used to evaluate the protective quality of metal oxide coated polymers (1-3). However, the projection of in-space durability of these materials based on thermal energy ground laboratory testing is not obvious as a result of differences in atomic oxygen arrival directions and energies.

If one can develop a quantitative understanding of differences between ground laboratory thermal energy atomic oxygen erosion and in-space high energy erosion, then low-cost large-area ground laboratory thermal energy plasma facilities could be used to estimate in-space durability with acceptable accuracy. Computational techniques have been developed with reasonable success in simulating the effects of atomic oxygen attack at defect sites on protected polymers (4-5). The Monte Carlo computational technique presented in this investigation is based on mechanistic interaction characteristics of atomic oxygen attack at defect sites in protected polymers.

2. Computational Procedure

The Monte Carlo computational program was designed using Microsoft Visual C++ for operation in the Windows° environment on an 80486 or Pentium platform.

2.1. SIMULATION GEOMETRY AND MATERIALS

The Monte Carlo computational model is a two-dimensional model consisting of an orthogonal array of square cells. The model simulates a protected polymer (on the exposed surface or both surfaces of the polymer) which has a defect (a non-protected area) in the protective coating. Model atoms are made to enter the defect and impinge upon the polymer cells which may be oxidized and thus removed, causing the erosion geometry to develop. The orthogonal array of cells can be up to a 500 x 500 cell array.

2.2. ATOMIC OXYGEN ENERGY

The location where the model atom enters the defect is randomly selected; however, the direction of arrival can be unidirectional, isotropic with a flux dependence on the cosine of the angle from normal incidence (to simulate an oxygen plasma environment), or angularly sweeping with a cosine flux dependence (to simulate atomic oxygen arrival on solar oriented surfaces).

In LEO, even with surfaces oriented perpendicular to the ram velocity direction, atomic oxygen arrival is angularly and energetically distributed as a consequence of thermal and orbital inclination velocity contributions (6-7). The arriving atomic oxygen is slightly more angularly distributed in a plane parallel to the Earth's horizon. The Monte Carlo model allows one to select a horizontally or vertically distributed (search light-like) arrival or simply paraxial fixed arrival.

The energy of arriving atoms can be chosen to replicate the distribution of energies from either a laboratory or in-space environment, including a fixed energy or a Maxwellian distribution (based on an assumed temperature), or combined distribution based on Maxwellian thermospheric atoms and taking into account the orbital velocity and the orbital inclination.

2.3. ATOMIC OXYGEN INTERACTION MODELING ASSUMPTIONS

Figure 1 shows the geometric configuration of the Monte Carlo model. Impact with a cell is based on an atom impinging upon a cross centered within each cell.

If an atomic oxygen atom impacts a cell (as determined by hitting a cross) then one of three events can occur: 1) reaction and removal of the cell, 2) scattering without reaction, or 3) recombination to form non-reactive O_2. The probability of reaction depends upon the specific material being impacted. For polymers the probability of an oxygen atom reacting with a cell is dependent upon the energy of the impacting atom and the angle of attack (based on slope averaging of the cell and its adjoining neighbors). Based on results of in-space measurements of the angular dependence of erosion yield of polymers, the angular dependence of the reaction probability is $(\cos \Theta)^{1/2}$ where Θ is the angle of impact relative to the normal of the slope averaged polymer cell surface (8-9). A variety of energy dependent reaction probability functions can be employed. If, through random processes, the atom does not react, then a random selection is made to determine if the atom recombines. If it does not recombine, then the atom is ejected with an ejection angle which can be specular, diffuse (using a cosine weighted random angle) or any user selectable gradation between. The atom can be made to lose a specific or random fraction of the difference between impacting energy and a randomly selected energy based on a Maxwellian energy distribution for thermally accommodated atoms for a user specified temperature.

2.4. COMPUTATIONAL SPEED UP TECHNIQUES

The Monte Carlo computational model employs two techniques for speed up computation when needed. These speed up techniques are useful when low reaction probabilities exist such as for thermal energy or thermally accommodated atoms.

The first technique limits the number of bounces an atom is allowed to undergo when it is almost trapped in an undercut cavity with a small defect opening. When an atom completes a user selected number of impacts (without reacting) and it has lost sufficient energy to be within a user prescribed proximity to atoms of thermal accommodation, then a decision is made as to whether it will be considered as reacted (with the next polymer impact), ejected out the defect opening or recombined.

The second technique which can be used to speed up the program for low reaction probability atoms is to let each atom represent a large number or bundle of atoms, thus increasing the probability of one atom in the bundle reacting.

3. Results and Discussion

3.1. DEMONSTRATION OF THEORETICAL EXPECTED RESULTS

If one assigns a probability of atomic oxygen reaction equal to one and probability of recombination of unreacted atoms equal to one, then predicted erosion geometries simulating an isotropic arrival plasma, fixed direction normal incidence and in-space horizontal search light normal incidence exposures shown in Figure 2 (a), (b), and (c) respectively are as one would expect. Ideally the Figure 2(a) undercut geometry should approximate a circle as a result of the cosine distributed angular arrival of the atomic oxygen flux. Figure 2(b) should be a perfect rectangle and Figure 2(c) is simply a more divergent undercut pattern of 2(b) resulting from the off normal oxygen velocity components caused by LEO thermal and orbital inclination velocity contributions.

If one assumes in-space atomic oxygen incidence at 38° from perpendicular and an atomic oxygen protective coating on both surfaces of the polymer, a reaction probability equal to one and the probability of recombination of unreacted atomic oxygen equal to one for atoms impacting the polymer and zero for atoms impacting the protective coating the undercut cavities shown in Figure 3 (a), (b), and (c) are predicted. Figure 3(a) assumes 100% of the non-reacted atoms are specularly scattered. One of the consequences of specular scattering is a portion of the incident atoms reflect off the sidewalls of the defect and react as well as shown in Figure 3(a). Figure 3(b) assumes 100% diffuse scattering of unreacted atoms leaving a protective coating surface. As one would expect there are undercut features approximating circles at the bottom of the undercut cavity and at top from scattering off the hole walls). Figure 3(c) is identical to 3(a) with the exception of assuming a horizontal search light exposure. As one would expect, it is a divergent version of 3(a).

3.2. COMPARISONS WITH EXPERIMENTALLY OBSERVED RESULTS

Two characteristics of atomic oxygen attack of polymers that can be used to assess how well the Monte Carlo computational model replicates experimentally observed results are the resulting texture of polymer surfaces and the shape of undercut cavities below protected polymers. Figure 4 compares the Monte Carlo predicted erosion caused by an isotropic thermal energy atomic oxygen plasma (Figure 4(a)) with a scanning electron microscope photograph of polyethylene (Figure 4(b)) after exposure to isotropic thermal energy atomic oxygen in an RF plasma asher to an effective fluence of 2.9×10^{21} atoms/cm^2 (10). Figure 4 should be compared with Figure 5 which shows the predicted erosion caused by in-space normal incident horizontal search light arrival (Figure 5(a)) and a scanning electron microscope photograph of chlorotrifluoroethylene after exposure to fixed direction arrival LEO atomic oxygen on the Long Duration Exposure Facility to a fluence of 5.77×10^{21} atoms/cm^2 (10). As can be seen by comparison of Figures 4(a) and 5(a) as well as 4(b) and 5(b), both the Monte Carlo predicted and experimentally observed surface roughness is much greater for the in-space directed atomic oxygen incidence than for the isotropic arrival.

A comparison of the Monte Carlo computational model predicted atomic oxygen undercut geometries with those experimentally observed can be seen in Figures 6 and 7. Figure 6 compares the Monte Carlo predicted and experimentally observed undercutting at the site of a small defect in the protective coating over a polymer exposed to an isotropic thermal energy plasma. Both Monte Carlo model (Fig. 6a) and the RF plasma asher exposed sample (Fig. 6b) produce undercut cavities much wider than the defect width or diameter.

Figure 6(b) is a scanning electron microscope photograph of an aluminized Kapton H polyimide sample after exposure in a plasma asher to an effective fluence of 7.15×10^{21} atoms/cm^2 and after chemical removal of the aluminum coating. As can be seen in this figure, there are two slightly overlapping undercut cavities as a result of two defects located near each other. The average diameter of the original larger defect in the aluminum protective coating (not shown) was found to be 0.66 μm as measured by scanning electron microscopy prior to its removal. In contrast, the undercut cavity is approximately 41.6 μm in diameter.

Figure 7 compares the Monte Carlo computational model atomic oxygen undercut geometry with an experimentally observed sample retrieved from LDEF after exposure to an atomic oxygen fluence of 7.15×10^{21} atoms/cm^2. The sample consisted of aluminized polyimide Kapton H having a crack or scratch defect. As can be seen by the model predictions in Figure 7(a) and experimentally observed results in Figure 7(b), there is an unsymmetrical character to the undercut cavity close to the protective coating surface. Atomic oxygen atoms that have entered from the upper left in both the model and experiment have a finite probability of scattering off the right walls of the scratch defect in protective coating thus enhancing the degree of undercutting on the left side of the undercut cavity. In Figure 7(b) the scanning electron microscope photograph was taken with the sample tilted to be able to view down the undercut cavity. Thus, the cavity actually is inclined and matches the Figure 7(a) model predictions.

4. Summary

A Monte Carlo computational model has been developed which simulates the erosion of polymers caused by atomic oxygen reaction with the polymers at defect sites in protective coatings on the polymers. The two-dimensional model is based on theoretical and experimentally observed mechanistic interactions of atomic oxygen with materials whose oxides are fully volatile. The model is capable of simulating a variety of atomic oxygen environments including thermal energy isotropic plasmas as well as fixed direction or sweeping arrival in-space exposure. The Monte Carlo model predicted atomic oxygen erosion patterns are in reasonable agreement with theoretical results expected if one assumes reaction probabilities of unity. The model also produces predicted undercut cavity results for both RF plasma asher exposure and in-space LEO exposure which reasonably approximate observed experimental results.

5. Acknowledgements

The authors gratefully acknowledge the assistance of Kim K. deGroh and Daniela Smith for their valuable assistance in providing scanning electron microscopy information for this paper.

6. References

1. Banks, B.A., Rutledge, S.K., de Groh, K.K., "Low Earth Orbital Atomic Oxygen, Micrometeoroid, and Debris Interactions with Photovoltaic Arrays," paper presented at the 11th Space Photovoltaic Research and Technology Conference (SPRAT XI), NASA Lewis Research Center, Cleveland, OH, May 7-9, 1991.

2. Rutledge, S.K., Olle, R. and Cooper, J., "Atomic Oxygen Effects on SiOx coated Kapton for photovoltaic arrays in low Earth orbit," paper presented at the IEEE Photovoltaic Specialists Conference, Las Vegas, NV, October 7-11, 1991.

3. Banks, B.A., Rutledge, S.K., de Groh, K.K., Stidham, C.R., Gebauer, L. and LaMoureaux, C., "Atomic Oxygen Durability Evaluation of Protected Polymers using Thermal Energy Plasma Systems," NASA TM 106855, presented at the International Conference on Plasma Synthesis and Processing of Materials, sponsored by the Metallurgical Society, Denver, CO, February 21-25, 1993.

4. Banks, B.A., Mirtich, M.J., Rutledge, S.K., and Swec, D.I., "Sputtered Coatings for Protection of Spacecraft Polymers," NASA TM 83706, presented at the 11th International Conference on Metallurgical Coatings, sponsored by the American Vacuum Society, San Diego, CA, April 9-13, 1984.

5. Banks, B.A., Rutledge, S.K., and Gebauer, L., "SiO$_x$ Coatings for Atomic Oxygen Protection of Polyimide Kapton in Low Earth Orbit," AIAA paper 92-2151, presented at the Coating Technologies for Aerospace Systems Materials Specialist Conference, Dallas, TX, April 16-17, 1992.

6. de Groh, K.K., and Banks, B.A., "Atomic Oxygen Undercutting of Long Duration Exposure Facility Aluminized Kapton Multi-layer Insulation," J. Spacecraft and Rockets, Vol. 31, No. 45, p. 656-664. August, 1994.

7. Banks, B.A., de Groh, K.K., Bucholz, J.L. and Cales, M.R., "Atomic Oxygen Interactions with Protected Organic Materials on the Long Duration Exposure Facility (LDEF)," Proceedings of the Third Long Duration Exposure Facility (LDEF) Conference held in Williamsburg, VA, Nov. 8-12, 1993.

8. Banks, B.A., Auer, B., and DiFilippo, F.J., "Atomic Oxygen Undercutting of Defects on SiO$_2$ Protected Polyimide Solar Array Blankets," in Materials Degradation in Low Earth Orbit (LEO), Srinivasan, V. and Banks, B.A., eds., Proceedings of Symposium sponsored by the TMS, ASM Joint Corrosion and Environmental Effects Committee, held at the 119th Annual Meeting of the Minerals, Metals and Materials Society, in Anaheim, CA, Feb. 17-22-1990.

9. Banks, B.A., Dever, J.A., Gebauer, L., and Hill, C.M., "Atomic Oxygen Interactions with FEP Teflon and Silicones on LDEF," Proceedings of the First LDEF Post-Retrieval Symposium, Kissimmee, FL, June 2-8, 1991, pp. 801-815.

10. Banks, B.A., Rutledge, S.K., Hunt, J.D., Drobotij, E., Cales, M.R., and Cantrell, G. "Atomic Oxygen Textured Polymers", NASA Technical Memorandum 106769, April, 1995.

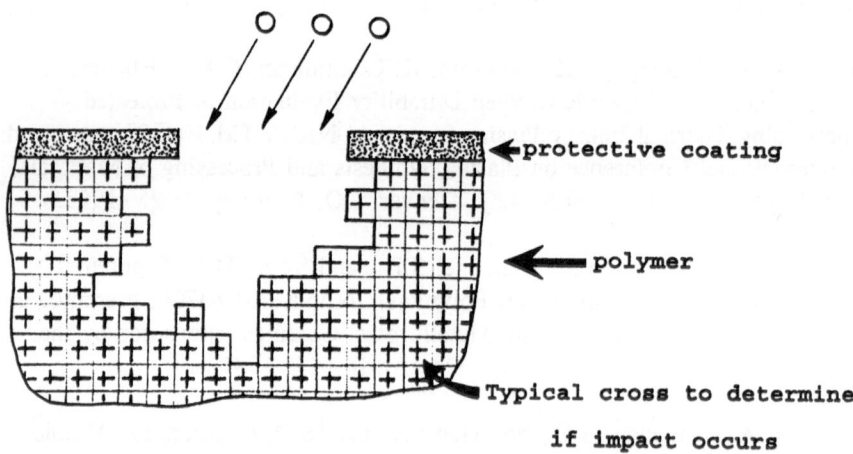

Figure 1. Geometric configuration of Monte Carlo model

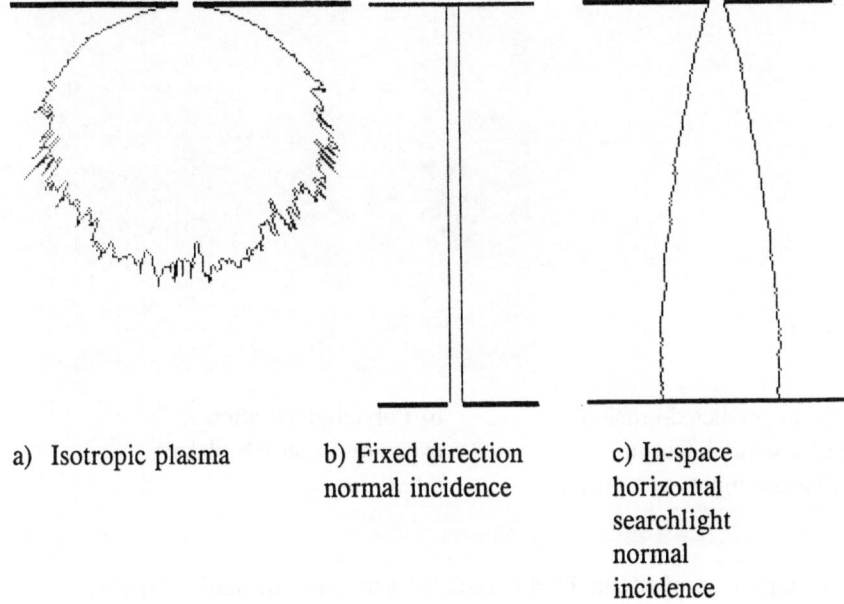

a) Isotropic plasma

b) Fixed direction
normal incidence

c) In-space
horizontal
searchlight
normal
incidence

Figure 2. Monte Carlo model predicted atomic oxygen erosion geometries assuming reaction probability and unreacted oxygen atomic recombination probability equal to one.

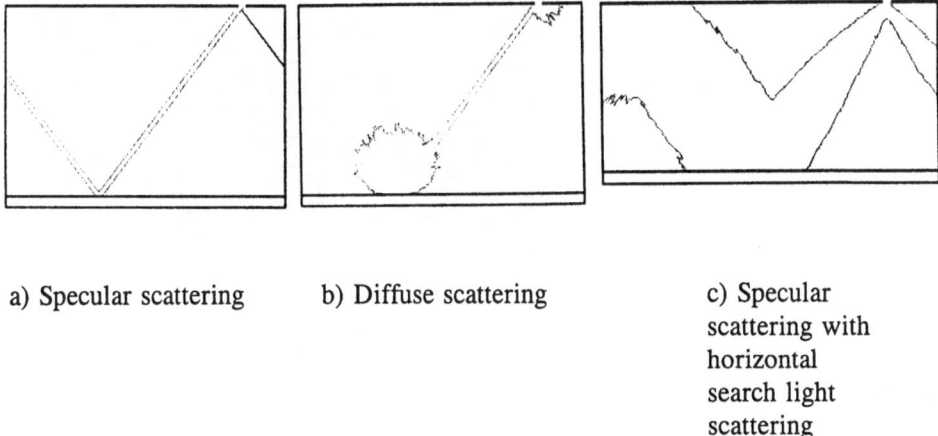

a) Specular scattering

b) Diffuse scattering

c) Specular
scattering with
horizontal
search light
scattering

Figure 3. Monte Carlo model predicted erosion for atomic oxygen incidence at 38° from normal, reaction probability equal to one, unreacted oxygen recombination probability equal to one for atoms leaving polymer surfaces and zero for atoms leaving protective coating surfaces.

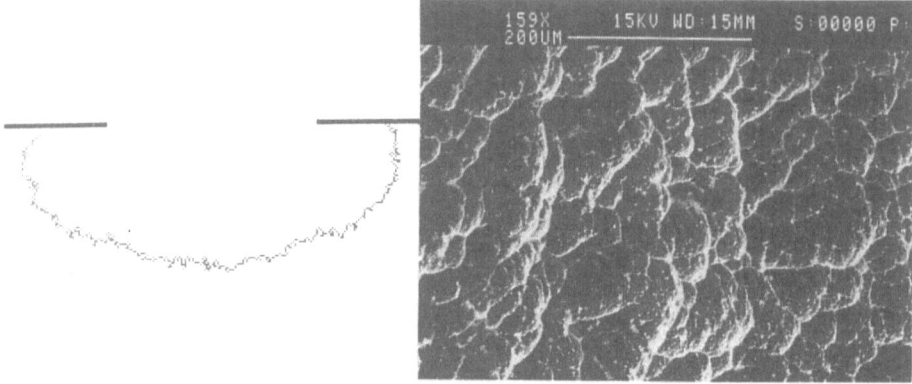

a) Monte Carlo predicted erosion
at the site of a wide defect in
the protective coating over a polymer

b) Polyethylene after
exposure in an RF plasma
asher.

Figure 4. Comparison of Monte Carlo predicted and experimentally observed atomic oxygen erosion resulting from exposure to an isotropic thermal energy plasma.

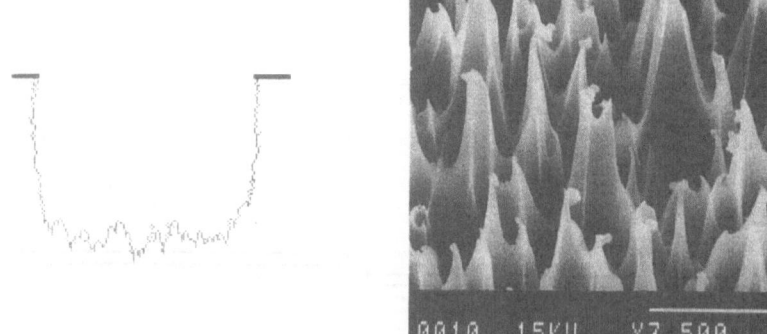

a) Monte Carlo predicted erosion
at the site of a wide defect in
the protective coating over a
polymer exposed to normal incident
horizontal search light atomic oxygen.

b) Chlorotrifluoroethylene
after in-space exposure
on LDEF.

Figure 5. Comparison of Monte Carlo predicted and experimentally observed atomic oxygen erosion resulting from exposure to in-space directed atomic oxygen arrival.

a) Monte Carlo model prediction

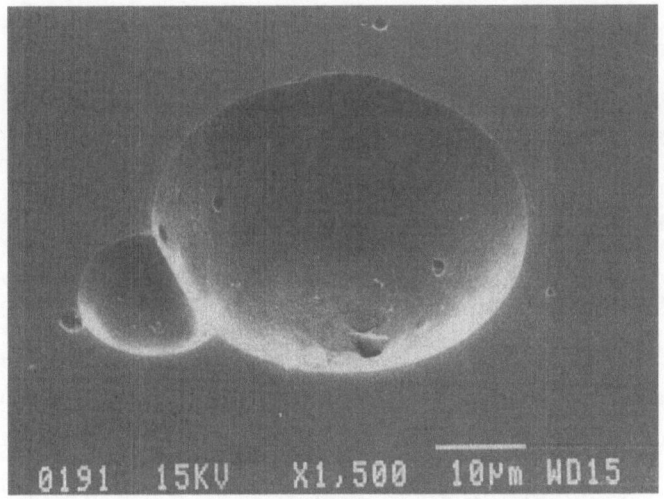

b) Undercut cavity at pin-window defect sites in an
aluminized Kapton H polyimide sample after exposure
in an RF plasma asher and chemical removal of the aluminum coating.

Figure 6. Comparison of Monte Carlo model predicted and experimentally
observed undercutting at the site of a small defect in a protective coating on a
polymer exposed to an isotropic thermal energy plasma.

a) Monte Carlo model prediction

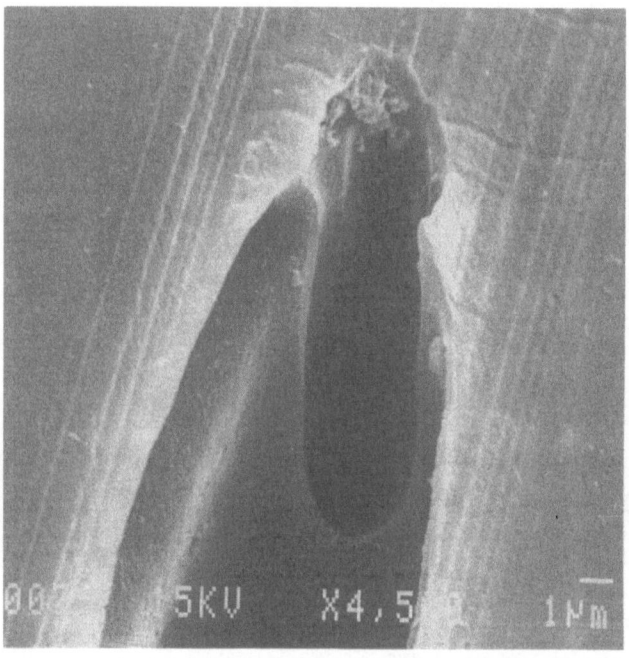

b) Undercut cavity at a defect in an aluminized Kapton H polyimide sample after in-space exposure on LDEF and chemical removal of the aluminum coating

Figure 7. Comparison of Monte Carlo model predicted and experimentally observed undercutting at the site of a crack or scratch defect in a protective coating on a polymer exposed to in-space atomic oxygen arriving at 38.1° from normal.

STRUCTURAL MODELS OF PHOTOSENSITIVITY OF POLYCRYSTAL FILMS

A.R. GOKHMAN, V.N. KADCHENKO, Z.B. POLYAK,
V.P. RZEPETSKIY
South Ukraine Pedagogical University, Department of Physics
26 Staroportofrankovskaya, Odessa 270020, Ukraine

Polycrystalline PbS-layers are advanced material for obtaining high-sensitive photoresistors in the IR field. Existing theories of PbS-films [1] confirm the essential significance of the grain sizes and structure to provide the optimum parameters of the IR devices. In particular, the films photoelectrical properties - average grain size extreme dependence has been determined.

The important characteristics of PbS-resistors are the value of dark resistance R_d, noise voltage U_n and stress photosignal U_p. It's established [2] that these characteristics essentially depend on the mechanical stresses arising in the films under strain. It's undoubtedly interesting to study an effect of the crystallographic texture and the nature of mechanical stresses in the polycrystalline PbS-layers on the photoresistor's quality in detail.

The physical PbS-layers from the same technological group that provided the proximity of studied film's topologies were investigated. Indeed, the identity of the grain's distribution by sizes in these films was established by special measurements. The values of U_n and R_d were measured by the standard methods. The grains' orientational distributions and residual mechanical phenomena were investigated by X-ray methods. The most complete description of the crystallographic texture is achieved when the three-dimensional orientation distribution function (ODF) of crystallites is used that determines a volume fraction of crystallites with given orientation relative to the coordinate system with the main axes of the polycrystal [3]. The properties of PbS-films are symmetrical with respect to the axis normal to a surface. It allows consideration of the crystallographic texture as axial, and to describe it as not the ODF, but a two-dimensional distribution function in the orientational space of the spherical angles (Θ, Φ) giving the orientation of a normal to an arbitrary crystallographic plane (hkl) coinciding with the crystal coordinate system such a function is known as the inverse pole figure (IPF) and it's represented on the stereographic triangle [4].The X-ray integral intensity for each crystallographic plane of investigated and a standard specimen was being measured in Cu

R. C. Tennyson and A. E. Kiv (eds.),
Computer Modelling of Electronic and Atomic Processes in Solids, 271–275.
© 1997 *Kluwer Academic Publishers.*

K$_\alpha$ -beam on the X-ray diffractometer and the IPF values were calculated by Morris' method [5]. The IPF of the PbS-films under investigation are presented on fig. 1.

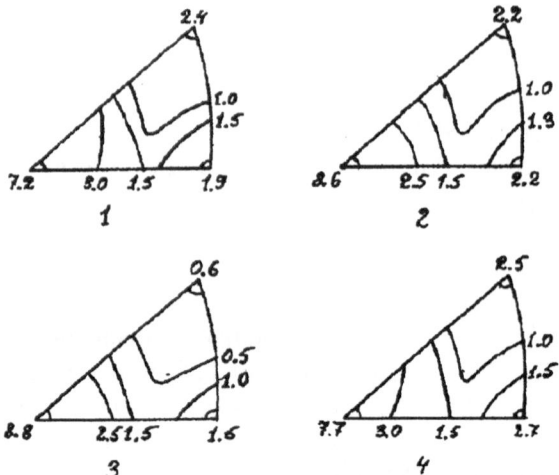

Figure 1. The inverse pole figures of PbS-films for direction normal to the coating plane : with U$_n$ = 0.5 mcv and R$_d$ = 530 ko (1); with U$_n$ = 8.3 mcv and R$_d$ = 343 ko (2); with U$_n$ = 7.0 mcv and R$_d$ = 185 ko (3); with U$_n$ = 8.0 mcv and R$_d$ = 470 ko (4).

The characteristic of the textural maximums on the IPF point out that the structural state of these films is the intermediate between random and monocrystalline. The component (200) is the preferential texture component: a fraction of grains with the plane (200) (r$_{200}$) coinciding with the plane of film made up of 30-40 % of the total grains number.

The value of R$_d$ is reduced with intensification of the texture component (200) but on especially sharp fall of the R$_d$ occurred under the r$_{200}$ excess of 38 % (fig. 2).

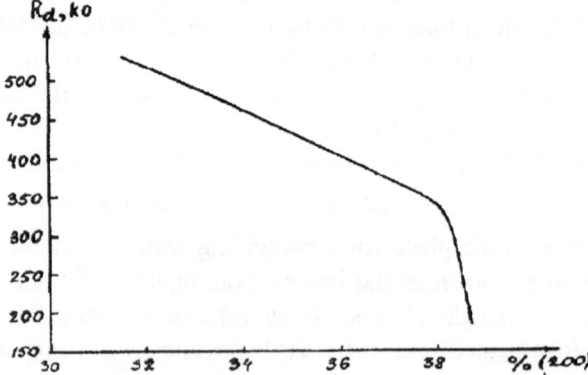

Figure 2. The dark resistance (R$_d$) versus a percentage of the PbS-films texture component (200).

The value of U_n depends on the r_{200} nonmonotonically, with a maximum for r_{200} in the 35.0 - 37.5 % interval (fig. 3).

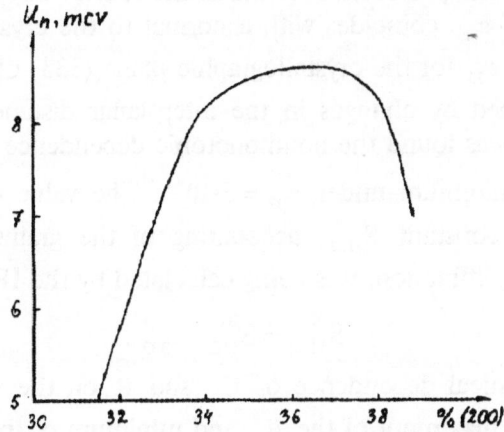

Figure 3. The noise voltage (U_n) versus a percentage of the PbS-films texture component (200).

Thus, a 37 - 38 % content of the structural component r_{200} corresponds to the worst parameters of PbS-films. The optimum parameters of these films has become apparent from the content of the component $r_{200} \leq 32\%$.

A technique to find the residual stresses (σ_{ik}) in isotropic materials by the residual strain (ε_{lm}) measured by the X-ray method is developed in [6]. An analogous method for the monocrystals is given in [7] and that for the poly-crystals with an arbitrary texture using the ODF is given in [8].

Let's determine the σ_{ik} by the ε_{lm} for thin films with an axial texture being described by the ODF. Hook's law for the OX_3 -direction normal to the poly-crystal plane is :

$$\varepsilon_{33} = -\frac{v}{E}\left(\sigma_{11} + \sigma_{22}\right) \tag{1}$$

where σ_{11} and σ_{22} are the stress tensor components in the polycrystal plane; $\sigma_{11} = \sigma_{22} = \sigma$ because of the sample symmetry; v and E are Poisson's and Young's coefficients accordingly.

A ratio of the material properties v and E is equal to the compliance tensor component S_{1122} in accordance with [8]. Accounting on the above, we find an expression for the σ -value from (1):

$$\sigma = \frac{\varepsilon_{33}}{2S_{1122}} \tag{2}$$

The component S_{1122} is considered in (2) not as the mechanical (S_{1122}^M) but as the X-ray elastic characteristic (S_{1122}^X) of the material. Unlike the mechanical characteristic the X-ray elastic one is being calculated as the average from only those grains of the polycrystal for which the reflex (hkl) being used for the X-ray measuring of ε_{lm} coincides with a normal to the crystallographic plane (hkl). The value of ε_{33} for the crystallographic plane (333) of the studied films was being determined by changes in the interplanar distance relative to the unstressed state. It was found the nonmonotonic dependence of the U_n on the ε_{33}-value had a minimum under $\varepsilon_{33} = 6 \cdot 10^{-4}$. The value of S_{1122}^X from the mechanical elastic constant S_{1122}, accounting of the grains' weight fraction (r_{333} taking part in diffraction, was being calculated by the IPF:

$$S_{1122}^X = S_{1122}^M \cdot r_{333} \qquad (3)$$

The nonmonotonical dependence of U_n and R_d on the value of S_{1122}^X has also been found : a maximum of the R_d and minimum of the U_n are reached under $S_{1122}^X = -0.189 \cdot 10^{-10}$ Pa^{-1}. U_n, R_d - σ dependence is calculated by (2), (3). It's observed that the R_d -maximum and U_n -minimum are reached under the $\sigma = -7.8$ MPa.

CONCLUSIONS

1. It's established for the first time that the photoelectrical parameters of PbS-films (physical layers) depend not on the grains' size and state of the lanarkit phase [9] only, but on the films' crystallographic texture. The values of texture parameters defining the films' optimum characteristics are found.
2. The level of PbS-photoresistors flicker-noise - to - so-called "X-ray elastic characteristics" correlations of films are found. A minimum noise level and maximum value of the dark resistance of the studied films corresponded to a percentage of the (r_{200})- component in the 37.0 - 38.0 % interval, residual strain $6 \cdot 10^{-4}$ and stresses ≈ -7.8 MPa. Thus, under applying of physical layers on the base-layer a problem is not that to take off fully the mechanical stresses in the film - base-layer system but to lead theirs to the definite optimum value [10].

REFERENCES

1. Neustroev, L.N. and Osipov,V.V., (1886), To a theory of the physical properties of photosensitive polycrystalline PbS-films., *Physics and Technique of Semiconductors*, **20**, 1, 59-65.

2. Polyak,Z.B., (1990), The radiactive effects in the photosensitive films of the lead sulphide., Thesis for a cand. of phys.-math. sciences., Odessa.
3. Bunge,H.J., (1965), Analysis by Orientation Distribution Function., *Z.Metallkunde* **2, 18,** 872-874.
4. Wasserman, G. and Grewen, J., (1962), *Texturen Metallischer Werkstoffe, Springer-Verlag.*, Berlin /Gottingen/ Heidelberg.
5. Morris, P.R., (1959), Reducing the effects of nonuniform pole distribution in inverse pole figures., *Appl. Phys.,* **30, 2,** 595-596.
6. Vasilyev, D.M. and Trofimov, V.V., (1984), A modern state of the X-ray measuring of macrostresses., *Industrial Laboratory* **50, 3,** 20-49.
7. Rybin, V.V. and Titovets, U.F., (1992), An applying of the X-ray parallel beam method for investigation of the elastic and plastic distortions in the monocrystals and separate grains of the large-crystalline specimens., *Industrial Laboratory.,* **58, 1,** 40-54.
8. Gokhman, A.R. and Reznik, L.I., (1991), To using of the texture integral characteristics in problems of the X-ray tensometry of the cubic and hexagonal polycrystals., *Industrial Laboratory.,* **57, 7,** 23-25.
9. Kadchenko, V.N., Kiv, A.E., Polyak, Z.B. and Trunov, A.V., (1994), Investigation of the adsorption-desorption processes of the water molecule in films by the spectra of low-friquency noise., *Surface. Physics, Chemistry, Mechanics.,* **5,** 126-128.
10. Vavilov, V.S., Gorin, B.M., Danilin, N.S., Kiv, A.E., Nurov, U.L. and Shakhovtsov, D.I., (1990), *The radiative methods in the solid-state electronics.*, Radio i Svyaz., Moscow.

PREDICTIVE MODELS OF EROSION PROCESSES IN LEO SPACE ENVIRONMENT: A BASIS FOR DEVELOPMENT OF AN ENGINEERING SOFTWARE

J. I. KLEIMAN[*,‡], Z. A. ISKANDEROVA[*,‡] , YU. I. GUDIMENKO [*,‡], V. LEMBERG[*], D. TALAS[*], AND R. C. TENNYSON[‡]

‡- University of Toronto Institute for Aerospace Studies;
*- Integrity Testing Laboratory Inc.;

Abstract

A great deal of effort has been made to understand the effects of the low Earth orbit (LEO) space environment on materials and structures. Among the various factors that are present in LEO space, the erosion of polymer-based materials by atomic oxygen presents one of the most important contributions to their rapid deterioration and consequent failure.

Based on a number of recently determined semi-empirical correlations between the erosion rates of carbon-based polymers and their chemical and physical parameters, an attempt was made to develop a practical engineering software guide for evaluation of erosion resistance of homopolymers, copolymers, polymer compositions, blends and alloys. The software package will provide basic data on the erosion behaviour of common, space-related polymer materials. What is more important, it is intended to provide estimated lifetimes for untested, newly synthesized or even hypothetical polymer-based materials of any degree of complexity in the harsh LEO environment, thus eliminating to a large extent the need for complicated, time-consuming, and expensive ground-based and/or flight experiments. Based on the estimated erosion yields and the environmental models of LEO, predictions can be made for the lifetimes of polymer-based materials in spacecraft applications.

1. Introduction

Polymeric materials undergo various types of degradation, the degree of which depends on the extent of the exposure, as well as on the nature of the chemical processes involved. The consequences of material degradation in space can be catastrophic and considerable effort is put towards selection and development of various protection schemes for materials. The description of degradation reactions under various environmental factors is generally a very complex problem.

Presently, there is a considerable amount of experimental evidence that in the interaction of materials with reactive neutral beams with few eV energies, various dynamic factors affect the course of chemical reaction rates at the surfaces, leading to enhanced oxidation, erosion, etching, etc., with high directionality. A very limited amount of

277

R. C. Tennyson and A. E. Kiv (eds.),
Computer Modelling of Electronic and Atomic Processes in Solids, 277–287.

theoretical work has been done, however, that can be used in their interpretation. The accelerated erosion of carbon- and polymer-based materials in LEO environment falls into this category and is mostly due to interaction with hyperthermal or fast (E=2-5 eV) atomic oxygen. While the major role of reactive, fast atomic oxygen $O(^3P)$ in erosion processes is well established, the mechanisms of surface erosion in LEO environment are only hypothetical at the present time. Development of a model for these processes remains an acute problem for fundamental research and, when accomplished, can serve as the basis for engineering predictions and selection of materials on the basis of their content and structural parameters.

2. Review of the Correlation Models Used in the Software Package Development

A number of correlation models were developed that take into account the chemical and physical aspects of the atomic oxygen (AO) interaction with polymer-based materials. A brief summary of the mechanisms and models that are at the cornerstone of the developed software package will be given below.

2.1 EROSION YIELD OF HYDROCARBON POLYMERS VERSUS THE CONTENT OF THEIR REPEAT UNIT

From the analysis of erosion of carbon, graphite and organic polymers in low Earth orbit , by FAO beams, and in different plasma environments, it was postulated [1,2] that both, atomic collisions with energy and momentum transfer, and chemical oxidative degradation processes are involved in the etching mechanisms.

It was found that in interactions of thermal atomic oxygen (TAO) with polymers, the surface chemical reaction rate is sensitive to the degree of aromaticity of the material. Reaction rates fall off sharply for polymers with higher degrees of aromaticity. In the interaction of hydrocarbon polymers with AO in LEO, the dependence of the erosion yield on degree of aromaticity $Re^{LEO}(\alpha)$ is much weaker, but still exists. It would appear collisionally-induced and enhanced processes are "smoothening" the structure - erosion resistance dependence of hydrocarbon polymers to AO attack.

It was suggested that erosion of polymer-based materials in LEO and by FAO beams can be viewed as a special case of a general physical-chemical phenomenon, which can be termed bombardment- induced and enhanced surface chemical etching (BESCE), with its very unique subthreshold low-temperature mechanisms [1,2]. Furthermore, it was conjectured that in the special case of carbon films and graphite, the erosion by FAO can be described as a subthreshold, low-temperature mechanism of chemical sputtering. In general, BESCE plays an important role in modern material science and industrial applications. This field covers etching and erosion of various materials in chemically active plasmas of different nature and content, etching and sputtering of different materials by reactive ion beams, and, some aspects of the LEO space environment interaction with materials.

Bombardment-induced and enhanced oxidative degradation constitutes the main destructive process for hydrocarbon polymers in LEO conditions and FAO beams. The content of hydrogen and volatile weakly- bonded hetero-atoms contribute to the bombardment-induced erosion due to hydrogen abstraction, weak bond excitation and rupture, with the "effective" mass density of carbon being the rate-limiting factor [1]. A procedure to calculate the number of "effective" carbon atoms in a material was established, taking into account that in oxygen-containing polymers, the intra-molecular oxygen can fully participate in the degradation and destruction processes together with the incoming fast atomic oxygen flux. A linear relationship was found between the erosion yield, R_e^{LEO} and an "effective carbon" chemical content factor, (γ), based on LEO experimental data for many hydrocarbon polymers

$$R_e^{LEO} \approx k\frac{\overline{M}}{\rho}\frac{N_T}{N_C - N_C^O} \approx k\frac{\overline{M}}{\rho}\gamma \approx k\gamma' \tag{1}$$

where M is the average atomic weight of the atoms in the polymer repeat unit, N_T is the total number of atoms in the unit, N_C is the number of carbon atoms in the unit, and N_C^O is the number of carbon atoms in the unit, bonded to oxygen atoms. The coefficient -k represents the average impact reaction probability for "in-space" atomic oxygen attack on hydrocarbon polymers. It is defined by the slope of the linear regression line in figure 1 that demonstrates the fit of the flight data from a large number of sources to the above correlation.

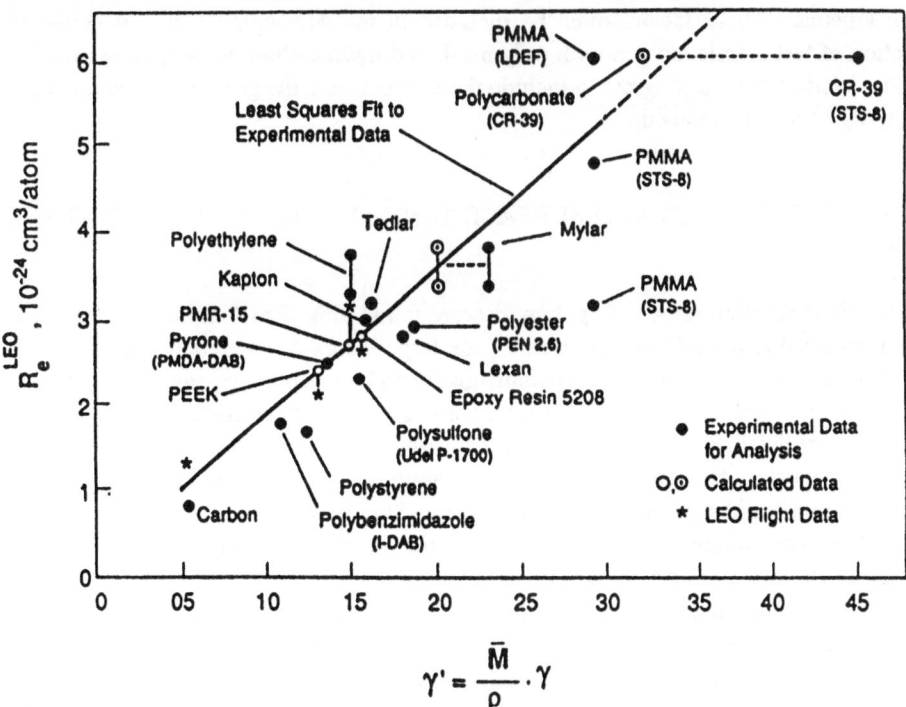

$$\gamma' = \frac{\overline{M}}{\rho} \cdot \gamma$$

Figure 1. Erosion yield versus the chemical content parameter for LEO experimental flight data. The solid line represents a linear correlation using a least squares regression ($R^2 = 0.905$) analysis. For details see [2].

Based on the data presented in figure 1, the reaction probability can be estimated as 0.11 - 0.12. It should be noted that a value of 0.138 was used in an earlier work to compute the undercutting effects of atomic oxygen on protected Kapton using a Monte-Carlo computer simulation [3]. It is also interesting to note that the reaction probability cited in [3] was found to be in close agreement with the space experimental results for carbon, assuming carbon monoxide being the reaction product, and a carbon-erosion yield "in-space" of $1.2 \cdot 10^{-24}$ cm^3/at [3]. The value derived from figure 1 for -k lends additional credence to the proposed model [1,2] where the "effective carbon content" is the limiting factor of accelerated erosion of hydrocarbon polymers in LEO space environment.

No such correlation as shown in figure 1 was found for TAO in flowing afterglow conditions. This relationship can take into account the atomic densities of the material and possible pathways of partial decomposition (decarbonilation and decarboxilation) for some oxygen-containing polymers due to bombardment-enhanced oxidative degradation. The $R_e^{LEO}(\gamma')$ correlation has important implications for predicting erosion yields of many hydrocarbon polymeric materials and composites.

For perfluorinated polymers, screening effects can be assumed to explain and describe their resistance to FAO erosion. Based on the data from synergistic effects for FAO/vacuum ultraviolet radiation combined experiments, photo- or ion-induced radical generation is likely the initiating step for the degradation and erosion sequences. There is growing evidence that both physical and chemical processes produce collisionally-induced distortion and rupture of chemical bonds and drive the bombardment-enhanced chemical etching kinetics. These factors must be included in the development of any model of interaction of fast atomic oxygen with polymer-based hydrocarbon or halogen-containing materials. Work is now in progress to include these effects into the general "erosion yield"-content, $R_e^{LEO}(\gamma')$ relationship.

2.2. PHYSICAL ASPECTS AND SURFACE TENSION EFFECTS AFFECTING THE EROSION RATES

Since the FAO reaction rates with polymers were found to be 3 to 4 orders of magnitude higher than for thermal AO oxidation rates, the large increase in reaction rates for FAO processes had to be explained. The relationships found between the erosion yields in FAO and the chemical content and structure of some hydrocarbon polymeric materials as described above [1,2] were based mostly on a "chemical approach", i.e. on analysis of the chemistry of the polymer repeat units. It would be beneficial, however, to evaluate the erosion yields of polymers from a "solid state physics" point of view. Such an attempt was made in [4].

The most important physical and kinetic consequences of polymer chemical reactivity relate to the kinetics of reactions in solid polymers which are sensitive to such structural and physical parameters of the polymer as crystallinity, mechanical strain, orientation, deformation, etc. Polymeric materials essentially consist of long, flexible chains which can be randomly entangled or laterally ordered in the form of very thin lamellae or extended fibrils. The response of such a system to external forces (including chemically active accelerated species) may depend not only on the strength of the individual macromolecular chains but also on the degree of the lateral cohesion between the chains.

Polymer strength depends upon the strength of its weakest links [5]. In relation to polymers, it was estimated that London forces contribute about 75 to 100% of the total cohesion, except for strongly polar molecules and when hydrogen bonding is taken into account. In general, these forces cannot be determined by any direct measurement. However, the Van der Waals forces between individual surface macromolecules represent the first order effect of the dispersion component of surface tension (γ_s^d), which is proportional to the power of -7 of the distance between molecules [6]. It was also found [7] that a direct proportionality exists between the surface and bulk properties for 18 of the 23 nonpolar polymers, i.e.:

$$\gamma_c \sim 0.44\delta^2 \tag{2}$$

where γ_c is the critical surface tension and represents the energy per unit area, and δ^2 is the cohesive energy density and represents the energy per unit volume. This linear correlation defines the relation between the free-energy of the surface and internal energy density of the bulk state. This correlation is independent of the physical state of the polymer material (i.e. elastomer, amorphous glass, or semicrystalline solid) and its density [7].

Based on the above principles, a critical analysis was conducted [4] to find possible correlations between the surface tension γ_s, the dispersion γ_s^d and the polar γ_s^p components of surface tension, the polarity x^p (representing the ratio of γ_s^p / γ_s), and the R_e^{LEO} for the interaction of FAO with various polymers exposed in LEO. A reasonably good correlation between R_e^{FAO} and γ_s^d for twelve materials was found and is presented in figure 2.

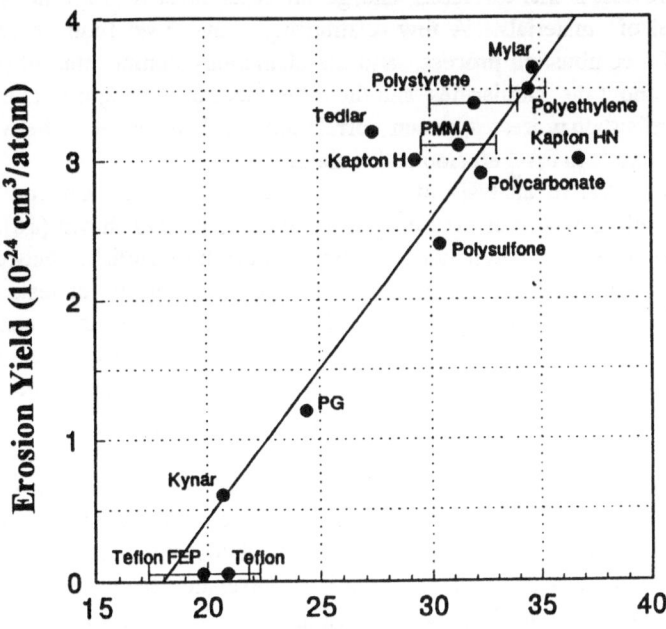

Dispersion Component of Surface Tension (dyn/cm)

Figure 2. Plots of erosion yield versus the dispersion component of surface tension for 14 polymer materials. A linear correlation using a least squares regression ($R^2=0.883$) is shown. For details see [4].

2.3 THE CORRELATION OF POLYMERS' EROSION RESISTANCE (EROSION YIELD) TO FAO WITH FLAME RESISTANCE (OXYGEN INDEX)

Certain analogies can be found between the erosion resistance of polymers to FAO and to plasma stripping and their combustion resistance characteristics, i.e. flammability [2,8]. For instance, phosphorus-containing organic polymers, being fairly resistant to flames, have also shown higher erosion resistance under FAO attack [8-10]. In all these processes, the final products are the simplest gaseous volatiles. The major products from erosion by FAO and in combustion processes of hydrocarbon polymers are, more or less, the same, i.e. water and carbon dioxide for complete combustion, and, predominantly, water, carbon oxide and carbon dioxide, in different ratios, in erosion by FAO [11-13]. Oxygen atoms, present in the content of a polymer repeat unit, may aid in erosion, etching or combustion processes [2,14,15]. It is reasonable, therefore, to assume that, in general, the more oxygen is present in the content of a polymer repeat unit, the less of an external supply of it will be required for the same erosion or complete combustion effect (at room temperature). Therefore, polymers with higher oxygen concentrations in the repeat unit, should have higher erosion yields and lower flame resistance, i.e. higher flammability.

A direct criterion of flame resistance of polymers is the Oxygen Index (OI) [15]. OI is defined as the minimum fraction of oxygen in the test atmosphere which will just support the combustion (after ignition) of a material. It means, that materials with smaller OI's are more flammable. At the present, the OI is probably one of the most valuable tests in fire research that provides numerical results, and, generally, is linearly proportional to the flame-retardant level of materials. A few relationships have been found between OI, the parameters of a combustion process, and the elementary composition of polymers [15]. Taking into account the similarities and the differences of the major final products in AO erosion and combustion processes, certain correlations could be expected between resistance of polymer materials to FAO erosion and to flame.

A comparative analysis of the available experimental data on OI for many hydrocarbon polymers and of their erosion yields in LEO had shown [16], with certain limitations, that an approximately linear relationship can be established between R_e^{LEO} and the inverse Oxygen Index as shown in Fig. 3. For the majority of tested hydrocarbon polymers (with Kapton and PEEK being the only exceptions), a clear trend can be found that can be used for making estimates and predictions. As can be seen from Fig.3 (dashed line), the highest values of R_e^{LEO}, that have been found in the LEO flight experiments are close to, or less than the numbers from the simplest expression, $R_e^{LEO}{}_{MAX}(cm^3/at) \leq 1/OI$. It means that this semi-empirical relationship can be used to estimate and to predict, to a certain extent, the upper limit of erosion yields of many polymers in LEO. It seems that the scatter of data in this basic correlation is due mostly to the difference in the final products of erosion and combustion. On the other hand, if one decides to choose the linear regression line (solid line, Fig.3), then, because of a larger than in γ-correlation scatter in data, its predictive power is not as good as in the first case (dashed line, Fig.3). It should, however, encompass a wider variety of polymers including the halogen-based ones.

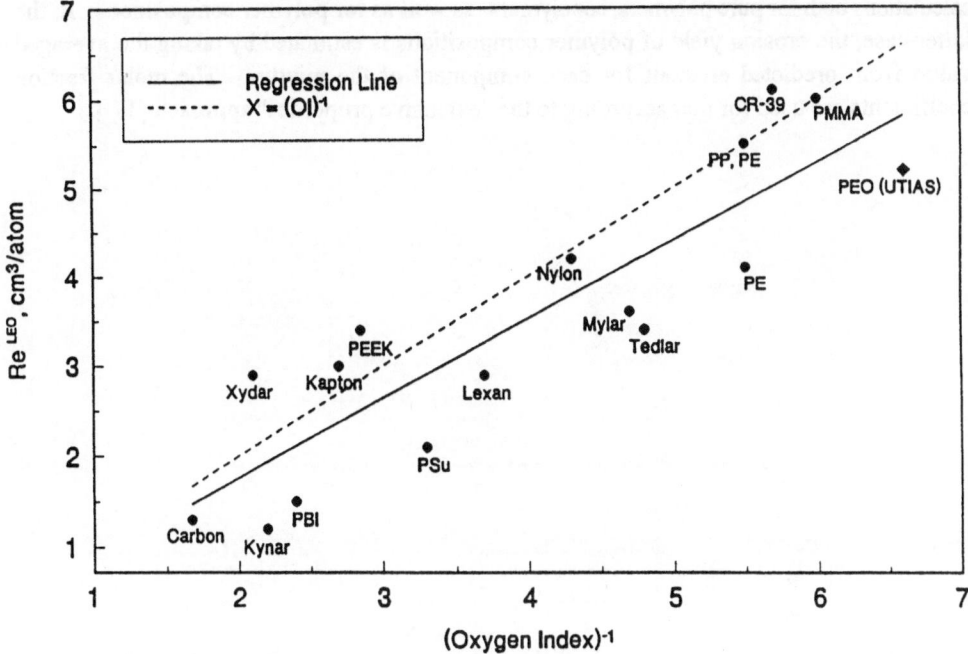

Figure 3. Erosion yield versus the inverse oxygen index for selected polymer materials. The solid line presents the result of a linear correlation using a least squares regression analysis. The dashed line presents the $Y = (OI)^{-1}$ relationship.

3.0 Predictive Erosion Resistance Software (PERS ™) Program

The correlations described above, when fully developed, can be used successfully as material selection guidelines in the design stages of systems and structures for LEO space missions, minimizing the very costly and time-consuming processes of material selection and testing. These correlations would make it possible to estimate the behaviour of new untested materials in promising applications without extensive testing. Since atomic oxygen is an integral part of many plasma sources, the application of the developed software can be extended to many terrestrial applications where the use of plasma is made.

An approach similar to that in designing the interactive computer program SYNTHIA [17] was taken in the present work. It should be noted that, in addition to being able to predict the erosion properties of existing and new polymers, the present software also presents an extensive database on various properties like the surface tension energies and their components, the oxygen indices, and, where such data is missing can calculate it, a process that can be very time consuming and not trivial in many cases.

3.1 THE SOFTWARE

The Predictive Erosion Resistance Software (PERS™) package has been specifically developed as an engineering tool for prediction of erosion yields of polymer-based materials by one or more of the mechanisms mentioned above. The system allows the performance of

284

calculations both for pure polymers, copolymers as well as for polymer compositions. In the latter case, the erosion yield of polymer compositions is estimated by taking the averaged value from predicted erosions for each component of the mixture. The molar fraction coefficients are used for that according to the "extensive properties" approach [17].

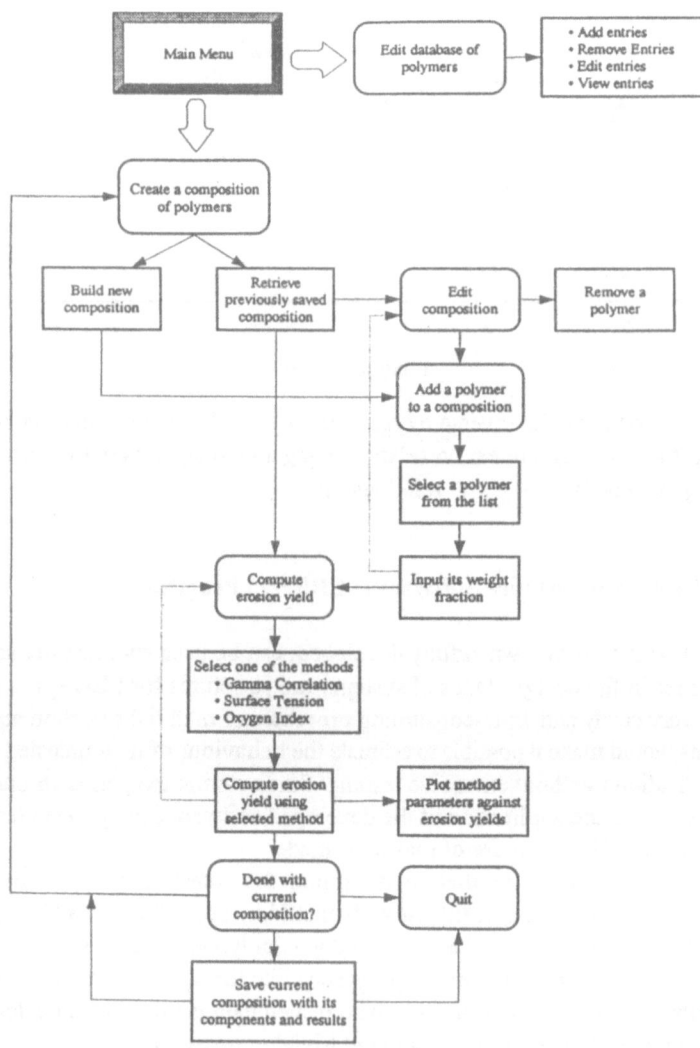

Figure 4. Flow chart of erosion yield calculation by the software package implementing the key correlations discussed in section 2 and shown in figures 1 - 3.

The General Algorithm that is used in the program is shown in Fig. 4 above. At the first stage, the polymer composition is specified by single or multiple selection from the list of available polymers using their generic names, chemical names, trade names, etc... The program retrieves the data on selected components from the built-in database and prompts the values for weight fractions (100% in the case of one component). The initialization of polymer composition is completed by converting the weight fractions to the molar ones. The existing data bases of atomic oxygen exposure effects on materials [18 - 20] will be also connected with the program. After the selection of appropriate Regression Analysis method ("Gamma-Correlation", "Surface Tension" or "Oxygen Index") has been done, the calculations are performed and the erosion yields for each component as well as for the polymer composition are estimated. The results are displayed and also could be added to the existing database for permanent storage. If estimates of the lifetime of a spacecraft component in a LEO mission are required, the erosion results from the present calculations can be combined with the environmental database models [20].

3.2 THE DATABASE

The program contains an extensive database of physical and chemical properties of polymers used frequently in LEO-flights. The database consists of a few sets of tables which describe the atomic content and structure of the Polymer Repeat Unit (PRU), volumetric, and other properties of polymers. Multi-field keys which correspond to the Generic Names, Chemical Symbols, Trade Names, Manufacturer Name and Density fields are used for the identification of each polymer in accordance with the classification given in [15]. A new polymer could be easily added to the existing database using the interactive mode of the program.

3.3 COMPUTATIONAL PROCEDURES

Computational procedures allow retrieval of the necessary data from the database and performance of the calculations in accordance with the selected algorithm. Each algorithm uses its own set of tables, for example, the data about atomic content of PRU are used by "Gamma-Correlation" and "Oxygen Index" methods and not used by "Surface Tension" approach. At the same time, the table with structure of PRU is mainly used by "Surface Tension" method.

A user-friendly interface is developed for easy data input and graphical representation of the results obtained. Special attention is given, where possible, to the comparison of theoretically predicted values of the erosion yields with those experimentally observed and used for the regression procedure. This allows us to check the correctness of proposed regression coefficients.

4. Conclusions

Based on a number of recently found semi-empirical correlations between the erosion rates of carbon-based polymers and their chemical and physical parameters, an attempt was made to develop a practical engineering guide for evaluation of existing simple polymers and blend polymer materials and for prediction of erosion properties of new materials. At the present stage three regression analysis methods, namely, "Gamma-Correlation", "Surface Tension" and "Oxygen Index" are used in the calculations of the erosion yields of many polymers. With further development of the PERS ™ software, new additional correlations could be implemented into the program. Combining the data on erosion behaviour from the described software with data accumulated from numerous space flights and presented in such packages as Environet by NASA [20], or the composite materials lifetime prediction nomograms [21] a much better understanding and prediction of polymer-based material behaviour in space can be achieved.

4. References

1. Iskanderova, Z., Kleiman, J.I., Gudimenko, Yu., and Tennyson, R.C. (1994) "Erosion of Polymers in LEO Space Environment: A Unified Approach to the Development of a Model", in the *Proc. of the Second International Space Forum* "Protection of Materials and Structures from the Low Earth Space Environment", Toronto, 1994, 22, Integrity Testing Laboratory; *LDEF Spaceflight Environmental Effects Newsletter,* (1994), **6**, No.6.

2. Iskanderova, Z., Kleiman, J.I., Gudimenko, Yu., and Tennyson, R.C. (1996) "The Influence of Content and Structure of Hydrocarbon Polymers on Erosion by Atomic Oxygen", *J. of Spacecraft and Rockets*, **32**, No.5, 878-884.

3. Banks, B.A.,Rutledge, Sh., Auer, B.M., and DiFilippo, F. (1990), in *Materials Degradation in Low Earth Orbit (LEO)*, Eds. Srinivasan, V. and Banks, B.A., TMS Publishers, 15.

4. Kleiman, J.I., Gudimenko, Yu., Iskanderova, Z., Tennyson, R.C., Morison, W.D., McIntyre, M.S., and Davidson, R. (1995) "Surface Structure and Properties of Polymers and Pyrolytic Graphite Irradiated with hyperthermal Atomic Oxygen", *Surface and Interface Analysis* **23**, 335-341.

5. Mark, H.F. (1979, July) *Adhesion Age,* 35; (1979 Sept), 45.

6. Kueble, D.H. (1971) *Physical Chemistry of Adhesion,* Wiley-Interscience, New York.

7. *Polymer Handbook,* 3rd Edition (1989).

8. Torre, L.P. and Pippin, H.G., (1986) "Structure-Property Relationship in Polymer Resistance to Atomic Oxygen", *Proc. of the 18-th International SAMPE Technical Conference* **18**, 1086.

9. Fewell, L. and Fenney, L. (1991) *Polymer Communications,* **32**, No.13, 303.

10. Koontz, S.L., Leger, L.J., Visentine, J., Hanton, D., Cross, J.B. and Hacks, Ch.L. (1995) *Journal of Spacecraft and Rockets,* **32**, No.3, 483.

11. Cross, J., Koontz, S.L., Gregory, J.C. and Edgell, M.J., (1990*) Materials Degradation in Low Earth Orbit (LEO)*, Eds. Srinivasan, V. And Banks, B.A., TMS Publishers, 1.

12. Brinza, D., (1993, June 22-23) *Proc. of the EOIM-3 BMDO Experiment Workshop,* JPL, Acadia, CA.

13. Minton, T.K. and Moore, T.A. (1995) "Molecular Beam Scattering From ^{13}C-Enriched Kapton and Correlation with the EOIM-3 Carousel Experiments", *LDEF-Third Post-Retrieval Symposium,* Ed. Levin, A., NASA, 1095.

14. Morea, W.M. (1988) *Semiconductor Lithography Practices and Materials,*

Plenum.

15. Van Krevelin, D.W. (1990) *Properties of Polymers*, 3-rd Edition, Elsevier, Amsterdam.

16.Kleiman, J.I., Iskanderova, Z, Gudimenko, Yu., Tennyson, R.C., Lemberg, V. and Talas, D. (1996) "Predictive Models of Erosion Processes in LEO Space Environment: A Basis for Development of an Engineering Software", *Proc. of the 3-rd International Space Conference, ICPMSE-3,* Toronto, 22-23, Integrity Testing Laboratory.

17. Bicerano, J. (1993) *Prediction of Polymer Properties*, Marcel Dekker, Inc.

18. *NASA MAPTIS Internet Database*

19. McCall, S.H.C.P., Pierre, J.E., Mabee, A., Tennyson, R.C., Morison, D., Kleiman, J. I., Iskanderova, Z., and Gudimenko, Yu. (1994) "The Atomic Oxygen Exposure Effects Module of the Database for the Properties of Black, White, Reflective and Transmissive Spectrally Selective Surfaces", in the *Proc. of SPIE International Symposium on Optics, Imaging and Instrumentation*, 24-29 July, 1994, San Diego, CA.

20. *NASA ENVIRONET Internet Database*

21. Tennyson, R.C. (1993) "Atomic Oxygen and its Effect on Materials", R.N. DeWitt et al (eds), in *The Behaviour of Systems in the Space Environment*, 233-257, Kluwer Academic Publishers

COMPUTER SIMULATION AND EXPERIMENTAL STUDIES OF ION IMPLANTATION IN POLYMERS FOR EROSION RESISTANCE IMPROVEMENT

Z. A. ISKANDEROVA*+ J. KLEIMAN*+ , YU. GUDIMENKO*+ ,
G. COOL* and R. C. TENNYSON*
University of Toronto Institute for Aerospace Studies
Integrity Testing Laboratory Inc. +
North York, Ontario, Canada

Introduction

Many polymeric materials degrade in low Earth orbit (LEO) due to the damaging environmental effects of atomic oxygen (AO), solar ultraviolet (UV) radiation, temperature cycling (TC) and micrometeoroid/debris impacts. Of these effects, atomic oxygen and ultraviolet radiation cause the most damage to the chemistry of polymeric surfaces and, therefore, have important implications for the durability of these materials [1, 2]. Oxidation of polymer surfaces resulting in enhanced erosion and mass loss has been observed for materials aboard shuttle missions such as STS-5, STS-8, and NASA's Long Duration Exposure Facility (LDEF), and in different ground testing facilities, including the unique AO facility at the University of Toronto Institute for Aerospace Studies (UTIAS) [1-3].

High-performance, thermal, stable polymers which are commonly considered for use on LEO missions include Kapton® (polyimide), Mylar® (polyethylene terephtalate), Teflon® FEP (fluorinated ethylene propylene) or PTFE Teflon® (polytetra-fluoroethylene). Advanced composite materials that consist of graphite fibers bonded with different kinds of epoxy resins, high-performance polyimides or polyetheretherketone (PEEK) polymer matrix materials are also commonly used in space. Kapton, epoxies and Mylar have an erosion yield $(2-4)*10-24$ cm3/atom [3] which makes them unsuitable for long term use in the LEO environment. Teflons are much more stable to hyperthermal, or fast (E = 2-5 eV) atomic oxygen (FAO) exposure, but in long duration missions and testing facilities, a synergistic effect of increased erosion rate in FAO and VUV radiation has been observed [2, 3].

Thin film coatings of transparent nonvolatile oxides, such as silicon or aluminum oxides, are viable candidates to provide protection of polymers in LEO. However, because of the differences in thermal expansion coefficients of plastics and oxide coatings, cracking or spalling occurs in these materials due to thermal cycling in LEO [4, 5]. Atomic oxygen gradually undercuts these and other defect sites in protective coatings, thus oxidizing the underlying polymer and forming a cavity underneath. That is why it was found to be impossible to provide high-quality protection with protective coatings in long duration missions in LEO.

289

R. C. Tennyson and A. E. Kiv (eds.),
Computer Modelling of Electronic and Atomic Processes in Solids, 289–299.
© 1997 *Kluwer Academic Publishers.*

This study presents, to the best of our knowledge, the first attempt to achieve high-quality protection of polymer materials from FAO, not by coatings, but by ion implantation. Computer simulation of high-dose, single or multiple implantation of specially-selected metal or semi-metal elements or their combinations with other selected species, were considered in the program for a few major polymers used in space. Based on these results, implantation in Kapton, PEEK and Teflon were per-formed in the energy range (10-100) keV, with subsequent Rutherford Backscattering Spectroscopy (RBS) analysis and comparison with computer modelling. Implanted materials were exposed to FAO in a unique ground-based UTIAS Accelerating Testing Facility, and showed excellent self-protection of the surface modified polymers, i.e., formation of highly-resistant surface structures. A few complementary surface analysis techniques, such as X-ray Photoelectron Spectroscopy (XPS), Secondary Ion Mass Spectroscopy (SIMS), and Scanning Electron Microscopy (SEM), were employed to investigate the content and structure of the modified surface layers.

Computer Simulation of Single and Binary Ion Implantation in Polymers

The modified erosion-resistant surface layers of polymers should meet numerous requirements, including FAO-resistance, abrasion resistance, UV tolerance and flexibility without modifying the polymer thermal and optical properties, a high degree of resistance to thermal cycling, thin, and strongly adherent to the polymer. Four chemical elements, Si, Al, Sm and Gd, which can form, at appropriate conditions, highly stable, protective oxide layers, transparent in the visual and UV regions, have been chosen for ion implantation, and used in the calculations, based on the well-known TRIM [6] computer code.

Another non-metal element, boron, was also considered for binary low or middle energy implantation with Si or Al. In the first case, the intention was to form a compound precursor which, after oxidation under FAO, can form a compound with content close to borosilicate glass. In the second option, a significant improvement in surface mechanical properties was expected. The rare Earth metals, Sm and Gd, were chosen because of their well-known ease of oxidation. Also, the thermal expansion coefficients of their oxides are a better match for organic polymers, than SiO_2 or Al_2O_3 [7, 8]. Computer simulation was carried out, based on the code TRIM 90.05, a well-known program that allows modelling of ion implantation in compound mataterials, including polymers [6]. Because of the comparatively large molecular volume for the polymers of interest, the special high-dose implantation effects have not been considered. The calculated depth distribution and concentration of implanted species served as a basis to make the appropriate choices and recommendations for implantation conditions.

Experiment

Kapton NH, PEEK and Teflon films, widely used in space and many terrestrial applications, were implanted with silicon, silicon + boron aluminum, and rare Earth metals Sm and Gd in the energy range 10-40 keV, with fluences recommended from TRIM computer simulation..The flux was (0.2-0.3) μA/cm2 for Sm and Ga, and ~(1-1.5) mA/cm2 for the light elements.

The equipment used for ion implantation was a Lintott 3A high current ion implanter, a cryo-pumped end station and beam line assured a clean implant. The beam deliberately implanted a larger area than necessary to eliminate the possibility of charge build-up on the polymer samples.

The ion-implanted Kapton, PEEK and Teflon specimens were analyzed using 2 MeV He RBS ($\theta = 110°$). The beam was collimated to an area of 1 mm^2 and the samples were analyzed using a beam current of 10-15 nA without rastering. An accumulated charge of 10 μC was used for obtaining the RBS spectrum, although visual inspection showed beam damage to the sample (dark spots). However, the spectra did not show any significant beam dose dependence.

X-ray photospectroscopy (XPS) was used to analyze the outermost 2-3 nm of the surfaces. All samples were analysed with a modified SSL SSX-100 X-ray spectrometer using monochromatized Al Kα exciting radiation. Survey scans of all samples were obtained using a 600 μm X-ray spot size and 150 eV pass energy.

Elemental compositions were determined using integrated photoelectron intensities corrected with Schofield cross-section and the electron mean free path. Binding energies were referenced to the hydrocarbon peak at 284.7 eV. A Gaussian-Lorentzian fitting program was used to curve-fit XPS spectra. A comparison was made between the expected atomic ratios and the corresponding measured atomic concentrations for each of the untreated polymeric films. High-resolution XPS spectra were used to determine the initial chemical environment contributions and follow them as a function of treatment conditions.

SIMS depth profiles were accumulated using a Cameca IMS 3f facility. An oxygen negative primary ion beam with E = 10 keV bombarded the sample in a raster of 250 μm^2 while monitoring positive secondary ions. These conditions provided good sensitivity for elements such as aluminum and silicon with adequate sensitivity for carbon and oxygen. Oxygen was monitored at mass 16 and mass 18 since the primary beam implants ^{16}O. ^{18}O is therefore representative of sample oxygen. The O$^-$ primary beam allows non-conducting samples to be profiled. The possibility of sample-charging was addressed in two ways. The sample was gold-coated to stabilize any charging. The profile was also collected using an auto charge compensation program that automatically compensates for charging conditions during the analysis. Depth calibration was done using a Dektak IIA surface profilometer to measure the SIMS crater.

The fast atomic oxygen exposure testing was conducted at UTIAS. Each sample was placed in the UTIAS Atomic Oxygen Beam Facility [2] in a holder that oriented the sample at 45° to the trajectory of the AO beam. A set of test samples were simultaneously exposed to atomic oxygen at an average flux of 1×10^{16} atoms/cm^{2}s (15 cm source to target distance) for a period of 5-6 hours and were held at a constant temperature for the duration of the test. The mass of each test sample was measured following exposure and the mass loss due to atomic oxygen erosion was computed using the control samples to account for adsorbed moisture.

Scanning electron microscopy (SEM), coupled with energy dispersive X-ray microanalysis (EDS), was employed to study the morphology and the elemental composition of the surfaces of all samples. SEM was employed in both the secondary electron (SE) mode and in the backscattered electron (BE) mode. While in BE mode, the BE detector was used in two different configurations, i.e., in compositional mode (BEC) and in topographical mode (BET). The BET mode is useful when topographical

Ion Implantation in Kapton
Aluminum @ 25 keV

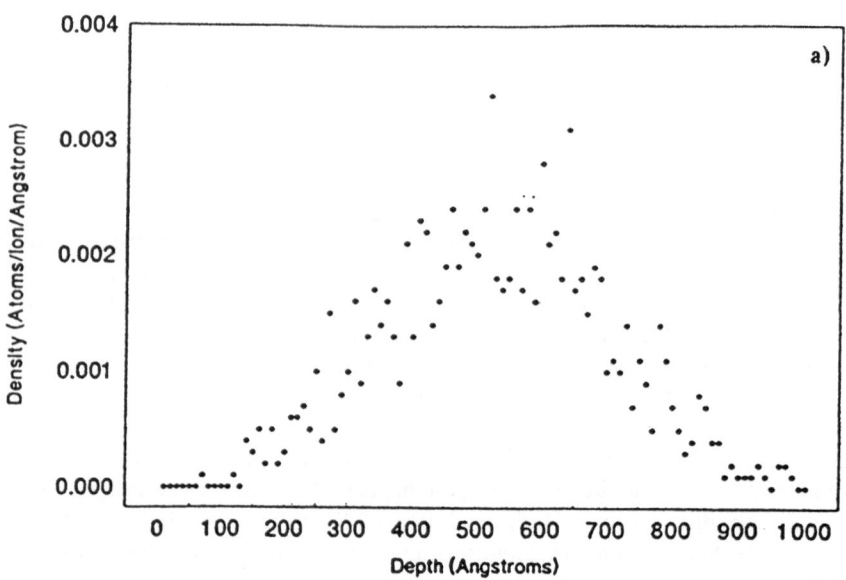

Ion Implantation in Kapton
Aluminum 40 keV

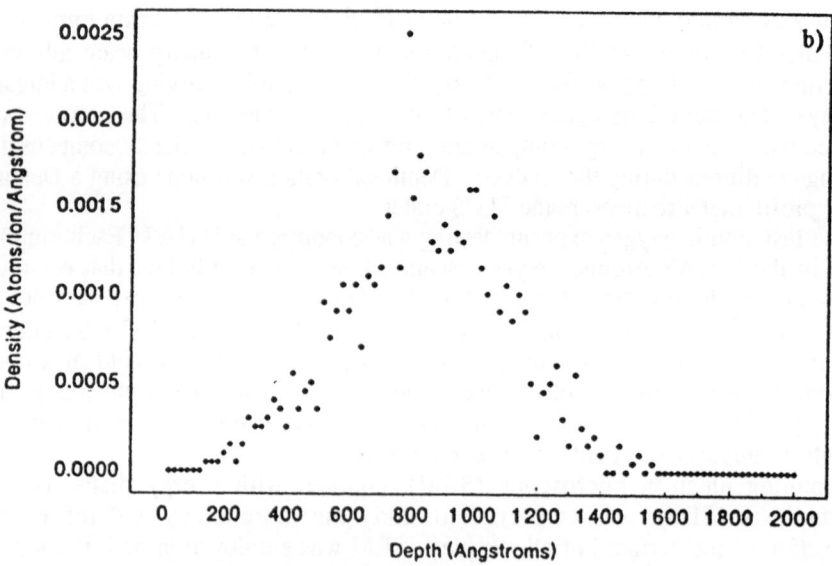

Figure 1.   Code TRIM computer simulation of depth distribution of Al, implanted in Kapton: (a) E = 25 keV; (b) E = 40 keV.

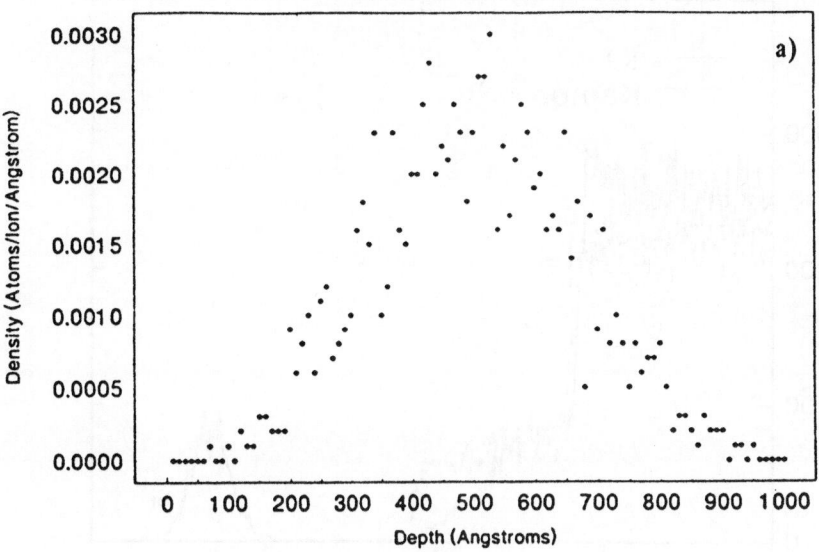

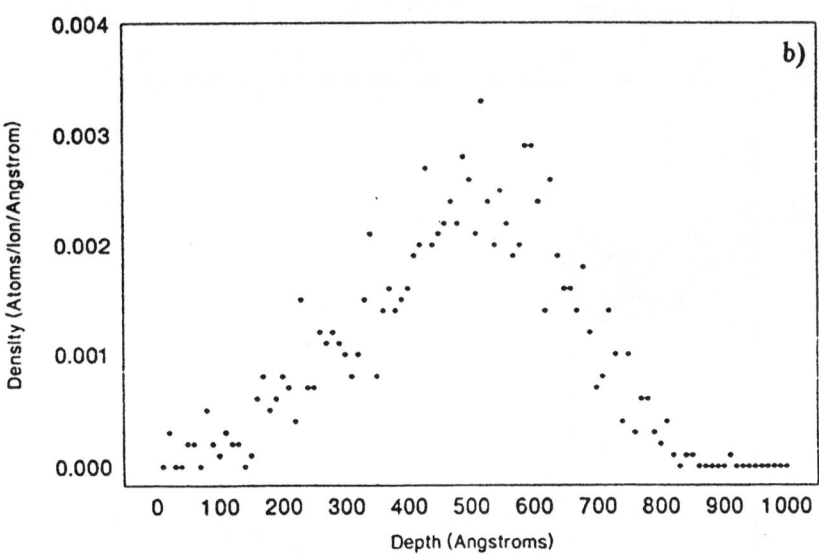

Figure 2.   Code TRIM computer simulation of depth distribution of Si, E = 25 keV
and B, E = 10 keV in double implantation in Kapton suitable to form a
borosilicate glass-like compound.

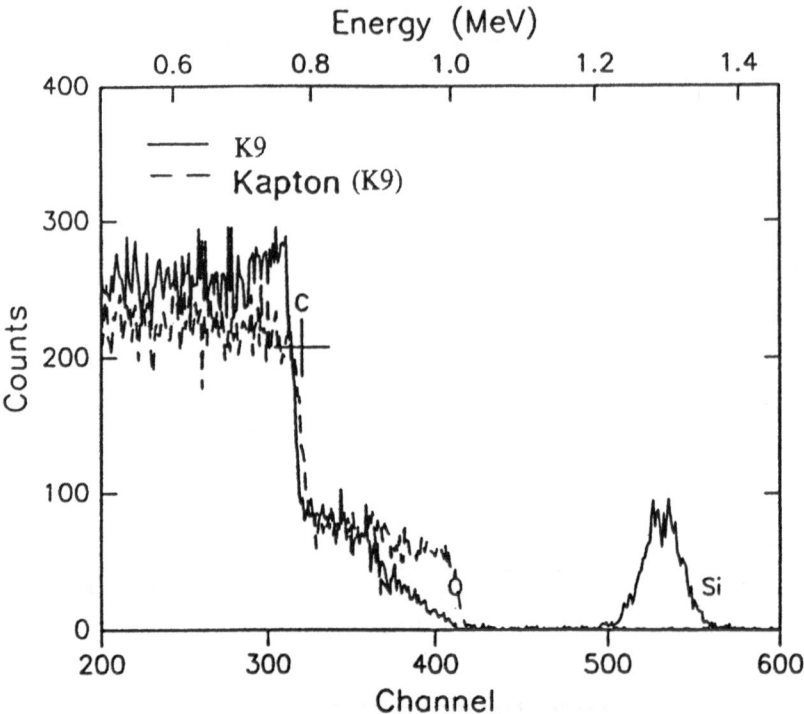

Figure 3a.   RBS comparative specta of the Si implanted sample K9, and the part under the holder (K9).

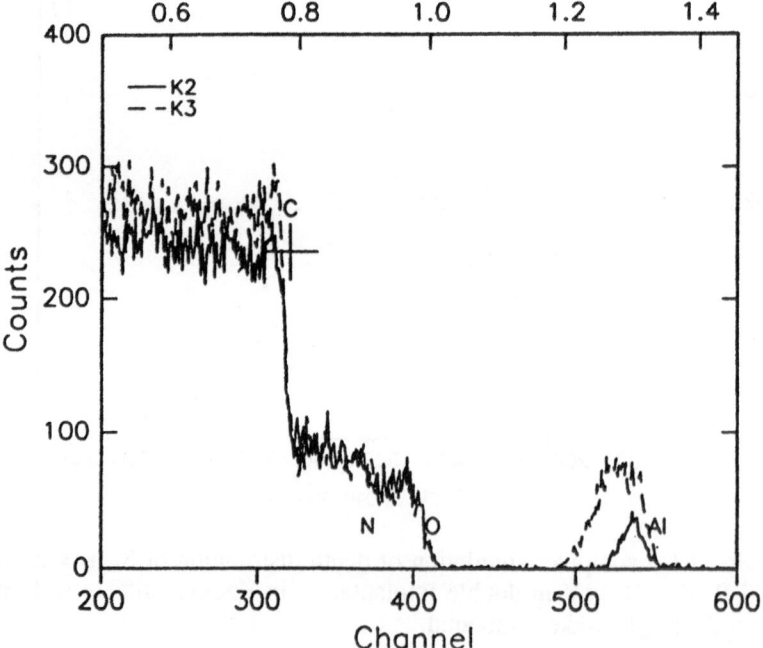

Figure 3b.   RBS comparative spectra of K2 and K3, low- and high-dose Al implanted samples.

information on the surface features is desirable. The surfaces of the polymer samples were coated with a thin carbon film before viewing to prevent charging.

Results and Discussion

The computer simulation results showed that implantation with Al and Si in Kapton, PEEK and Teflon is preferential in the energy range (25-40) keV, if the modified layer is expected to be ~(500-1000) Å in depth (Fig. 1). Very thin implanted layers, with a depth $d \sim (150\text{-}250)$ Å, and narrow sharp distribution of implanted species should be expected with implantation of heavy metals, Sm and Gd. A reasonable selection of energies and fluences can be made for binary implantation to have similar depth distributions and required ratios for different elements. An example for (Si+B) in the low-energy range, with $E_{Si} = 35$ keV for Si and $E_B = 10$ keV for B, is presented in Fig. 2. To imitate the content of the "precursor" for further formation of the compound close to borosilicate glass [6], the ratio of the implanted elements, i.e., the fluences, should be $\varphi_B/\varphi_{Si} \approx 1/3$. It was also shown that in another case of (Si+B) or (Al+B), where B is considered for mechanical properties improvement, B implantation with $E_B \geq 100$ keV can be recommended for the modification of the surface layer with depth $d \geq 0.5$ μm. Based on the calculated depth distribution and atomic concentrations of implanted elements, and using the atomic content of stoichiometric oxides or, in the case of Si+B, the content of borosilicate glass, as an upper limit, a reasonable estimate has been done for the fluences of Si, B and Al, and also for Sm or Gd, to be implanted in the experiment.

The appearance of all polymers after implantation was changed by the well-known carbonization effect [9-11]. Darkening of all implanted Kapton and PEEK samples was visually indicated with the naked eye, when compared to the pristine materials, and was stronger for higher doses and heavy elements. A lustrous graphite-like shade was visible on high-dose implanted Kapton and PEEK. Teflon became opaque, and a white-grey "ash" developed on the surface. This specific behaviour of Teflon can be explained by the destruction (decomposition) of this polymer under ion bombardment, as with implantation of light elements at higher energies [12]. The absence (or very low, less than ~13% concentration) of the irradiated ions in Teflon samples was confirmed later by RBS. The experimental RBS data, presented in Fig. 3, show that Al and Si were successfully implanted in Kapton and PEEK, with the depth of ion implantation being in reasonable agreement with TRIM computer simulations (see Figs. 1 and 2). Note that the theoretical ranges obtained by TRIM, are quite approximate in the case of high-dose implantation of polymers, because the composition and density of the irradiated layer can change dramatically due to depletion of light, volatile gaseous species [9-12]. Note also that the light element B cannot be detected by RBS. The retained dose was shown to be ~77% for Al and ~69% for Si - both "shrinking" effects due to depletion of volatile elements at ion bombardment [2, 12], and thermal or radiation-enhanced diffusion, confirmed for both elements by SIMS, can be responsible for the retention decrease, measured by RBS. RBS results, not presented here, showed that well-defined thin (~100-150Å) layers were implanted in Kapton and PEEK with Sm and Gd.

For comparison of Kapton composition before and after implantation, the spectrum obtained from the part of the sample under the holder was used as a reference in Fig. 3. The results show the depletion of oxygen, and partly nitrogen (N-peak is too low to make a quantitative decision), and an enrichment in carbon, i.e., carbonization, that is in good agreement with [9-12]. The comparatively fewer changes in Al-implanted samples, as detected by RBS, can be explained by increased retention of oxygen, "freed" in ion bombardment, due to reaction with highly reactive implanted Al. This possibility was strongly supported by the results of SIMS depth profile analysis, which show an increase in oxygen concentration in the region of maximum concentration of the implanted aluminum (details are published elsewhere). The results of XPS analysis of the "as implanted" samples (Table I) are also in good agreement with both computer simulation and RBS data. XPS did not show the implanted species in the thin (30-50Å) top surface layer (see also [10]). This is in agreement with TRIM results (Figs. 1 and 2), which show the implanted species to be located at a depth of more than ~100Å. The depletion of light volatiles, O and N, and an increase in carbon at the surface are also clearly confirmed.

TABLE I. Surface content of pristine and implanted Kapton by XPS analysis, at %

Kapton Samples	Elements			
	C	O	N	Implanted
Pristine (theoretical)	75.9	17.2	6.9	—
Pristine (experimental)	79.6	14.3	5.6	—
Implanted	83-88	12-8	2.2-0	Not detected

FAO exposure and testing have been conducted on ion implanted samples. The samples have been weighed, stored in the vacuum chamber for 2 hours for degassing, weighed again, and then exposed to FAO flux for 5 hours, and weighed for the last time. For some of the samples, intermediate weighing has also been conducted. The total mass loss was compared with the value for control Kapton for the same time of exposure, and this data allowed for rough estimates of the stabilization effects. It is important to note that some mass loss is to be expected due to organic etching from the implanted layer by FAO. This is partly compensated for by the oxygen uptake due to oxidation of the implanted species, depending on the depth of the formed oxide-based layer. The usual calculations of an average value for erosion yield R^{FAO}, based on the assumptions of linear etching kinetics, are not appropriate in this special case. After the formation of the stable erosion-resistant top layer, the erosion process should halt on the protected surface. The time for this experiment was ~1-2 hours and seems to be approximately equal to the time of etching of the carbonized layer, depending on the real distribution of the implanted element. For the remainder of the testing time, the average erosion rate of the surface dropped very significantly, almost to a value not measurable by weighing. Some of the kinetic trends have been demonstrated by intermediate weighting of the samples, for instance, after every hour of FAO exposure.

After finishing the FAO testing with a fluence $\varphi_{FAO} \approx (1\text{-}2)\times10^{20}$ at/cm^2, the implanted samples appeared clear, transparent, with a glass-like shiny surface, and without any visual surface features. This appearance was confirmed by SEM at different magnifications and was in strong contrast to the heavily eroded, matte control samples (Fig. 4). A strong increase in the concentration of Si or Al, and also O, with a following decrease of the carbon content, has been shown at the surface of the samples by XPS (Table II). High-resolution XPS also confirmed the formation of oxide-like surface layers, in good agreement with the results of improved resistivity of the surface-modified materials.

TABLE II. XPS surface composition of implanted polymers after FAO testing

Sample No.	Element Detected, at %				
	C	O	N	Si	Al
K contr.	79.6	14.3	5.4	—	—
K₃/AO	28	45	2.7	—	24
K₉/AO	17	54	—	29	—

Conclusions

Computer simulation of ion implantation in thermally stable, high-performance polymers was used to predict the concentration and depth distribution of light and heavy ions, including binary implantation with specially-selected species, in the energy range (10-100) keV. The simulation approach, based on binary atomic collisions, is often useful to predict the final results, even for high-dose implantation. For instance, Kapton and PEEK implantation with Al, Si, B, Sm, Gd, was successfully performed in the conditions recommended from the computer simulation, and the results of RBS analysis showed good agreement with theoretical calculations. At the same time, from the comparison with the experiment, it was shown that Teflon, the most chemically stable polymer, is unstable under ion implantation.

High-dose ion implantation with metal and semi-metal elements, or their special combinations, has proven itself as a new method to form highly-stable, erosion-resistant, protective surface layers on advanced polymer materials, widely used in space and many terrestrial applications. Exposure and testing of this polymer in the FAO testing facility at UTIAS clearly demonstrated that with the high-dose ion implantation of properly-selected species and/or their combination, polymer materials resistance to FAO can be highly improved. The surface-modified materials are expected to have superior performance in space in long duration missions.

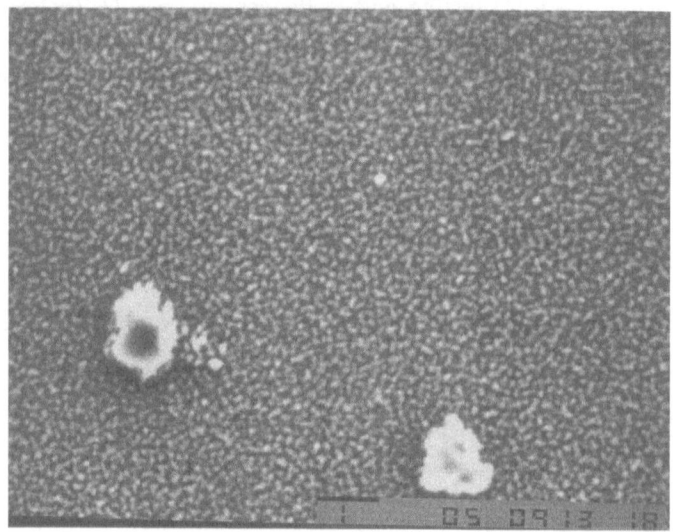

(a) Control

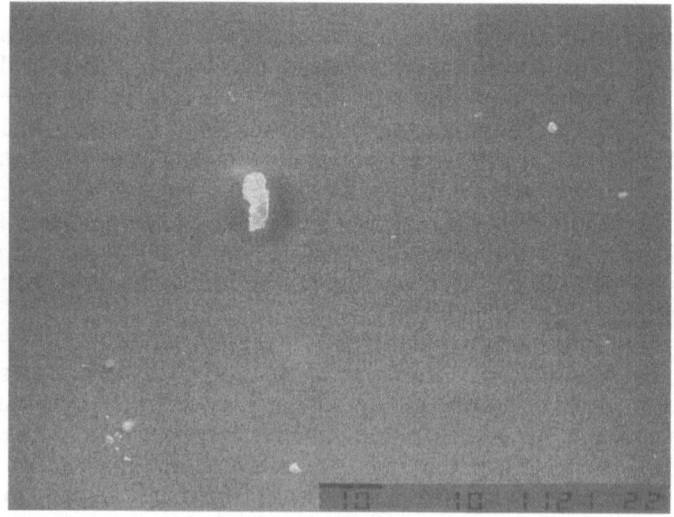

(b) Modified

Figure 4. Scanning electron microscopy of kapton samples, exposed to FAO for 5 hours.
(a) control pristine sample, x10,000; (b) (Si+B) implanted sample (K8), x 1,000

References

1. Banks, B. A., "Atomic Oxygen," LDEF Materials Data Analysis Workshop, compiled by B. A. Stein and Ph. R. Young, NASA CP 10046, 1990, p. 191.
2. Tennyson, R. C., Canadian Journal of Physics 69 (1991), 1190; Tennyson, R. C., "Atomic Oxygen and Its Effects on Materials," in The Behaviour of Systems in the Space Environment, Eds. R. N. DeWitt et al, Kluwer Acad. Publ., 1993, p. 233.
3. Iskanderova, Z. A., Gudimenko, Yu. I., Kleiman, J., and Tennyson, R. C., J. Spacecraft and Rockets 32, No. 5, 1995, pp. 878-884.
4. Banks, B. A., Rutledge, S. K., Auer, B. M., and DiFilippo, F., in Materials Degradation in Low Earth Orbit (LEO), Eds. V. Srinivasan and B. A. Banks, publ. of TMS, Pennsylvania, 1990, p. 15.
5. Kleiman, J., Iskanderova, Z. A., Perez, F. J., and Tennyson, R. C., Surface and Coatings Technology 76-77, 1995, p. 827.
6. Ziegler, J. F., TRIM-90, "Transport of Ions in Matter," Int. Business Machine Corp., 1990.
7. Physical-Chemical Properties of Oxides, Ed. G. V. Samsonov, Moscow, "Metallurgy," 1988.
8. Handbook of Tables for Applied Engineering Science, 2nd Ed., Ed. R. E. Bobz, G. L. Tuve, CRC, 1989.
9. Venkatesan, T., Calcagno, L., Elman, B. S., and Foti, G., in Ion Beam Modification of Insulators, Eds. P. Mazzoldi and G. Arnold (Elsevier, Amsterdam, 1987), p. 301.
10. Davenas, J., Xu, X. L., Boiteux, G., and Sage, D., Nucl. Instr. and Meth. B39, 1989, p. 754.
11. Davenas, J., Boiteaux, G., Xu, X. L., and Adem, E., Nucl. Instr. and Meth. B32, 1988, p. 136.
12. Lee, E. H., Lewis, M. B., Blau, P. J., and Mansur, L. K., J. Mater. Res., Vol. 6, No. 3, 1991, pp. 610-628; Ochsner, R., Kluge, A., Zechel-Malonn, S., Gong, L., and Ryssel, H., Nucl. Instr. and Meth. B80/81, 1993, pp. 1050-1054.

COMPUTER MODELLING OF THE ANOMALOUS ULTRASOUND ATTENUATION IN GLASSES

E.P. NIKONOVA and V.N. SOLOVJEV
State Pedagogical Institute
Department of Computer Sciences
Krivoy Rog, 54 Gagarin Avenue
324086 Ukraine

Abstract
The first wide investigation of ultrasound anomalous attenuation in temperature interval is reported.

This investigation is based on numerical calculations of energetical spectra of soft atomic potentials. The processes of resonant and relaxational absorption are taken into the consideration. Results of the calculations allow us to describe low-temperature peaks, so as high-temperature peaks of attenuation.

1. Introduction

The elastic and dielectric behaviour of amorphous solids at low temperatures differs completely from those of crystalline solids [1]. For example, the acoustic and dielectric absorption is strongly enhanced, compared with crystals and a large absorption peak is found around liquid nitrogen temperature (See Fig. 1).

With increasing frequency the peak moves slowly up to the higher temperature. The height of this maximum increases linearly with frequency and quadratic frequency dependence of absorption is found on its high temperature side.

On the low-temperature side, a second peak or shoulder appears at temperatures below 20 K. The position of this low-temperature absorption peak depends more strongly on frequency than the high-temperature peak. Below the maximum, acoustic attenuation drops rapidly as T^3 and becomes independent of frequency.

Below helium temperature, a number of anomalous effects are observed, which can be understood within the framework of a phenomenological model of two-level systems (TLS's) [2,3].

R. C. Tennyson and A. E. Kiv (eds.),
Computer Modelling of Electronic and Atomic Processes in Solids, 301–307.
© 1997 *Kluwer Academic Publishers.*

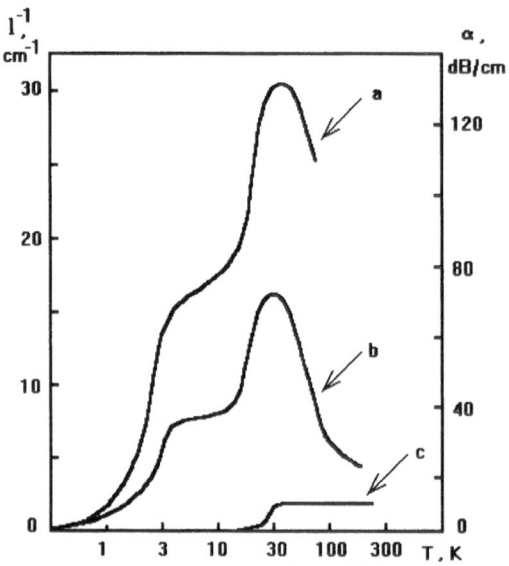

Figure 1. Experimental evidence of the temperature dependence of the longitudinal ultrasonic absorption α in vitreous silica. Glass (a) 930, (b) 507 MHz. Quartz crystal : (c) 1000 MHz. l^{-1} inverse mean free path ; T - temperature .

Although in many cases it is clear that the TLS model is very helpful, in others the simplification introduced by its use is inadequate for detailed understanding. This model is inconsistent with a number of thermodynamical and transport properties of glasses at temperatures that are higher than helium [1]. As for ultrasound attenuation, the TLS model does not explain the existence of a high temperature peak within a single approach to describe and interpret the curves in Fig. 1.

In this paper, we present a model that fits both the low- and high-temperature attenuation data to about 300 K.

The paper is organised as follows. We shall first examine the TLS models and a new approach, named "soft potential model". Second is the original methodology of computer modelling of the relaxational absorption. Temperature and frequency dependence are studied here. Last is the analysis of results achieved and comparison with experiment.

2. TLS Model and the Soft Atomic Potentials

A two-level system model has been introduced by Anderson, Halperin, and Varma [2], and independently by Phillips [3]. According to the TLS model, an atom or a group of atoms are associated with a TLS which resides in a double-well potential $V(x)$ in regard to some generalised coordinate x (See Fig. 2).

We assume that the space between the two lowest levels (E) of such a system is much smaller than the distance to the third level. In such a case we have a two-level systems of energy

$$E = \left(\Delta^2 + \Delta_0^2\right)^{1/2} \qquad (2.1)$$

$\Delta_0 = \hbar\omega_0 \exp(-\lambda)$ being the tunnelling parameter of the two-level system; Δ describes the asymmetry of the potential $V(x)$. Here, ω_0 is the order of the vibrational frequency in a well. The asymmetry Δ and the overlap parameter λ are supposed to be randomly and uniformly distributed over comparatively wide intervals of their values so that the corresponding density of states $N(\lambda, \Delta) = N_0$ is constant.

Recently a new approach appeared in the theory of glasses that includes a TLS model as a limiting case. This is the "soft potential model" (SP) [4]. This approach accounts for the existence of the two-well potentials in glasses by fluctuations of microscopic structure parameters. As a result, there is a finite probability for soft atomic potentials to occur, i.e. for a quasi-elastic constant to be small or even negative for at least one of the local modes. For this mode, according to the SP model, the local potential $V(x)$ can be represented as:

$$V(x) = V_0\left[\eta(x/a)^2 + t(x/a)^3 + (x/a)^4\right], \qquad (2.2)$$

where $a \cong 1 \overset{0}{A}$ is a characteristic interatomic distance, $V_0 \cong 30$ eV is characteristic atomic energy. Coefficients η and t are assumed to be small, random and uncorrelated quantities ($|\eta|, |t| \ll 1$). If $\eta < 0$ the potentials (2.2) are of the two-well type and the TLS's may occur.

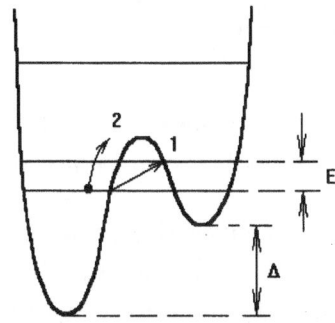

Figure 2. Double-well potential and the three lowest energy levels. Transition 1 corresponds to tunnelling process, transition 2 is thermally activated.

The computer simulations of real amorphous systems (α-Si, α-Ge, α-SiO$_2$) [5] have shown that distributions of parameters η and t have the form depicted in Fig. 3.

The SP model reproduces all the results of the TLS model, and also leads to a number of new conclusions concerning properties of glasses in the region of intermediate temperatures [6].

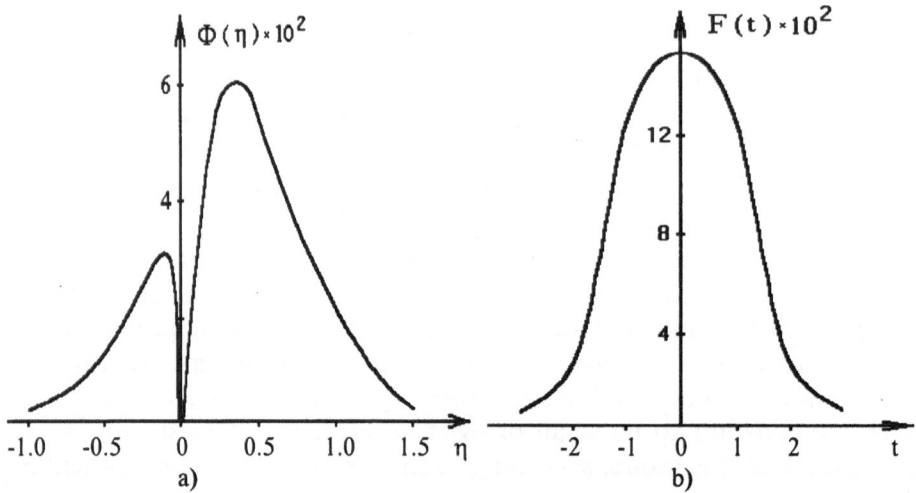

Figure 3. The distribution functions of the parameters η and t of potential (2.2).

3. Basic equations

At low temperatures, the most likely interaction of a phonon and a TLS is the resonant scattering where the energy splitting E coincides with the phonon energy $\hbar\omega$. It leads to absorption (resonant absorption α_{res}). For an individual TLS, it is given by;

$$\alpha_{res} = \frac{\pi \gamma^2 \Delta_0^2}{\rho \upsilon^3 E^2} \varpi th\left(\frac{E}{2T}\right),$$
(3.1)

where ρ is the mass density of the glass, υ is the sound velocity, and γ is the mean deformation potential of the constant describing the TLS - strain interaction.

The relaxation of the TLS gives rise to a second absorption process (nonresonant or relaxational α_{rel}). This absorption is due to the modulation of the energy E of the TLS by the periodic strain of the acoustic wave. As a result, the level population is displaced from its equilibrium value. Such a shift results in energy dissipation and in sound attenuation. This attenuation for one TLS is given by

$$\alpha_{rel} = \frac{\gamma^2}{\rho v^3 T^2} \frac{\Delta_0^2}{E^2} \frac{1}{ch^2(E/2T)} \frac{\varpi^2 \tau}{1+(\varpi\tau)^2},$$ (3.2)

where

$$\tau^{-1} = \frac{3}{2\pi \hbar^4 \rho v^3} E \Delta^2 \gamma^2 coth(E/2T)$$ (3.3)

is the relaxation rate of TLS with energy splitting E. Formally, the relaxation process can be described by a particle moving in double-well potential, and the origin of the absorption lies in a tunnelling type process (transition 1 on Fig. 2) or a thermally activated one (transition 2 on Fig. 2)

The calculation of energy levels for potential (2.2) is fulfilled by a variational method [7]. The Schrödinger equation is

$$-\frac{\hbar^2}{2m}\frac{d^2}{dx^2}\Psi_n + V(x)\Psi_n = E\Psi_n.$$ (3.4)

To simplify the computations, it is convenient to scale the problem, by introducing the following non-dimensional parameters;

$$y = \frac{x}{r_0 \eta_L^{1/2}} \quad , \quad \alpha = \eta/\eta_L, \quad \beta = \frac{t}{\eta_L^{1/2}} \quad , $$ (3.5)

where

$$\eta_L = \left(\frac{\hbar^2}{2mV_0 r_0}\right)^{1/2} \cong 10^{-2}.$$

This transformation reduces the Schrödinger equation to,

$$\frac{d^2}{dy^2}\tilde{\Psi}_n + \tilde{V}(x)\tilde{\Psi}_n = \tilde{E}\tilde{\Psi}_n$$ (3.6)

where $\tilde{V}(y) = w^{-1} V(r_0 y\eta_L^{1/2}) = \alpha y^2 + \beta y^3 + y^4$,
and the characteristic energy w is,

$$w = V_0 \eta_L{}^2 \cong 10 \text{ K} \qquad (3.7)$$

The energy levels E_n of Eq. (3.4) are connected with $\tilde{E}_n$ by the relation,

$$E_n = w \, \tilde{E}_n \qquad (3.8)$$

The linear variation method was used within the framework of the harmonic oscillator representation and with standard matrix elements. The matrix diagonalization routine was based on the Jacobi method [8].

Having an energy spectrum (3.8), the problem is to determine the average of (3.1) and (3.2), according to the energetic splits with $E(\alpha,\beta)$ based on the distribution functions α,β (or η,t), asshown in Fig. 3.

4. Numerical results and discussion

The results of the calculations on vitreous silica are presented in Fig. 4. It can be seen that the soft-potential model successfully exhibits the major features of both the low- and high-temperature peaks of attenuation. Additionally, our results allow us to interpret the nature of high-temperature peaks. For a thermally activated relaxation process, the relaxation time τ usually varies with temperature according to the Arrhenius relationship,

$$\tau = \tau_0 \exp(V/T) \qquad (4.1)$$

where τ_0 is a constant and V is the activation energy for the process [1].

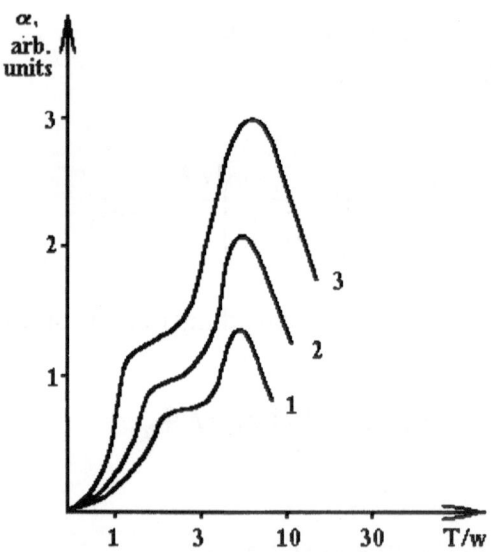

Figure 4. Calculated temperature dependence of the attenuation of the sound waves in vitreous silica. (1) 1000, (2) 500, (3) 250 MHz.

The validity of Eq.(4.1) can therefore be tested by considering the frequency dependence of the temperature T_m at which the peak in the absorption is observed. A straight line of the logarithm of the angular frequency log ω as a function of T_m^{-1} was observed, thus indicating that Eq.(4.1) is valid. From the slope of this line, the mean value of the activation energy can be deduced and a value of roughly 0.05eV is found for vitreous silica. The experiment with the peak shows that a single relaxation time approximation is not sufficient. For the formal description of the activation energy distribution, relaxation distribution times have to be introduced. If the particle is assumed to vibrate with the frequency τ_0^{-1} in one of the two wells in double-well potential, the distribution in the activation energy is equivalent to a distribution in the barrier height.

In our units, the barrier height is $V_\beta = w/4(\eta/\eta_L)^2$ which must correspond to the peak in density of the distribution of parameter η in figure 3, i.e. $|\eta| = \eta_L(4V_\beta/w)^{1/2} \cong 0.15$. As a consequence of figure 3a, the achieved result also can be regarded as satisfactory and thus explains the nature of a high-temperature peak

References

1. Hunklinger, S. and Raychaudhuri, A.K. (1986) Thermal and elastic anomalies in glasses at low temperatures, *J.Low Temp. Phys* **9**, 1-56.
2. Anderson, P.W., Halperin, B.I., and Varma, C.M. (1972) Anomalous low-temperature thermal properties of glasses and spin glasses, *Phil.Mag.* **25**, 1-9.
3. Phillips, W.A. (1972) Tunnelling states in amorphous solids, *J.Low Temp. Phys.* **7**, 351-357.
4. Ilyin, M.A., Karpov, V.G., and Parshin, D.A. (1987) Parametri myagkich atomnich potencialov v steklach, *JETF* **91**, 291-296.
5. Dyadina, G.A., Karpov, V.G., and Solovjev, V.N. (1989) Fluctuatcii lokalnich atomnich potencialov v amorfnich vestchestvach, *FTT* **31**, 148-155.
6. Galperin, Yu.M., Karpov, V.G., and Solovjev, V.N. (1988) Plotnost kolebatelnich sostoyaniy v steklach, *JETF* **94**, 373-384.
7. Galperin, Yu.M., Karpov, V.G., and Solovjev, V.N. (1988) O nizkotemperaturnoy teploemkosti amorfnich vestchestv, *FTT* **30**, 3636-3642.
8. Sunney, I.Chan, David Stelman, and Larry E.Thompson. (1964) Quartic oscillator as a basic for energy level calculations of some anharmonic oscillators, *J.Chem. Phys.* **41**, 2828-2835.

THE FRACTAL MODELS OF DEFECTS GROWTH IN SOLIDS

M. RYBACZUK
TECHNICAL UNIVERSITY OF WROCŁAW
INSTITUTE OF MATERIALS SCIENCE
AND APPLIED MECHANICS
UL. SMOLUCHOWSKIEGO 25,
50-370 WROCŁAW, POLAND

1. MICRO, MESO, and MACRO SCALES OF LENGTH

The fractal models of fatigue defects growth constitute an essential topic of this paper, however the proposed methods are completely general and applicable in many other problems.

All physical processes in a material may take place at any distinct length scales. At macroscopic scale we have variables related to whole samples. At the opposite limit we meet microscopic objects as particles or atoms and their degrees of freedom. However there is also an intermediate, mesoscopic range of scales. Essentially structure defects as grain boundaries, dislocations, small cracks appear at this scale. Note that all meso–objects are well localized in space comparing to makro range. In the continual medium approximation we neglect short range scales comparable to distances between particles. On the other hand many important processes (like fatigue one for example) take place or correspond to other range of length.

Under external load applied to a sample all degrees of freedom at any scale range become excited. The input energy measured in terms of hysteresis loop flows down from macroscopic scale to deeper levels (see Fig. 1). Finally at micro–level we obtain some heat outflow. As we know heat corresponds to oscillations of atoms.

Not all input energy flows out as a heat. Some part, called cold work, becomes stored at defects at mesoscopic scale. That entails the fatigue process related to an external loading. The most important theoretical problem is to find the correspondence between macro and meso scopic picture of fatigue.

R. C. Tennyson and A. E. Kiv (eds.),
Computer Modelling of Electronic and Atomic Processes in Solids, 309–319.
© 1997 *Kluwer Academic Publishers.*

310

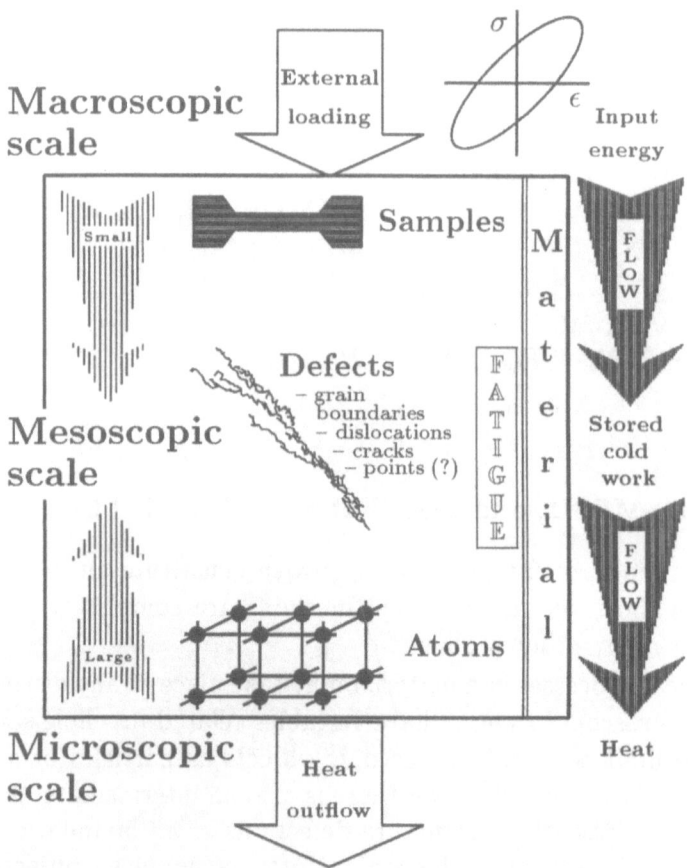

Figure 1. The energy flow and energy accumulation at different length scales in a material. The fatigue defect born at mesoscopic level grows up to a macroscale.

Defects at any stage of evolution are modeled by means of fractals with fixed fractal measure and dimension. The cold work $\mathcal{E}$ stored at defect is assumed to be linear in fractal measure ν_D but the proportionality factor $a(D)$ depends on fractal dimension D. Note that the fractal measure has been understood as and represented by suitable projective quantity (see appendix A). Next representing the projective quantity by suitable powers one obtains:

$$\mathcal{E} = a(D)\nu_D. \tag{1}$$

In such way we have generalized the usual formula for the dislocation core energy or surface energy. This is also the state equation for fractal defects linking energy, measure and dimension.

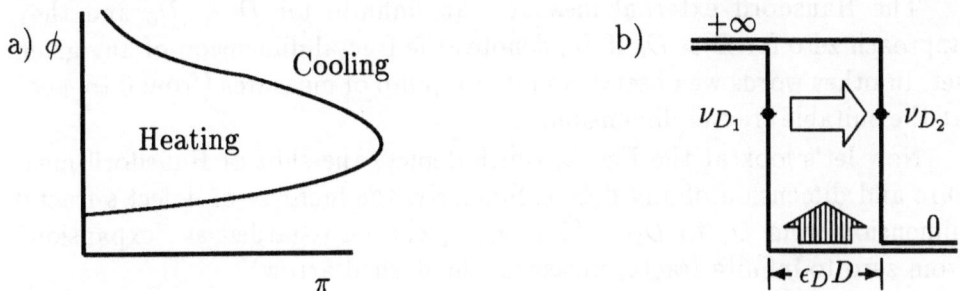

Figure 2. The inversion curve for the van der Waals gas a). In the area inside the curve the gas warms itself during expansion, outside of this area the gas cools itself. The growth of fractal dimension interpreted as expansion from 0 to ∞ of fractal measures from some interval b).

As was said at the beginning the fractal defects grow at the mesoscopic scale. Thus we need the mesoscopic characteristics of a material. They come from the factor $a(D)$ in the state equation for fractal defects. Suppose that we have the small relative shifts of all variables. Such shifts are given by small, real numbers ϵ. Then the state equation can be written as equation for epsilons:

$$\frac{\Delta\mathcal{E}}{\mathcal{E}} = \epsilon_{\mathcal{E}} = D\epsilon_\nu + \epsilon_D \ln\left(\frac{\nu_D}{\mathcal{A}(D)}\right), \quad \mathcal{A}(D) = \left(\frac{\pi\, a(D)}{\pi\, D}\right)^{-1}, \quad (2)$$

where $\mathcal{E}' = \mathcal{E}+\Delta\mathcal{E} = (1+\epsilon_{\mathcal{E}})\mathcal{E}$, $\nu'_D = (1+D\epsilon_\nu)\nu_D$ and $D' = (1+\epsilon_D)D$. Instead of the factor $a(D)$ we have the characteristic fractal measure $\mathcal{A}(D)$ (defined by the multiplicative derivative (3)).

$$f : x \ni \mathcal{R} \longrightarrow f(x) \in \mathcal{R}_+,$$

$$\frac{\pi f(x)}{\pi x} = \lim_{\epsilon\to 0}\left\{\frac{f((1+\epsilon)x)}{f(x)}\right\}^{\frac{1}{\epsilon}}, \quad f((1+\epsilon)x) \simeq f(x)\left(\frac{\pi f(x)}{\pi x}\right)^{\epsilon}. \quad (3)$$

The last formula describes local (in the vicinity of x) approximation of $f(x)$ in terms of $\pi f(x)/\pi x$ in close analogy to the well–known additive approximation $f(x + \Delta x) \simeq f(x) + f'(x)\Delta x$. The quantity $\mathcal{A}(D)$ has physical dimension $(length)^D$.

The curve $\mathcal{A}(D)$ is the inversion curve comparing to the van der Waals gas [1]. The p, V–plane (or equivalently reduced π, ϕ) is divided into the two distinct regions as shown in the Fig. 2. During the Joule–Thomson process the temperature of gas increases inside the bounded area (heating) and decreases (cooling) outside of it. The border line is called the inversion curve for any real gas.

The Hausdorff external measures are infinite for $D < D_0$ and they approach zero for $D > D_0$ if D_0 denotes the fractal dimension of any given set. In other words we observe an infinite jump of measures (from 0 to $+\infty$) at the suitable fractal dimension.

Now let's look at the Fig. 2, which depicts the shift of Hausdorff measure and dimension of any defect. Similarly, the increase of defect's fractal dimension from D_1 to $D_2 = (1 + \epsilon_D)D_1$ can be regarded as "expansion" from zero to infinite fractal measure (the dashed arrow).

Above $\mathcal{A}(D)$ (i.e., for $\nu_D > \mathcal{A}(D)$) the defect energy increases with growing fractal dimension, whereas below $\mathcal{A}(D)$ we observe opposite behavior. The irreversible, fatigue evolution runs according to $\mathcal{A}(D)$. Note that $\mathcal{A}(D)$ describes the mesoscopic level of any material and similarly as for the van der Waals gas, many different thermodynamical processes are possible for the same $\mathcal{A}(D)$.

In principle there are many different objects with intermediate size. Thus in material many mesoscopic scales may simultaneously exist. In this paper we assume that there is one well defined mesoscopic level. This simplification strongly reduces amount of necessary material characteristics but the resulting simple models appear to be still applicable in some problems. The position of mesoscopic scale becomes then one of the most important material characteristics.

In brittle materials mesoscopic level is close to macroscale. Any attempt to accumulate energy in material will yield a macroscopic crack. At limit configuration, when meso and macro levels coincide we have cracking without fatigue. For deep mesoscopic level close to the shortest microscale, any defect has long way (in terms of length) to cover and large amount of energy can be accumulated in a material. That configuration is typical for plastic materials. At the limit case, when meso and micro scales coincide we have neither fatigue nor cracks. The whole input energy flows out as a heat. Such infinitely plastic material behaves similarly to a liquid. However there is one exclusion from that last possibility. Suppose that the shortest scale corresponds to objects with no kinetic degrees of freedom. Then there is no heat and fatigue process becomes possible. Suppose that energy is stored at broken fibres in composite but there are no smaller objects then fibres. In turn fibres do not move or oscillate as a whole. From theoretical point of view such ideal material, or may be rather some mathematical model, is very interesting because we can observe the fatigue defects growth over all possible length scales. The last model was applied to test fractal description of defects growth [2].

At present time there is no experimental technique allowing for continuous observation of defects and their transformations in bulk materials. We can examine surfaces only. Of course one may stop the fatigue test and make cross-section of a sample. However then we can observe some frozen structure of defects solely without any preceding or future evolution. Computer experiments, simulations make possible continuous monitoring of defects transformations. Unfortunately at present time only simplified models can be studied in this way. In effect there are no direct evidence for theoretical models of fatigue based on defects evolution.

Mathematics says that any ball with finite, nonzero radius may include even infinite fractal with lower fractal dimension [3]. That is not possible in real physical systems because of discrete structure at microscopic, atomic scale. The fractal measure of maximal defect (with fractal dimension D_0) included into ball with radius ρ_0 is proportional to the suitable power $\rho_0^{D_0}$. Let's write the characteristic measure $\mathcal{A}(D_0)$ in the same way, i.e., $\mathcal{A}(D_0) = L_0^{D_0}$. We consider the evolution of an isolated defect (i.e., $\epsilon_{\mathcal{E}} = 0$) growing with respect to measure and dimension only. The general evolution equation (2) gives:

$$0 = D_0 \epsilon_\nu + \epsilon_D \ln \left(\frac{\rho_0^{D_0}}{\mathbf{L}^{D_0}} \right), \qquad \epsilon_\nu = \frac{\rho' - \rho_0}{\rho_0}, \qquad \epsilon_D = \frac{D' - D_0}{D_0} = \frac{\Delta D}{D_0},$$

(4)

where ρ' and D' denote the suitable ball radius and fractal dimension after small transformation of defect. Simple calculations yield:

$$\rho' = \rho_0 \left(1 + \frac{\Delta D}{D_0} \ln \frac{\mathbf{L}_0}{\rho_0} \right).$$

(5)

Assuming $\Delta D > 0$ (the increase of fractal dimension, more complicated defect) and $\rho_0 > \mathbf{L}_0$ (large defect) we obtain $\rho' < \rho_0$. It means that defect becomes localized in smaller ball. In general the characteristic length $\mathbf{L}$ can be different for different dimensions D. The effect described by the equation (5) resembles the fractal sieve. Only close packed defects with fractal measure equal $\mathcal{A}(D)$ pass through sieve without any change in spatial localization ($\rho' = \rho_0$). The above mechanism gives very weak, logarithmic localization only.

Let $\mathbf{L}$ does not vary due to labile D. In other words we assume $a(D)$ in the form $a(D) = E/\mathbf{L}^D$, where E is some fixed energy. Integrating (5) for defects growth from D to D' we obtain:

$$\frac{\rho_{D'}}{\mathbf{L}} = \left(\frac{\rho_D}{\mathbf{L}} \right)^{\frac{D}{D'}},$$

(6)

where ρ_D denotes the covering ball radius for fractal defect with fractal dimension D. For large shifts of dimensions $D \ll D'$, one gets $\rho_{D'} = \mathbf{L}$ independently of the initial value of ρ_D. Therefore isolated defect (i.e., without incoming energy) will reach inversion curve $\mathcal{A}(D)$ if the range of D–shifts is large enough. In other words, the new defect with structure depicted by D' will be born at length scale given by $\mathcal{A}(D')$ ($\mathcal{A}(D)$ attracts defects). That is our idea of <u>mesoscopic length</u>. Therefore $\mathcal{A}(D)$ defines the mesoscopic scale of length in a material.

Any growth process is then equivalent to some thermodynamical transformation of system with state equation (1). Even for the same equation (1) many distinct thermodynamical processes are possible. Note also some specific property of relative differentials from (2). The relative shifts given by means of epsilons do not differ themselves for extensive and intrinsic thermodynamical quantities. That makes fractal growth possible.

The mesoscopic length scale may also vary during defects evolution. The state equation (1) can be treated as first order energy expansion. Taking into account the second order term, describing interaction between separate fractal defects one obtains:

$$\mathcal{E} = a(D)\nu_D + b(D)\nu_D^2. \tag{7}$$

After simple calculations relative shifts give:

$$\epsilon_{\mathcal{E}} = D\epsilon_\nu \frac{1 + 2\dfrac{b(D)}{a(D)}\nu_D}{1 + \dfrac{b(D)}{a(D)}\nu_D} + \epsilon_D \frac{\ln\left(\dfrac{\nu_D}{\mathcal{A}(D)}\right) + 2\dfrac{b(D)}{a(D)}\nu_D \ln\left(\dfrac{\nu_D}{\mathcal{B}(D)}\right)}{1 + \dfrac{b(D)}{a(D)}\nu_D}, \tag{8}$$

where $\mathcal{B}(D)^{-2} = (\pi b(D)/\pi D)$ defines the new characteristic length corresponding to interaction radius. For small defects (low interaction energy) $a(D) \gg b(D)\nu_D$ the evolution equation coincides with (2). In the opposite limit, for large defect with great interaction energy $a(D) \ll b(D)\nu_D$, the new characteristic measures $\mathcal{B}(D)$ plays the role of $\mathcal{A}(D)$ comparing to (2):

$$\epsilon_{\mathcal{E}} = 2D\epsilon_\nu + 2\epsilon_D \ln\left(\frac{\nu_D}{\mathcal{B}(D)}\right). \tag{9}$$

Now the defects evolution runs according to $\mathcal{B}(D)$ and all previous arguments (an inversion, localization) apply to this new mesocopic scale. In the computer simulations of fibre breaking process in composites such rapid jumps of defects clusters are observed when the separate defects density approaches some characteristic value corresponding to the stress concentration radius [2]. In subsequent part we confine ourselves to the two quite different models of fractal defects evolution. The other interesting problems like entropy shifts and mesoscopic scale dependence upon external stress for example are beyond the scope of this paper.

2. THE MODEL OF LOW CYCLE FATIGUE IN METALS

At macroscopic level we measure the external load energy in terms of hysteresis loop $\Delta \mathcal{E}$. Let's assume that $\Delta \mathcal{E}$ remains constant during fatigue test. From experimental observations of LCF it follows that for many metals $\Delta \mathcal{E}$ is proportional to some power σ_a^α of the applied stress amplitude σ_a:

$$\Delta \mathcal{E} \propto \sigma_a^\alpha. \tag{10}$$

If the temperature of fatigue sample remains approximately constant (the heat outflow does vary in time) then the cold work per cycle also becomes constant. In effect, after N cycles the energy accumulated at defects ΔW is proportional to $N\Delta \mathcal{E}$. Combining with (10) one obtains power dependence $\Delta W \propto N\sigma_a^\alpha$. If energetical fatigue criterion holds, i.e., the final damage appears when accumulated energy equals some critical (for sample) value W_c one get power low linking fatigue lifetime with stress amplitude. It is known as <u>Morrow equation</u> and commonly written in the form:

$$\sigma_a = \sigma_f \left(2N_f\right)^b, \tag{11}$$

where N_f denotes the number of half–cycles, and σ_f is some constant. Energetical criterion relates equations (10) and (11) giving $b = -1/\alpha$. However α and b can be measured independently. The experimental values for α and b violate relation $b = -1/\alpha$.

After some algebra the evolution equation (2) can be rewritten in integral form:

$$\frac{\mathcal{E}_{end}}{\mathcal{E}_0} = \exp\left\{ \int_{D_0}^{D_{end}} \left[\Phi(D, \mathcal{E}) + \ln\left(\frac{\mathcal{E}}{a(D)\mathcal{A}(D)}\right) \right] \frac{dD}{D} \right\}, \tag{12}$$

under the same assumption as above. $\mathcal{E}$ denotes the current energy, $\mathcal{E}_{end} = \mathcal{E}_0 + N\Delta\mathcal{E}$ ($N = 2N_f$), where $\mathcal{E}_0$ is the initial energy. In similar way D_0 and D_{end} are fractal dimensions at the beginning and at the end of the irreversible fatigue process suitably. The direction of thermodynamical process is fixed by the functional $\Phi(D, \mathcal{E})$:

$$\Phi(D, \mathcal{E}) = D\frac{\epsilon_\nu}{\epsilon_D}. \tag{13}$$

We assume small constant energy increase during one cycle $\mathcal{E}_0 \gg \Delta\mathcal{E}$ but after enough number of cycles $N\Delta\mathcal{E} \gg \mathcal{E}_0$ (the large rearrangement of defect). Combining (13) with (10) one gets:

$$\sigma_a^\alpha = \kappa\mathcal{E}_0 N^{-1} \exp\left\{ \int_{D_0}^{D_{end}} \left[\Phi(D, \mathcal{E}) + \ln\left(\frac{\mathcal{E}}{a(D)\mathcal{A}(D)}\right) \right] \frac{dD}{D} \right\}, \tag{14}$$

where κ is some constant. Clearly the term N^{-1} corresponds to relation $b = -1/\alpha$ whereas the exponential term carries corrections coming from fractal degrees of freedom. The concrete form of evolution is given by $\Phi(D, \mathcal{E})$ and according to the universal form of Wöhler curve we look for suitably general Φ.

For defect growing along some curve $\mathcal{F}(D)$ (fractal measure versus dimension) the direction $\Phi = \ln[(\mathcal{F}(D)^{-1}\mathcal{F}'(D)]$, where $\mathcal{F}'(D) = \pi\mathcal{F}(D)/\pi D$ (the logarithm is characteristic). Moreover if the position at evolution curve depends on energy only, then: $\Phi = \ln(\phi(\mathcal{E}/\mathcal{E}_0))$ ($\mathcal{E}$ is the current energy). The correction to Φ should be linear in n (n current number of cycles) at the beginning whereas close to final damage we expect corrections linear in n/N (the system becomes sensitive to upper limitation for lifetime). In effect Φ has the following form:

$$\Phi(D, \mathcal{E}_0 + n\Delta\mathcal{E}) = \Phi_0 + \ln\frac{N}{1 + n\dfrac{\Delta\mathcal{E}}{\mathcal{E}_0}}, \tag{15}$$

where Φ_0 dependence on n can be neglected. The logarithm of energy exactly cancels similar term in evolution equation and simple integration gives:

$$b = \left(\ln\frac{D_{end}}{D_0} - 1\right)\frac{1}{\alpha}. \tag{16}$$

Cracks correspond to $D \geq 2$. On the other hand in crystals or polycrystals the linear defects ($D = 1$), dislocations always exist. Consequently $1 \leq D_0 \leq 2$. The opposite value D_{end} can be taken from examinations of fracture surfaces.

The corrections in (16) coming from fractal excitations were verified in fatigue test for medium–carbon steel (0.45% C). The suitable power exponents are (the constant total strain amplitude test): $\alpha = 11$, $\beta = -0.059$, $D_{end} \simeq 2.2$. That yields the value of dimension $D_{0steel} \simeq 1.4 - 1.5$. The initial dimension D_0 shows how complicated defects should be to evolve in irreversible manner. On the other hand the start point of fatigue defects growth is supposed to be known in such metal. The fatigue process begins with formation of slip bands in ferrite.

The derived initial value of D_0 is very close to the known, exactly solvable model from dislocation theory. Suppose that we examine the condensation of identical straight, parallel dislocation lines in a plane. Before an obstacle dislocations collapse with (singular) density $\rho \propto x^{-\frac{1}{2}}$, where x denotes distance to an obstacle. The fractal formed from such condensation points solely will have fractal dimension equal 1.5. On the other hand slip bands can also come from condensing dislocations.

The same experiment was repeated for α–brass to verify if the initial value od D_0 exceeds fatigue threshold for steel as it should be for more plastic material. Experimental value $D_{0brass} \simeq 1.7$. However we have no such simple intuition for fatigue threshold in brass as in the case of steel. In both metals the derived values for the initial structure of defects fractal dimension at fatigue threshold are reasonable but the direct observation and experimental evidence is still missing. The arguments to support (15) are rather heuristic ones. On the other hand many roentgenographic observations suggests linear changes (of mean interatomic distance) at the begining and close to the end of fatigue process with large platou for the intermediate times (or numbers of cycles).

3. INTER PORES CRACK GROWTH IN SINTERED STEELS

The sintered steels are very brittle material with initial structure of pores involved by powder metallurgy. During fatigue process we observe the cracks growing between pores. Once more the initial structure of defects evolves to final transparent crack in a sample. We have to model both: pore structure and fatigue crack.

At first we look for the suitable range of magnifications. For each picture we estimate (box–counting) fractal dimension for observed structure of pores. Next computer finds contours of all pores and once more we evaluate fractal dimension for contours solely. We seek for range of magnifications in which the above two fractal dimensions coincide. Then fractal dimension will depend on linear size of pores and their distribution solely but not on the internal structure of separate pores. Since details of individual pore form are not important we can model pores by points but the distribution of points should have the same fractal dimension as real structure.

Once fractal model of pore structure is found we may generate crack linking pores. At next step we compare fractal dimensions of generated model cracks and cracks observed in sintered steel. For proper model both dimensions should coincide.

Sintered steels are produced from powders and during technological process high pressures are applied. Therefore structure of grains, being dense packed, should be locally close to hexagonal one. In turn pores originate predominantly at surfaces of adjoint powder grains. In effect we expect the hexagonal structure to be visible also in spatial distribution of pores. At large macroscopic scale the pore distribution becomes uniform and hexagonal order is missing. In effect the fractal modeling pore structure should be composed with hexagonal cells. Each cell contains a fractal with dimension close to value obtained from experimental observations.

To prepare the needed mathematical model we have adopted the ordi-

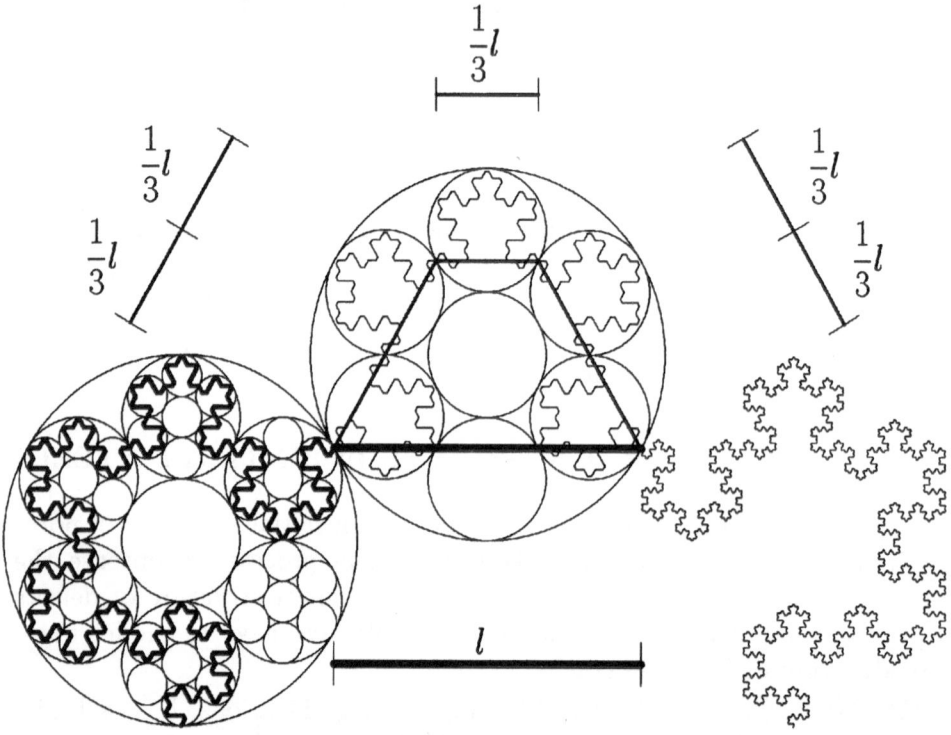

Figure 3. The crack recurrent construction applied to the fractal model of pore distribution. The pore structure fractal model has dimension 1.63 quite close to experimental value 1.68. The fractal dimension of the model crack equals 1.47 whereas experimental value is slightly greater 1.51.

nary construction of middle third <u>Cantor dust</u>. In the traditional method the middle third part of a segment is removed at each step of recurrence. However that is the same as intersection of shells with decreasing (by factor 1/3) radiuses. A <u>shell</u> with center x and radiuses R_{min}, R_{max} is a collection of all points y with distance $d(x,y)$ limited by radiuses: $R_{min} \leq d(x,y) \leq R_{max}$. Replacing one dimensional shells by two dimensional ones we construct the suitable Cantor dust with dimension 1.63 and hexagonal symmetry. The fractal dimension of real pores equals 1.68.

Next we generate a crack according to simple geometrical rule, which does not favorize any length scale. Suppose that we have two clouds of defects with a single common point. Then a crack should run through this common point and inside a cloud of defects we approximate crack by straight segment. Under current resolution we treat the cloud of defects as uniform defect. Increasing magnification we notice that initial cloud divides itself into smaller ones and we once more apply above crack form approximation. The model crack (in fact formed from <u>von Koch type curve</u> shown in the Fig. 3) has fractal dimension 1.47. At the same time fractal dimension of real crack equals 1.51. For both models: pore distribution and crack form, the model and experimental values of fractal dimensions nearly coincide. Model values are slightly smaller because our construction doesn't employ any noise. The Cantor dusts filling cells are very regular. According to experimental observations cells have linear size $\simeq 0.4mm$, equivalent to few diameters of powder grains. This range of length constitutes the mesoscopic level in sintered steels. Therefore meso and macro scales are quite close as it should be in brittle materials. Let us notice a difference between the current model and fractal model from previous sections. The fractal dimension of defect in sintered steels decreases $(D_{pores} > D_{crack})$.

The constructed models of pore distribution and cracks may be applied to study other important characteristics like stress intensity factor K_I for example. The energy conserving cascade process examined in [4], [5] gives $K_I \propto \sigma l^{\alpha}$, where σ denotes stress and l is fractal crack linear size. The power α depends on crack form and its fractal dimension D. For cracks with fractal surface $(2 \leq D \leq 3)$ $\alpha = \frac{1}{2}(3 - D)$, whereas for smooth crack with fractal contour of dimension D $(1 \leq D \leq 2)$ one gets $\alpha = \frac{1}{2}(2 - D)$. That agrees with numerical FEM calculations.

Acknowledgments Financial support under KBN grant nr 7 T07A 01909 is gratefully acknowledged.

References

1. Rybaczuk M., (1992), The fatigue evolution of fractal defects in metals, in: K. T. Rie, proceedings of *Third International Conference On Low Cycle Fatigue And Elasto-Plastic Behavior Of Materials*, Berlin FRG September 7–11, 1992, Elsevier London and New York 1992.
2. Stoppel P. and Rybaczuk M., (1996), Simulations of random fractal fibre breaking in composites, to be presented at *Ninth International Conference on Fracture*, Sydney, April 1–5, 1997.
3. Falconer K., (1990) *Fractal Geometry*, John Wiley & Sons, Chichester and New York.
4. Goldsztein R. W., Mosolov A. B., (1991) Cracks with fractal surfaces, *WAN SSSR*, vol. **319** Nr 4, (in Russian).
5. Goldsztein R. W., Mosolov A. B., (1992) Fractal cracks, *PMM*. vol. **56**, (in Russian).

COMPUTER MODEL FOR M/OD IMPACT DAMAGE ASSESSMENT ON SPACECRAFT MATERIALS

R . C . TENNYSON, G . SHORTLIFFE
University Of Toronto Institute for Aerospace Studies
North York , Ontario, Canada

1. Introduction

Spacecraft in low earth orbit (LEO) are vulnerable to impact damage resulting from collisions with micrometeoroids and orbital debris (M/OD). Micrometeoroids originate naturally from planetary or asteroidal collisions and cometary ejecta. Although there is no particular direction in which micrometeoroids approach the earth, spacecraft experience a bias in the direction of travel (known as the RAM direction). This can readily be seen from the impact distributions measured on the NASA Long Duration Exposure Facility (LDEF) which spent almost six years in space. Figure 1 presents this data for impacts arising from the M/OD environment and it is quite apparent that the RAM direction recorded significantly more impacts, although it is noteworthy that even the back surface was hit as well. Artificial space debris consists of everything from spent satellites and rockets to aluminum oxide fuel particles, paint chips and fragmentation objects from collisions of these bodies in orbit. A plot of the M/OD environment as a function of altitude and particle size is shown in Fig.2. Although micrometeoroids have been a design issue since the 1960's, concern about the growing amount of orbital debris did not receive serious attention until the mid-1980's. The problem has grown to such an extent that the probability of impact by space debris exceeds that due to micrometeoroids in some cases. There is currently an astounding 2.5×10^6 kg of debris in LEO. Since first entering space in 1957, almost 20,000 objects have been launched with approximately 7,000 still in orbit.

Many studies have been reported on impact damage from hypervelocity particles on metallic materials (see [1] and [2] for example), which provide design information on how to calculate damage areas and penetration depths. However, very little data has been published on such materials as polymer-based composites. These latter materials are used extensively in spacecraft structures and components , such as antenna struts and panels. Composites provide significant advantages over metals because of their high specific stiffness and strength (i.e.; "specific" refers to a property divided by material density) and low coefficients of thermal expansion. On the other

R. C. Tennyson and A. E. Kiv (eds.),
Computer Modelling of Electronic and Atomic Processes in Solids, 321–330.
© 1997 *Kluwer Academic Publishers.*

hand, they also pose serious problems in the space environment such as outgassing in vacuum, erosion due to atomic oxygen and VUV degradation. Protective measures have been developed such as coatings ,surface modifications and improved resin formulations. However, the problem of surface and structural damage due to M/OD impacts has not been resolved. Not only can M/OD cause local fracture and delamination damage of composites, but penetration of protective coatings can occur and thus open sites for erosion of the substrate.

At this point in time, sufficient information exists to describe the space M/OD environment as a function of altitude and satellite inclination (Fig.2). Based on LDEF data (Fig.1) , a model can be constructed for impact distributions around an orbiting satellite, although it is strictly valid only for the LDEF orbital parameters. Using impact damage models for materials, one can then proceed to estimate the total accumulated damage for a given operational lifetime. Together with probability estimates of significant hits, a corresponding risk assessment for the satellite can be made. This report describes the contents of a computer model required to yield these assessments. Graphs are presented illustrating the elements that must be included in the model , together with a case study example for a composite boom structure.

2. Computer Model Elements

There are seven elements that constitute the computer model required to assess M/OD effects on an orbiting satellite. These can be summarized as follows;

- M/OD Space Environment
- Satellite Orbital Parameters
- Structural Component
- Impact Probability
- Damage Assessment
- Failure Criteria
- Risk Assessment

Figure 3 presents a flow chart of the interaction of these elements to form the overall computer model. The following sections describe each component of the model.

2.1 M/OD SPACE ENVIRONMENT

The M/OD space environment consists of micrometeoroids and orbital debris. The micrometeoroid environment is essentially a function of particle diameter , with only a small variation with altitude . On the other hand, orbital debris is a function of particle size, altitude, inclination and a growth function which takes into account the increasing population density with time. The occurrence of solar flux activity and periodic meteor storms can also enhance the population of the M/OD environment. One computer model that takes into account most of these factors, including the structural configuration parameters, is the NASA Environet which can be accessed on the Internet at http://envnet.gsfc.nasa.gov.

2.2 SATELLITE ORBITAL PARAMETERS

The satellite orbital parameters that need to be input are; **altitude, inclination, lifetime operation (years) and the start date.** This information is essential in accounting for future solar and storm activities. The desired lifetime for the mission determines the debris enhancement due to population growth.

2.3 STRUCTURAL COMPONENT

For a given structural component on an orbiting satellite, to estimate the M/OD impact flux it is necessary to know the following parameters; **orientation relative to RAM , exposed area and the material type.** Additional information on material properties are also required, such as **density , modulus of elasticity and thickness.** These properties are required for the impact damage models that vary with material type. For metallic materials, damage models are reasonably well defined (see [1]). Relatively little information exists for polymers and composites.

2.4 IMPACT PROBABILITY

The probability of impact (POI) can be calculated from the Environet computer model once the above information is input. The output is given as a function of particle size for a tumbling surface. In other words, the spacecraft component is assumed to be tumbling in orbit, thus exposing the whole surface area to the M/OD environment. This is a reasonable assumption since particles are bombarding the structure from all directions. On the other hand, if the surface is protected from exposure in certain directions, then a correction for directional dependence must be taken into account. Because this is an impact flux model, no velocity dependence is predicted. The flux calculations represent the sum of both micrometeoroids and orbital debris , and define the total number of impacts of a specific diameter and larger / unit area / year. Calculation of the POI first requires calculating the probability of no impact (PNI) which is given by the function,

$$PNI = e^{-fAt} \tag{1}$$

f = flux , A = exposed area, t = time. To obtain the total POI over the prescribed lifetime in orbit, one must calculate the PNI for each year. Thus, for N years, one obtains,

$$\{PNI\}_N = \{PNI\}_1 * \{PNI\}_2 * \{PNI\}_3 * \dots \{PNI\}_N \tag{2}$$

where t = 1 year in Eq. (1). The corresponding POI is then given by,

$$\{ POI \}_N = 1 - \{ PNI \}_N \tag{3}$$

2.5 DAMAGE ASSESSMENT

Based on the type of material being studied, and its associated properties, one must select a damage model that provides information on the craters formed on impact, penetration depth , the spallation damage and secondary debris impacts. Once the local damage has been assessed, a structural damage model may be required depending on the component application. For example, if the component is a structural member subject to inertial loads and thermal cycling, then it may be necessary to evaluate fracture toughness, fatigue and buckling (for thin walled members). Clearly, if penetration prevention is a design consideration, then component thickness is a critical parameter. Thus the need for damage criteria that include thickness in the model.

2.6 FAILURE CRITERIA

Once the damage has been assessed, one needs to translate this information into failure modes. As noted above, failure criteria can be described in terms of wall penetration, loss of stiffness and/or strength, reduction in buckling capacity, fracture and fatigue strength, or in some defined operational performance criteria.

2.7 RISK ASSESSMENT

Depending on the criticality of the component to the spacecraft mission and its operational requirements , one must then compare the POI evaluated over the design lifetime and the associated degradation in component performance. This translates into comparing the POI of a particle impact for a specified particle size that is known to violate the above failure criteria. Clearly if this POI exceeds the prescribed acceptability limit, then modifications to the component design or the implementation of a protection system are required.

3. CASE STUDY - A COMPOSITE ROBOT ARM IN SPACE

3.1 DEFINITION OF THE M/OD ENVIRONMENT

Assume the composite structure is located at approximately 400 km altitude, 52^0 inclination, to be launched in 1998 for a 10-year lifetime. The exposed area is 15 m^2 . The debris growth rate from the Environet model is 2% per year. No storm activity is considered. Once these parameters are input, the POI as a function of particle size was determined and plotted in Fig.4.

3.2 DAMAGE MODELS

If the composite structure is considered to be a 33 cm diameter cylinder with a wall thickness of 0.27 cm , then the radius / thickness ratio indicates that one of the damage criteria that needs to be factored into the analysis is that of loss in structural stability under inertial loading. To assess this effect, one must know the crater hole size and the subsequent damage to the rear wall due to secondary debris impacts. It must be postulated that significant stiffness reductions can occur if large areas of the shell surface suffer impact damage and cracking.

From hypervelocity impact tests on composite flat plates and cylinders (including tests by the authors), data have been obtained on crater diameter as a function of an energy parameter given by $(Et/Dp)^{1/3}$, where E = energy, t = thickness, and Dp = particle diameter. These results are presented in Fig. 5. From other studies by Tennyson, it was possible to assess the buckling strength of cylinders as a function of hole size for axial compression , torsion and bending (Fig. 6). Thus, it is necessary to estimate hole size from initial impact and secondary impacts on the cylinder wall.

It was found from these impact tests that when a projectile of 9 mm diameter traveling at 7 km/s penetrated the wall, the debris plume caused a fracture zone of about 20 cm diameter consisting of multiple holes from the debris particles. As the particle size was reduced, it was observed that a 6mm particle was capable of producing similar cracking over the same area and thus represents the lower bound for design purposes. Using this hole size and the cylinder parameters, it can be seen in Fig. 6 that this damage will result in a 50% buckling strength reduction. From a design viewpoint, this constitutes a structural failure.

Using the POI data from Fig. 4 and the energy curve in Fig. 5, a failure curve can be constructed for the composite structure as a function of particle size and velocity (Fig . 7) .

3.3 RISK ASSESSMENT

From Fig.7 it can be seen that as the velocity increases, the critical particle size required to cause structural failure decreases. This increases the probability of impact and thus increases the risk of failure. Based on these results, it is estimated that the overall probability of failure over a ten year space mission is about 1% ~ 2% , assuming impact velocities in the range of 14km/s and higher.

4. ACKNOWLEDGMENTS

The authors wish to acknowledge the financial support for this project from the Institute for Space and Terrestrial Science, an Ontario Centre of Excellence, and the Natural Sciences and Engineering Research Council of Canada.

5. REFERENCES

[1] McDonnell, J.A.M. (1992) *Hypervelocity Impacts in Space,* University of Kent, Canterbury, U.K.
[2] National Research Council (US) (1995) *Orbital Debris - A Technical Assessment,* National Academy Press, Washington

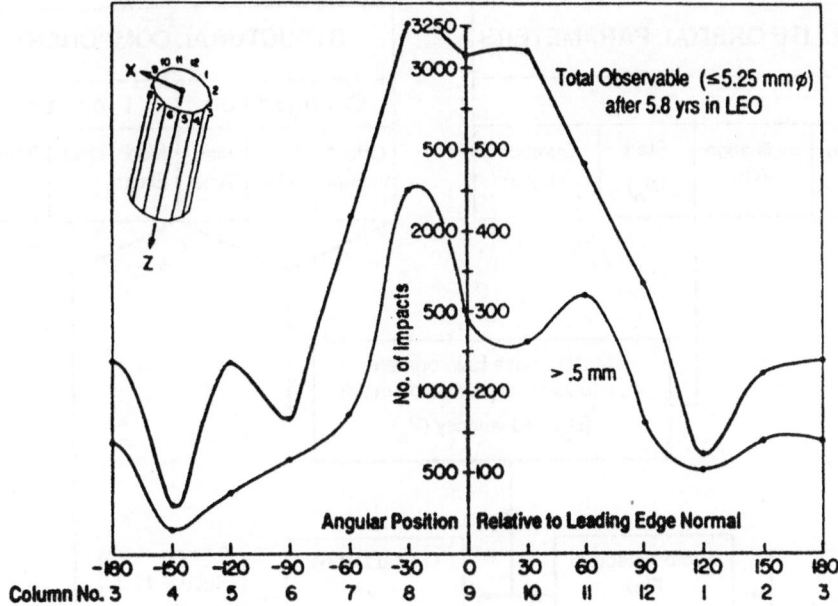

Fig. 1 Circumferential Distribution of Micrometeoroid/Debris
Impacts on LDEF (NASA M & D/SIG Rep't; Aug. 90)

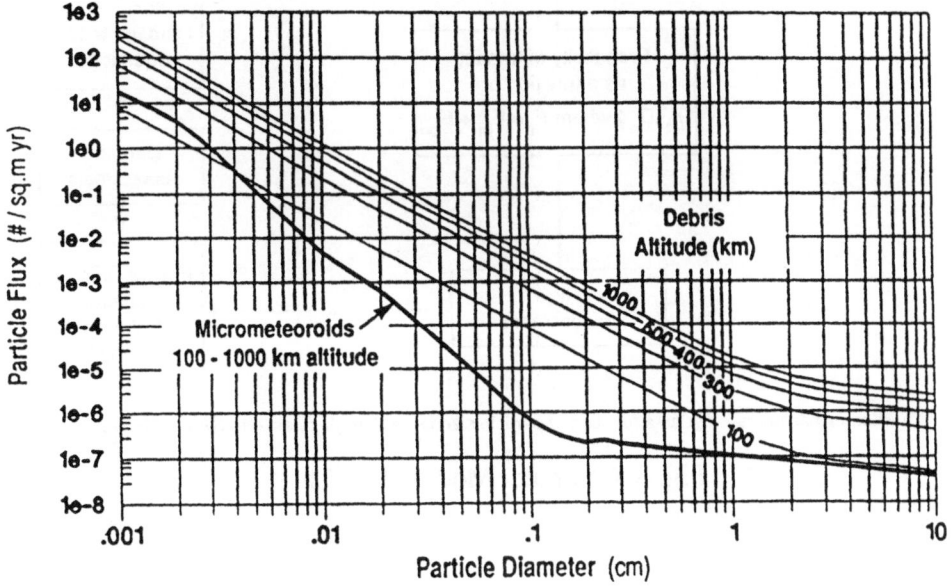

Fig. 2 Micrometeoroid and Orbital Debris Flux vs Particle Size as a
Function of Altitude (NASA CR#BB000883A; Jan. 1991)

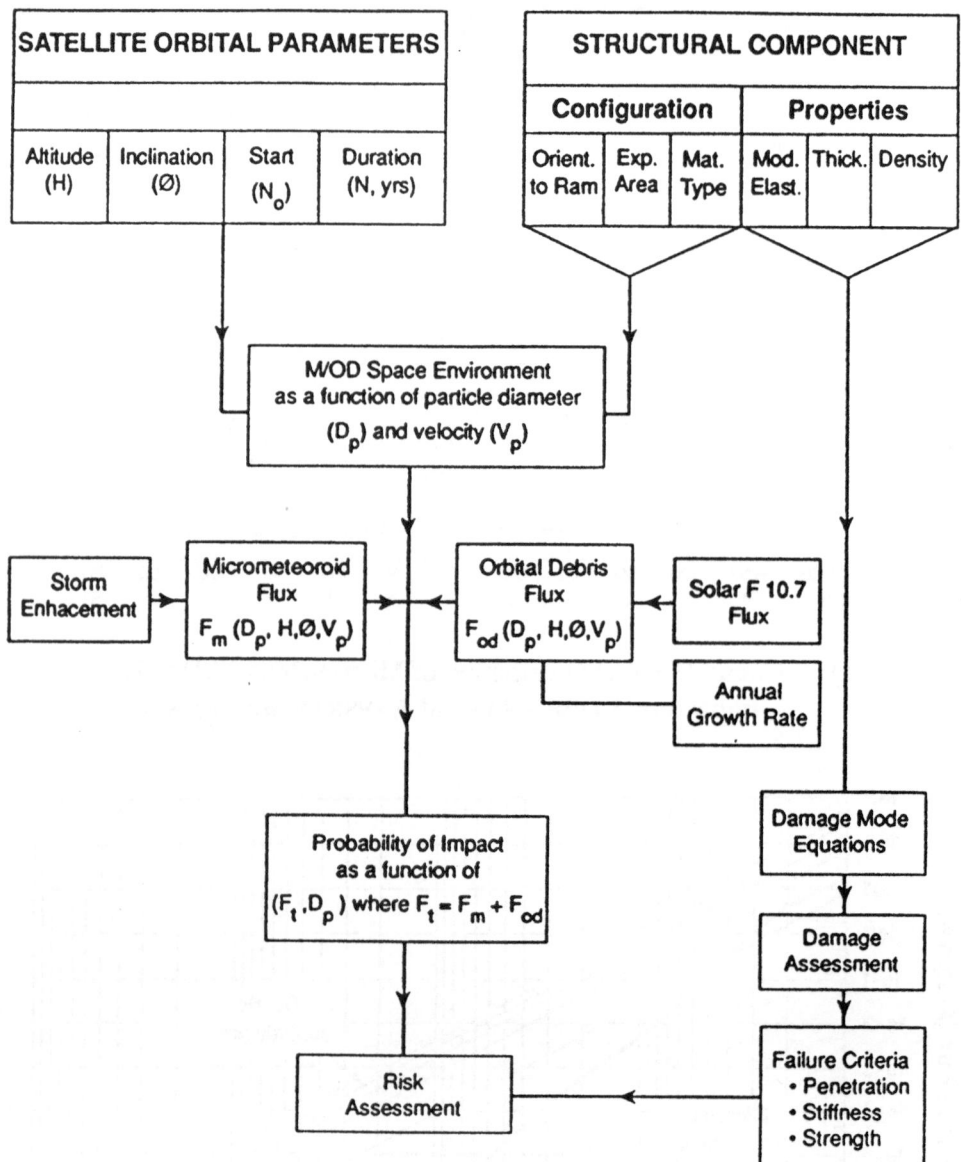

Flux = # hits/unit area/yr as a function of particle size D_p and angle relative to RAM for N years

Fig. 3 Computer Model Flow Chart

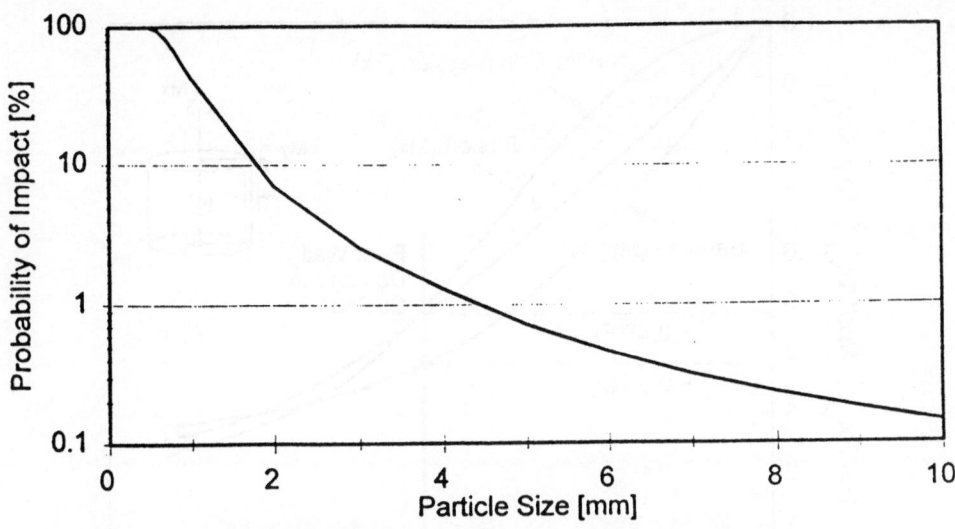

Fig. 4 Probability of Impact (10 yr. lifetime)

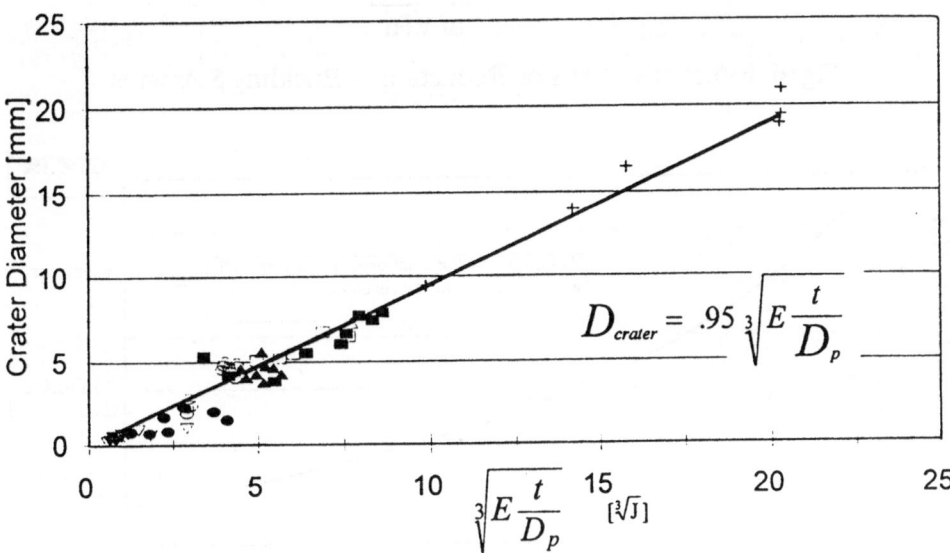

$$D_{crater} = .95 \sqrt[3]{E \frac{t}{D_p}}$$

Fig. 5 Effect of Energy Parameter on Impact Crater Size

330

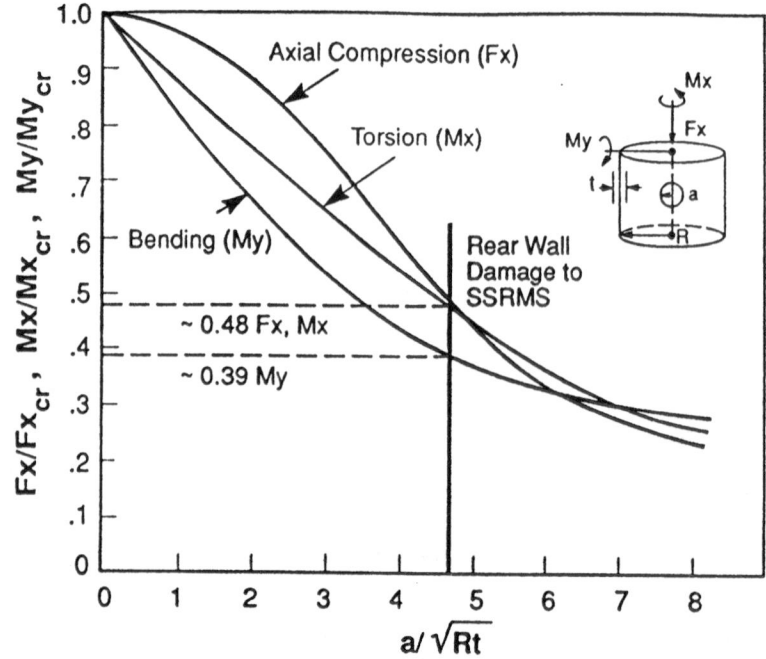

Fig. 6 Effect of Cutouts on Reduction of Buckling Strengths

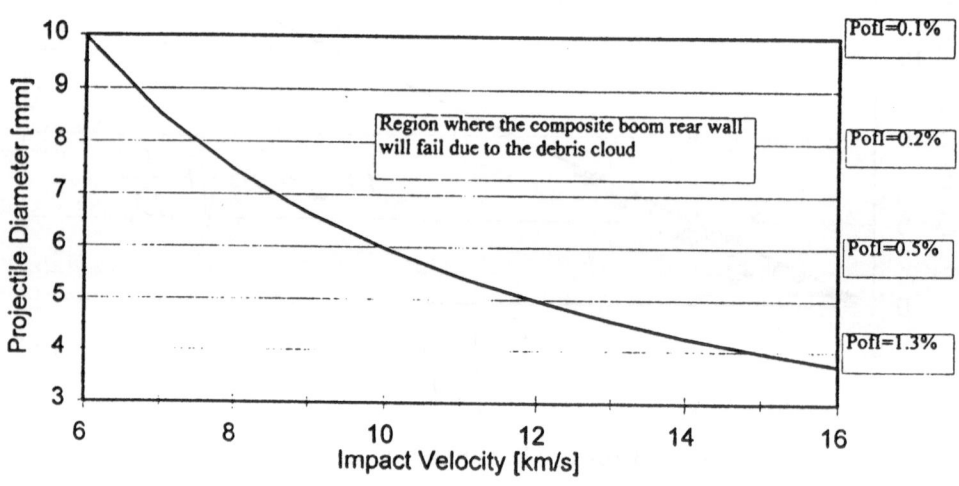

Fig. 7 Failure Envelope for Composite Booms

APPENDIX A
ORGANIZING COMMITTEES

International Organizing Committee

Co-Chairs:

R. C. Tennyson, Workshop Director - NATO countries
University of Toronto Institute for Aerospace Studies,
4925 Dufferin St., Toronto, Ontario, Canada M3H 5T6
Tel (416) 667-7710
Fax. (416) 667-7799
E-mail: rcten@utias.utoronto.ca

A. E. Kiv, Workshop Director - East European countries
Department of Theoretical Physics,
South Ukrainian Pedagogical University
26 Staroportofrankovskaya, Odessa, 270020 Ukraine
Tel. +380 (0482) 32 51 07
Fax. +380 (0482) 23 10 63
E-mail: akiv@tm.odessa.ua

Members:

J. Kleiman, Vice-Director - NATO countries
Integrity Testing Laboratory Inc.,
4925 Dufferin St., Toronto, Ontario, Canada M3H 5T6
Tel. (416) 667-7700
Fax. (416) 667-7799
E-Mail: jkleiman@utias.utoronto.ca

Ya. Roizin, Vice-Director - East European countries
Odessa State University,
2, P. Velikogo St., 270100, Odessa, Ukraine
E-mail: roizin@dtp.odessa.ua

E. Parilis
Califomia Institute of Technology,
Department of Physics 200 - 36,
Pasadena, CA 91125, USA
Fax. (310) 634-7417
E-mail: parilis@romeo.caltech.edu

L. Jacak
Institute of Physics,
Technical University of Wroclaw
27 Wybrzeze Wyspianskiego
50-370 Wroclaw, Poland

D. Maric
Swiss Center for Scientific Computing (SCSC-CSCS)
Via Cantonale, CH-6928 Manno, Switzerland
        fax. +41-91-50.67.11
        E-mail: dmaric@cscs.ch

Local Organizing Committee

E. Rysiakiewicz-Pasek, Chair
        Institute of Physics, Technical University of Wroclaw
        27 Wybrzeze Wyspianskiego
        50-370 Wroclaw, Poland
        Tel. +48(71) 20 36 14
        Fax. +48 (71) 22 96 96
        E-mail: ewar@rainbow.if.pwr.wroc.pl

K. Marczuk
        Institute of Physics, Technical University of Wroclaw
        27 Wybrzeze Wyspianskiego
        50-370 Wroclaw, Poland

K. Pater
        Institute of Physics, Technical University of Wroclaw
        27 Wybrzeze Wyspianskiego
        50-370 Wroclaw, Poland

APPENDIX B
WORKSHOP LECTURERS

I. Abarenkov,
Physics Department, St. Petersburg State University,
Ulyanovskaya 1, Petrodvorets, St. Petersburg 198904, Russia

R. M. Balabay,
Department of Computer Sciences & Applied Mathematics,
State Pedagogical Institute
Gagarin Avenue 54, Krivoy Rog, 324086, Ukraine

B. A. Banks,
NASA Lewis Research Center,
21000 Brookpark Rd., M.S. 302 - 1, Cleveland, Ohio 44135

E. P. Britavskaya,
South Ukrainian Pedagogical University,
26 Staroportofrankovskaya Str. 270020, Odessa, Ukraine

R. Catlow,
The Royal Institution of Great Britain,
21 Albemarle St., London WIX 4BS, UK

A. Gokhman,
South Ukrainian Pedagogical University,
26, Staroportofrankovskaya Street, Odessa 270020, Ukraine

A. Harker,
Centre for Materials Research, Department of Physics and Astronomy,
University College, Gower Street, London, WC1E 6BT

Z. Iskanderova,
Institute for Aerospace Studies, University of Toronto,
4925 Dufferin St., Toronto, Ontario Canada M3H 5T6

V.V. Kirsanov,
Tver State Technical University,
170026, Tver, Nab. A. Nikitin, 22, Russia

Z. M. Khakimov,
Institute of Nuclear Physics of Uzbekistan Academy of Sciences
Ulughbek, 702132, Tashkent, Uzbekistan

A. E. Kiv,
Department of Theoretical Physics,
South Ukrainian Pedagogical University
26 Staroportofrankovskaya, Odessa 270020 Ukraine

J. Kleiman,
Institute for Aerospace Studies, University of Toronto,
4925 Dufferin St., Toronto, Ontario Canada M3H 5T6

E. Kotomin,
Institute of Solid State Physics, Latvia University,
8, Kengaraga Street, Riga 226053, Latvia

V.V. Kovalchuk,
South Ukrainian Pedagogical University ,
26, Staroportofrankovskaya Str., Odessa 270020 Ukraine

A. I. Melker,
Department of Metal Physics, Physics Mechanics Faculty,
St. Petersburg State Technical University,
Polytekhnicheskaya str. 29, St. Petersburg, 195251, Russia

E. P. Nikonova,
Department of Informatics & Applied Mathematics,
State Pedagogical Institute,
Gagarin Avenue 53A, Krivoy Rog, 324086, Ukraine

V.V. Novikov,
Odessa Polytechni University,
270044 Odessa, Ukraine

E. Parilis,
Physics Department 200-36
California Institute of Technology, Pasadena, California 91125

V. E. Puchin,
Institute of Chemical Physics, University of Latvia,
Rainis Blvd. 19, Riga, LV-1586, Latvia

A.B. Roitsin,
Institute of Semi-conductor Physics, Academy of Sciences of Ukraine,
Prospect Nauki, 45, Kiev, 28, 252650, Ukraine

Ya. O. Roizin,
        Odessa State University,
        2, P. Velikogo Str., 270100, Odessa, Ukraine

M. Rybaczuk,
        Institute Of Materials Science And Applied Mechanics ,
        Technical University Of Wroclaw ,
        Wybrzeze Wyspianskiego 27, 50-370 Wroclaw, Poland

A. L. Shluger,
        Department of Physics, University College,
        Gower Street, London, WC1E 6 BT UK

Yu. N. Shunin,
        Aviation University of Riga, Department of Physics,
        Lomonosova Street 1, LV-1019, Riga 226010, Latvia

A. Sokol,
        Department of Computer Sciences & Mathematics,
        State Pedagogical Institute,
        Gagarin Avenue 54, Krivoy Rog, 324086 Ukraine

V. N. Soloviev,
        Department of Computer Sciences & Applied Mathematics,
        State Pedagogical Institute,
        Gagarin Avenue 54, Krivoy Rog, 324086, Ukraine

V.A. Telezhkin,
        Donetsk Institute for Physics and Technology,
        72 Luxembourg Str., Donetsk, Ukraine 340114

R. C. Tennyson,
        Institute for Aerospace Studies, University of Toronto,
        4925 Dufferin St., Toronto, Ontario Canada M3H 5T6

F.T. Umarova,
        Institute of Nuclear Physics, Uzbekistan Academy of Sciences,
        Ulughbek, 702132, Tashkent, Uzbekistan

V.S. Znamenski,
        Kabardin-Balkerian State University,
        P.B. 46, Nalchik-04, KBR 360004, Russia

Index